Student Solutions
Manual FOR Stewart's

MULTIVARIABLE
CALCULUS

Concepts AND Contexts

dan clegg
Palomar College

BROOKS/COLE PUBLISHING COMPANY

I(T)P ® *An International Thomson Publishing Company*

Pacific Grove • Albany • Belmont • Bonn • Boston • Cincinnati • Detroit
Johannesburg • London • Madrid • Melbourne • Mexico City • New York
Paris • Singapore • Tokyo • Toronto • Washington

 **A GARY W. OSTEDT BOOK**

Project Development Editor: *Beth Wilbur*
Editorial Associate: *Carol Benedict*
Marketing Team: *Caroline Croley, Christine Davis*
Ancillaries Coordinator: *Dorothy Bell*

Typesetting/Production Coordination: *Andrew Bulman-Fleming*
Cover Design: *Vernon T. Boes*
Cover Photograph: *David Bruce Johnson, Ian Sabell*
Printing and Binding: *Webcom Limited*

For more information, contact:

BROOKS/COLE PUBLISHING COMPANY
511 Forest Lodge Road
Pacific Grove, CA 93950
USA

International Thomson Publishing Europe
Berkshire House 168-173
High Holborn
London WC1V 7AA
England

Thomas Nelson Australia
102 Dodds Street
South Melbourne, 3205
Victoria, Australia

Nelson Canada
1120 Birchmount Road
Scarborough, Ontario
Canada M1K 5G4

International Thomson Editores
Seneca 53
Col. Polanco
11560 México, D. F., México

International Thomson Publishing GmbH
Königswinterer Strasse 418
53227 Bonn
Germany

International Thomson Publishing Asia
60 Albert Street
#15-01 Albert Complex
Singapore 189969

International Thomson Publishing Japan
Hirakawacho Kyowa Building, 3F
2-2-1 Hirakawacho
Chiyoda-ku, Tokyo 102
Japan

Printed in Canada

10 9 8 7 6 5 4 3 2 1

ISBN 0-534-34436-4

Preface

This *Student Solutions Manual* contains detailed solutions to selected exercises in the text *Multivariable Calculus: Concepts and Contexts* (Chapters 8–13 of *Calculus: Concepts and Contexts*) by James Stewart. Specifically, it includes solutions to the odd-numbered exercises in each chapter section, review section, True-False Quiz, and Focus on Problem Solving section. Also included are all solutions to the Concept Check questions.

Each solution is presented in the context of the corresponding section of the text. In general, solutions to the initial exercises involving a new concept illustrate that concept in more detail; this knowledge is then utilized in subsequent solutions. Thus, while the intermediate steps of a solution are given, you may need to refer back to earlier exercises in the section or prior sections for additional explanation of the concepts involved. Note that, in many cases, different routes to an answer may exist which are equally valid; also, answers can be expressed in different but equivalent forms. Thus, the goal of this manual is not to give the definitive solution to each exercise, but rather to assist you as a student in understanding the concepts of the text and learning how to apply them to the challenge of solving a problem.

I would like to thank James Stewart for entrusting me with the writing of this manual, for reviewing the solutions and making suggestions, and for allowing me the opportunity to contribute to aspects of his text. Andy Bulman-Fleming not only typeset and produced this manual, he also prepared solutions for comparison of accuracy and style. His assistance and suggestions were much appreciated. I thank Brian Betsill and Kathi Townes of TECH-arts for creating and coordinating the illustrations in this manual. Finally, I would like to thank Gary W. Ostedt and Beth Wilbur of Brooks/Cole Publishing Company for their trust, assistance, and patience.

dan clegg

Contents

Contents

Chapter 8   Infinite Sequences and Series

Section 8.1   Sequences

1. (a) A sequence is an ordered list of numbers. It can also be defined as a function whose domain is the set of positive integers.

(b) The terms a_n approach 8 as n becomes large. In fact, we can make a_n as close to 8 as we like by taking n sufficiently large.

(c) The terms a_n become large as n becomes large.

3. The first six terms of $a_n = \dfrac{n}{2n+1}$ are: $\dfrac{1}{3}, \dfrac{2}{5}, \dfrac{3}{7}, \dfrac{4}{9}, \dfrac{5}{11}, \dfrac{6}{13}$. It appears that the sequence is approaching $\dfrac{1}{2}$. $\displaystyle\lim_{n\to\infty}\frac{n}{2n+1} = \lim_{n\to\infty}\frac{1}{2+1/n} = \frac{1}{2}$

5. The numerators are all 1 and the denominators are powers of 2, so $a_n = 1/2^n$.

7. In each term, the numerator is 2 greater than the number of the term, and the denominator is the square of the number that is 3 greater than the term number. So $a_n = \dfrac{n+2}{(n+3)^2}$.

9. $\displaystyle\lim_{n\to\infty}\frac{1}{5^n} = \lim_{n\to\infty}\left(\frac{1}{5}\right)^n = 0$ by Equation 6 with $r = \dfrac{1}{5}$. Convergent

11. $\displaystyle\lim_{n\to\infty}\frac{n^2-1}{n^2+1} = \lim_{n\to\infty}\frac{1-1/n^2}{1+1/n^2} = 1$. Convergent

13. $\{a_n\}$ diverges since $\dfrac{n^2}{n+1} = \dfrac{n}{1+1/n} \to \infty$ as $n \to \infty$.

15. $\{a_n\} = \left\{\cos\frac{\pi}{2}, \cos\pi, \cos\frac{3\pi}{2}, \cos 2\pi, \cos\frac{5\pi}{2}, \cos 3\pi, \cos\frac{7\pi}{2}, \cos 4\pi, \ldots\right\}$
$= \{0, -1, 0, 1, 0, -1, 0, 1, \ldots\}$

This sequence oscillates among 0, -1, and 1 and so diverges.

17. $a_n = \dfrac{\pi^n}{3^n} = \left(\dfrac{\pi}{3}\right)^n$, so $\{a_n\}$ diverges by Equation 6 with $r = \dfrac{\pi}{3} > 1$.

19. $\displaystyle\lim_{x\to\infty}\frac{\ln\left(x^2\right)}{x} = \lim_{x\to\infty}\frac{2\ln x}{x} \overset{\text{H}}{=} \lim_{x\to\infty}\frac{2/x}{1} = 0$, so by Theorem 2, $\left\{\dfrac{\ln\left(n^2\right)}{n}\right\}$ converges to 0.

21. $b_n = \sqrt{n+2}-\sqrt{n} = \left(\sqrt{n+2}-\sqrt{n}\right)\dfrac{\sqrt{n+2}+\sqrt{n}}{\sqrt{n+2}+\sqrt{n}} = \dfrac{2}{\sqrt{n+2}+\sqrt{n}} < \dfrac{2}{2\sqrt{n}} = \dfrac{1}{\sqrt{n}} \to 0$ as

$n \to \infty$. So by the Squeeze Theorem with $a_n = 0$ and $c_n = 1/\sqrt{n}$, $\left\{\sqrt{n+2}-\sqrt{n}\right\}$ converges to 0.

23. $\displaystyle\lim_{x\to\infty}\frac{x}{2^x} \overset{\text{H}}{=} \lim_{x\to\infty}\frac{1}{(\ln 2)\,2^x} = 0$, so by Theorem 2, $\{n2^{-n}\}$ converges to 0.

25. $0 \le \dfrac{\cos^2 n}{2^n} \le \dfrac{1}{2^n}$ [since $0 \le \cos^2 n \le 1$], so since $\displaystyle\lim_{n\to\infty}\frac{1}{2^n} = 0$, $\left\{\dfrac{\cos^2 n}{2^n}\right\}$ converges to 0 by the Squeeze Theorem.

1

27.

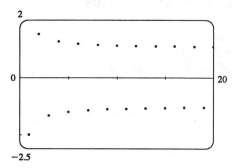

From the graph, we see that the sequence $\left\{(-1)^n \dfrac{n+1}{n}\right\}$ is divergent, since it oscillates between 1 and -1 (approximately).

29.

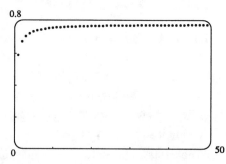

From the graph, it appears that the sequence converges to about 0.78.

$$\lim_{n\to\infty} \frac{2n}{2n+1} = \lim_{n\to\infty} \frac{2}{2+1/n} = 1, \text{ so}$$

$$\lim_{n\to\infty} \arctan\left(\frac{2n}{2n+1}\right) = \arctan 1 = \frac{\pi}{4}.$$

31.

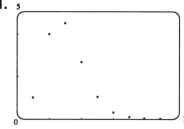

From the graph, it appears that the sequence converges to 0.

$$0 < a_n = \frac{n^3}{n!} = \frac{n}{n} \cdot \frac{n}{(n-1)} \cdot \frac{n}{(n-2)} \cdot \frac{1}{(n-3)} \cdots\cdots \frac{1}{3} \cdot \frac{1}{2} \cdot \frac{1}{1}$$

$$\leq \frac{n^2}{(n-1)(n-2)(n-3)} \quad \text{(for } n \geq 4\text{)}$$

$$= \frac{1/n}{(1-1/n)(1-2/n)(1-3/n)} \to 0 \text{ as } n \to \infty$$

So by the Squeeze Theorem, $\{n^3/n!\}$ converges to 0.

33. (a) $a_1 = 1$, $a_2 = 4 - a_1 = 4 - 1 = 3$, $a_3 = 4 - a_2 = 4 - 3 = 1$, $a_4 = 4 - a_3 = 4 - 1 = 3$, $a_5 = 4 - a_4 = 4 - 3 = 1$. Since the terms of the sequence alternate between 1 and 3, the sequence is divergent.

(b) $a_1 = 2$, $a_2 = 4 - a_1 = 4 - 2 = 2$, $a_3 = 4 - a_2 = 4 - 2 = 2$. Since all of the terms are 2, $\lim\limits_{n\to\infty} a_n = 2$ and hence, the sequence is convergent.

35. (a) Let a_n be the number of rabbit pairs in the nth month. Clearly $a_1 = 1 = a_2$. In the nth month, each pair that is 2 or more months old (that is, a_{n-2} pairs) will have a pair of children to add to the a_{n-1} pairs already present. Thus, $a_n = a_{n-1} + a_{n-2}$, so that $\{a_n\} = \{f_n\}$, the Fibonacci sequence.

(b) $a_n = \dfrac{f_{n+1}}{f_n} \;\Rightarrow\; a_{n-1} = \dfrac{f_n}{f_{n-1}} = \dfrac{f_{n-1} + f_{n-2}}{f_{n-1}} = 1 + \dfrac{f_{n-2}}{f_{n-1}} = 1 + \dfrac{1}{f_{n-1}/f_{n-2}} = 1 + \dfrac{1}{a_{n-2}}$. If

$L = \lim\limits_{n\to\infty} a_n$, then $L = \lim\limits_{n\to\infty} a_{n-1}$ and $L = \lim\limits_{n\to\infty} a_{n-2}$, so L must satisfy $L = 1 + \dfrac{1}{L} \;\Rightarrow\;$

$L^2 - L - 1 = 0 \;\Rightarrow\; L = \frac{1+\sqrt{5}}{2}$ (since L must be positive).

37. $3(n+1)+5 > 3n+5$ so $\dfrac{1}{3(n+1)+5} < \dfrac{1}{3n+5}$ $\Leftrightarrow$ $a_{n+1} < a_n$, so $\{a_n\}$ is decreasing.

39. $\left\{\dfrac{n-2}{n+2}\right\}$ is increasing since $a_n < a_{n+1}$ $\Leftrightarrow$ $\dfrac{n-2}{n+2} < \dfrac{(n+1)-2}{(n+1)+2}$ $\Leftrightarrow$

$(n-2)(n+3) < (n+2)(n-1)$ $\Leftrightarrow$ $n^2 + n - 6 < n^2 + n - 2$ $\Leftrightarrow$ $-6 < -2$, which is true.

41. Since $\{a_n\}$ is a decreasing sequence, $a_n > a_{n+1}$ for all $n \geq 1$. Because all of its terms lie between 5 and 8, $\{a_n\}$ is a bounded sequence. By the Monotonic Sequence Theorem, $\{a_n\}$ is convergent, that is, $\{a_n\}$ has a limit L. L must be less than 8 since $\{a_n\}$ is decreasing, so $5 \leq L < 8$.

43. We show by induction that $\{a_n\}$ is increasing and bounded above by 3.

Let P_n be the proposition that $a_{n+1} > a_n$ and $0 < a_n < 3$. Clearly P_1 is true. Assume that P_n is true.

Then $a_{n+1} > a_n$ $\Rightarrow$ $\dfrac{1}{a_{n+1}} < \dfrac{1}{a_n}$ $\Rightarrow$ $-\dfrac{1}{a_{n+1}} > -\dfrac{1}{a_n}$.

Now $a_{n+2} = 3 - \dfrac{1}{a_{n+1}} > 3 - \dfrac{1}{a_n} = a_{n+1}$ $\Leftrightarrow$ P_{n+1}. This proves that $\{a_n\}$ is increasing and bounded above by 3, so $1 = a_1 < a_n < 3$, that is, $\{a_n\}$ is bounded, and hence convergent by the Monotonic Sequence Theorem. If $L = \lim\limits_{n\to\infty} a_n$, then $\lim\limits_{n\to\infty} a_{n+1} = L$ also, so L must satisfy

$L = 3 - 1/L$ $\Rightarrow$ $L^2 - 3L + 1 = 0$ $\Rightarrow$ $L = \frac{3 \pm \sqrt{5}}{2}$. But $L > 1$, so $L = \frac{3+\sqrt{5}}{2}$.

45. $(0.8)^n < 0.000001$ $\Rightarrow$ $\ln(0.8)^n < \ln(0.000001)$ $\Rightarrow$ $n\ln(0.8) < \ln(0.000001)$ $\Rightarrow$

$n > \dfrac{\ln(0.000001)}{\ln(0.8)}$ $\Rightarrow$ $n > 61.9$, so n must be at least 62 to satisfy the given inequality.

47. (a) First we show that $a > a_1 > b_1 > b$.

$a_1 - b_1 = \frac{1}{2}(a+b) - \sqrt{ab} = \frac{1}{2}\left(a - 2\sqrt{ab} + b\right) = \frac{1}{2}\left(\sqrt{a} - \sqrt{b}\right)^2 > 0$ (since $a > b$)

$\Rightarrow$ $a_1 > b_1$. Also $a - a_1 = a - \frac{1}{2}(a+b) = \frac{1}{2}(a-b) > 0$ and

$b - b_1 = b - \sqrt{ab} = \sqrt{b}\left(\sqrt{b} - \sqrt{a}\right) < 0$, so $a > a_1 > b_1 > b$. In the same way we can show that

$a_1 > a_2 > b_2 > b_1$ and so the given assertion is true for $n = 1$. Suppose it is true for $n = k$, that is,

$a_k > a_{k+1} > b_{k+1} > b_k$. Then

$a_{k+2} - b_{k+2} = \frac{1}{2}(a_{k+1} + b_{k+1}) - \sqrt{a_{k+1}b_{k+1}} = \frac{1}{2}\left(a_{k+1} - 2\sqrt{a_{k+1}b_{k+1}} + b_{k+1}\right)$
$= \frac{1}{2}\left(\sqrt{a_{k+1}} - \sqrt{b_{k+1}}\right)^2 > 0,$

$a_{k+1} - a_{k+2} = a_{k+1} - \frac{1}{2}(a_{k+1} + b_{k+1}) = \frac{1}{2}(a_{k+1} - b_{k+1}) > 0,$

and $b_{k+1} - b_{k+2} = b_{k+1} - \sqrt{a_{k+1}b_{k+1}} = \sqrt{b_{k+1}}\left(\sqrt{b_{k+1}} - \sqrt{a_{k+1}}\right) < 0$ $\Rightarrow$

$a_{k+1} > a_{k+2} > b_{k+2} > b_{k+1}$, so the assertion is true for $n = k+1$. Thus, it is true for all n by mathematical induction.

(b) From part (a) we have $a > a_n > a_{n+1} > b_{n+1} > b_n > b$, which shows that both sequences, $\{a_n\}$ and $\{b_n\}$, are monotonic and bounded. So they are both convergent by the Monotonic Sequence Theorem.

(c) Let $\lim\limits_{n\to\infty} a_n = \alpha$ and $\lim\limits_{n\to\infty} b_n = \beta$. Then $\lim\limits_{n\to\infty} a_{n+1} = \lim\limits_{n\to\infty} \dfrac{a_n + b_n}{2}$ $\Rightarrow$ $\alpha = \dfrac{\alpha + \beta}{2}$ $\Rightarrow$

$2\alpha = \alpha + \beta$ $\Rightarrow$ $\alpha = \beta$

Section 8.2   Series

1. (a) A sequence is an ordered list of numbers whereas a series is the *sum* of a list of numbers.

(b) A series is convergent if the sequence of partial sums is a convergent sequence. A series is divergent if it is not convergent.

3.

n	s_n
1	3.33333
2	4.44444
3	4.81481
4	4.93827
5	4.97942
6	4.99314
7	4.99771
8	4.99924
9	4.99975
10	4.99992
11	4.99997
12	4.99999

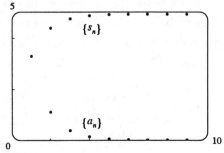

From the graph, it seems that the series converges. In fact, it is a geometric series with $a = \frac{10}{3}$ and $r = \frac{1}{3}$, so its sum is

$$\sum_{n=1}^{\infty} \frac{10}{3^n} = \frac{10/3}{1 - 1/3} = 5.$$ Note that the dot corresponding to $n = 1$ is part of both $\{a_n\}$ and $\{s_n\}$.

5.

n	s_n
1	0.50000
2	1.16667
3	1.91667
4	2.71667
5	3.55000
6	4.40714
7	5.28214
8	6.17103
9	7.07103
10	7.98012

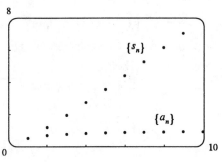

The series diverges, since its terms do not approach 0.

4

7.

n	s_n
1	0.64645
2	0.80755
3	0.87500
4	0.91056
5	0.93196
6	0.94601
7	0.95581
8	0.96296
9	0.96838
10	0.97259

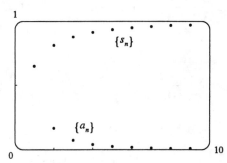

From the graph, it seems that the series converges to 1. To find the sum, we write

$$s_n = \sum_{i=1}^{n}\left(\frac{1}{i^{1.5}} - \frac{1}{(i+1)^{1.5}}\right) = \left(1 - \frac{1}{2^{1.5}}\right) + \left(\frac{1}{2^{1.5}} - \frac{1}{3^{1.5}}\right)$$

$$+ \left(\frac{1}{3^{1.5}} - \frac{1}{4^{1.5}}\right) + \cdots + \left(\frac{1}{n^{1.5}} - \frac{1}{(n+1)^{1.5}}\right) = 1 - \frac{1}{(n+1)^{1.5}}.$$

So the sum is $\lim_{n\to\infty} s_n = 1$.

9. (a) $\lim_{n\to\infty} a_n = \lim_{n\to\infty} \frac{2n}{3n+1} = \frac{2}{3}$, so the *sequence* $\{a_n\}$ is convergent by (8.1.1).

(b) Since $\lim_{n\to\infty} a_n = \frac{2}{3} \neq 0$, the *series* $\sum_{n=1}^{\infty} a_n$ is divergent by the Test for Divergence (7).

11. $4 + \frac{8}{5} + \frac{16}{25} + \frac{32}{125} + \cdots$ is a geometric series with $a = 4$ and $r = \frac{2}{5}$. Since $|r| = \frac{2}{5} < 1$, the series

converges to $\frac{4}{1-2/5} = \frac{4}{3/5} = \frac{20}{3}$.

13. $\sum_{n=1}^{\infty} \frac{(-3)^{n-1}}{4^n} = \frac{1}{4}\sum_{n=1}^{\infty}\left(-\frac{3}{4}\right)^{n-1}$. The latter series is geometric with $a = 1$ and $r = -\frac{3}{4}$. Since

$|r| = \frac{3}{4} < 1$, it converges to $\frac{1}{1-(-3/4)} = \frac{4}{7}$. Thus, the given series converges to $\left(\frac{1}{4}\right)\left(\frac{4}{7}\right) = \frac{1}{7}$.

15. For $\sum_{n=1}^{\infty} 3^{-n}8^{n+1} = \sum_{n=1}^{\infty} 8\left(\frac{8}{3}\right)^{n}$, $a = \frac{64}{3}$ and $r = \frac{8}{3} > 1$, so the series diverges.

17. $\sum_{n=1}^{\infty}[2(0.1)^n + (0.2)^n] = 2\sum_{n=1}^{\infty}(0.1)^n + \sum_{n=1}^{\infty}(0.2)^n$. These are convergent geometric series and so by

Theorem 8, the sum is also convergent. $2\left(\frac{0.1}{1-0.1}\right) + \frac{0.2}{1-0.2} = \frac{2}{9} + \frac{1}{4} = \frac{17}{36}$

5

19. $\lim\limits_{n\to\infty} a_n = \lim\limits_{n\to\infty} \dfrac{n}{\sqrt{1+n^2}} = \lim\limits_{n\to\infty} \dfrac{1}{\sqrt{1+1/n^2}} = 1 \neq 0$, so the series diverges by the Test for

Divergence.

21. Converges. $s_n = \sum\limits_{i=1}^{n} \dfrac{1}{i\,(i+2)} = \sum\limits_{i=1}^{n} \left(\dfrac{1/2}{i} - \dfrac{1/2}{i+2} \right)$ (using partial fractions) $= \dfrac{1}{2} \sum\limits_{i=1}^{n} \left(\dfrac{1}{i} - \dfrac{1}{i+2} \right)$.

The latter sum is

$$\left(1 - \dfrac{1}{3}\right) + \left(\dfrac{1}{2} - \dfrac{1}{4}\right) + \left(\dfrac{1}{3} - \dfrac{1}{5}\right) + \cdots + \left(\dfrac{1}{n-1} - \dfrac{1}{n+1}\right) + \left(\dfrac{1}{n} - \dfrac{1}{n+2}\right)$$

$$= 1 + \dfrac{1}{2} - \dfrac{1}{n+1} - \dfrac{1}{n+2} \text{(telescoping series).}$$

Thus, $\sum\limits_{n=1}^{\infty} \dfrac{1}{n\,(n+2)} = \dfrac{1}{2} \lim\limits_{n\to\infty} \left(1 + \dfrac{1}{2} - \dfrac{1}{n+1} - \dfrac{1}{n+2} \right) = \dfrac{1}{2}\left(1 + \dfrac{1}{2}\right) = \dfrac{3}{4}$.

23. Converges. $\sum\limits_{n=1}^{\infty} \dfrac{3^n + 2^n}{6^n} = \sum\limits_{n=1}^{\infty} \left(\dfrac{3^n}{6^n} + \dfrac{2^n}{6^n} \right) = \sum\limits_{n=1}^{\infty} \left[\left(\tfrac{1}{2}\right)^n + \left(\tfrac{1}{3}\right)^n \right] = \dfrac{1/2}{1-1/2} + \dfrac{1/3}{1-1/3} = 1 + \dfrac{1}{2} = \dfrac{3}{2}$

25. Converges. $s_n = \left(\sin 1 - \sin \tfrac{1}{2}\right) + \left(\sin \tfrac{1}{2} - \sin \tfrac{1}{3}\right) + \cdots + \left(\sin \dfrac{1}{n} - \sin \dfrac{1}{n+1} \right) = \sin 1 - \sin \dfrac{1}{n+1}$,

so $\sum\limits_{n=1}^{\infty} \left(\sin \dfrac{1}{n} - \sin \dfrac{1}{n+1} \right) = \lim\limits_{n\to\infty} s_n = \sin 1 - \sin 0 = \sin 1$.

27. $\lim\limits_{n\to\infty} a_n = \lim\limits_{n\to\infty} \arctan n = \dfrac{\pi}{2} \neq 0$, so the series diverges by the Test for Divergence.

29. $0.\overline{5} = 0.5 + 0.05 + 0.005 + \cdots = \dfrac{0.5}{1-0.1} = \dfrac{5}{9}$

31. $0.\overline{307} = 0.307 + 0.000307 + 0.000000307 + \cdots = \dfrac{0.307}{1-0.001} = \dfrac{307}{999}$

33. $\sum\limits_{n=0}^{\infty} (x-3)^n$ is a geometric series with $r = x - 3$, so it converges whenever $|x - 3| < 1 \quad \Leftrightarrow$

$-1 < x - 3 < 1 \quad \Leftrightarrow \quad 2 < x < 4$, and the sum is $\dfrac{1}{1 - (x-3)} = \dfrac{1}{4 - x}$.

35. $\sum\limits_{n=0}^{\infty} \left(\dfrac{1}{x}\right)^n$ is geometric with $r = \dfrac{1}{x}$, so it converges whenever $\left|\dfrac{1}{x}\right| < 1 \quad \Leftrightarrow \quad |x| > 1 \quad \Leftrightarrow \quad x > 1$ or

$x < -1$, and the sum is $\dfrac{1}{1 - 1/x} = \dfrac{x}{x - 1}$.

37. After defining f, We use `convert(f,parfrac);` in Maple, `Apart` in Mathematica, or `Expand Rational` and `Simplify` in Derive to find that the general term is

$$\frac{1}{(4n+1)(4n-3)} = -\frac{1/4}{4n+1} + \frac{1/4}{4n-3}.$$ So the nth partial sum is

$$s_n = \sum_{k=1}^{n}\left(-\frac{1/4}{4k+1} + \frac{1/4}{4k-3}\right) = \frac{1}{4}\left(\frac{1}{4k-3} - \frac{1}{4k+1}\right)$$

$$= \frac{1}{4}\left[\left(1-\frac{1}{5}\right) + \left(\frac{1}{5}-\frac{1}{9}\right) + \left(\frac{1}{9}-\frac{1}{13}\right) + \cdots + \left(\frac{1}{4n-3} - \frac{1}{4n+1}\right)\right] = \frac{1}{4}\left(1 - \frac{1}{4n+1}\right)$$

The series converges to $\lim\limits_{n\to\infty} s_n = \frac{1}{4}$. This can be confirmed by directly computing the sum using `sum(f,1..infinity);` (in Maple), `Sum[f,{n,1,Infinity}]` (in Mathematica), or Calculus Sum (from 1 to ∞) and `Simplify` (in Derive).

39. For $n=1$, $a_1 = 0$ since $s_1 = 0$. For $n > 1$,

$$a_n = s_n - s_{n-1} = \frac{n-1}{n+1} - \frac{(n-1)-1}{(n-1)+1} = \frac{(n-1)n - (n+1)(n-2)}{(n+1)n} = \frac{2}{n(n+1)}.$$

Also, $\displaystyle\sum_{n=1}^{\infty} a_n = \lim_{n\to\infty} s_n = \lim_{n\to\infty}\frac{1-1/n}{1+1/n} = 1.$

41. (a) The first step in the chain occurs when the local government spends D dollars. The people who receive it spend a fraction c of those D dollars, that is, Dc dollars. Those who receive the Dc dollars spend a fraction c of it, that is, Dc^2 dollars. Continuing in this way, we see that the total spending after n transactions is $S_n = D + Dc + Dc^2 + \cdots + Dc^{n-1} = \dfrac{D(1-c^n)}{1-c}$ by (3).

(b) $\displaystyle\lim_{n\to\infty} S_n = \lim_{n\to\infty}\frac{D(1-c^n)}{1-c} = \frac{D}{1-c}\lim_{n\to\infty}(1-c^n)$

$$= \frac{D}{1-c} \text{ (since } 0 < c < 1 \;\Rightarrow\; \lim_{n\to\infty} c^n = 0)$$

$$= \frac{D}{s} \text{ (since } c + s = 1) \quad = kD \quad \text{(since } k = 1/s).$$

If $c = 0.8$, then $s = 1 - c = 0.2$ and the multiplier is $k = 1/s = 5$.

43. $\displaystyle\sum_{n=2}^{\infty}(1+c)^{-n}$ is a geometric series with $a = (1+c)^{-2}$ and $r = (1+c)^{-1}$, so the series converges when $\left|(1+c)^{-1}\right| < 1 \;\Rightarrow\; |1+c| > 1 \;\Rightarrow\; 1+c > 1$ or $1+c < -1 \;\Rightarrow\; c > 0$ or $c < -2$. We calculate the sum of the series and set it equal to 2: $\dfrac{(1+c)^{-2}}{1-(1+c)^{-1}} = 2 \;\Leftrightarrow\;$

$$\left(\frac{1}{1+c}\right)^2 = 2 - 2\left(\frac{1}{1+c}\right) \;\Leftrightarrow\; 1 = 2(1+c)^2 - 2(1+c) = 0 \;\Leftrightarrow\; 2c^2 + 2c - 1 = 0 \;\Leftrightarrow\;$$

$c = \dfrac{-2\pm\sqrt{12}}{4} = \dfrac{\pm\sqrt{3}-1}{2}$. However, the negative root is inadmissible because $-2 < \dfrac{-\sqrt{3}-1}{2} < 0$. So $c = \dfrac{\sqrt{3}-1}{2}$.

7

45.

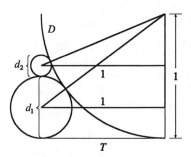

Let d_n be the diameter of C_n. We draw lines from the centers of the C_i to the center of D (or C), and using the Pythagorean Theorem, we can write $1^2 + \left(1 - \frac{1}{2}d_1\right)^2 = \left(1 + \frac{1}{2}d_1\right)^2$ $\Leftrightarrow$
$1 = \left(1 + \frac{1}{2}d_1\right)^2 - \left(1 - \frac{1}{2}d_1\right)^2 = 2d_1$(difference of squares) $\Rightarrow$ $d_1 = \frac{1}{2}$. Similarly,
$1 = \left(1 + \frac{1}{2}d_2\right)^2 - \left(1 - d_1 - \frac{1}{2}d_2\right)^2 = 2d_2 + 2d_1 - d_1^2 - d_1 d_2 = (2 - d_1)(d_1 + d_2)$ $\Leftrightarrow$
$d_2 = \dfrac{1}{2 - d_1} - d_1 = \dfrac{(1 - d_1)^2}{2 - d_1}$, $1 = \left(1 + \frac{1}{2}d_3\right)^2 - \left(1 - d_1 - d_2 - \frac{1}{2}d_3\right)^2$ $\Leftrightarrow$

$d_3 = \dfrac{[1 - (d_1 + d_2)]^2}{2 - (d_1 + d_2)}$, and in general, $d_{n+1} = \dfrac{\left(1 - \sum\limits_{i=1}^{n} d_i\right)^2}{2 - \sum\limits_{i=1}^{n} d_i}$. If we actually calculate d_2 and d_3

from the formulas above, we find that they are $\dfrac{1}{6} = \dfrac{1}{2 \cdot 3}$ and $\dfrac{1}{12} = \dfrac{1}{3 \cdot 4}$ respectively, so we suspect that

in general, $d_n = \dfrac{1}{n(n+1)}$. To prove this, we use induction: assume that for all $k \le n$,

$d_k = \dfrac{1}{k(k+1)} = \dfrac{1}{k} - \dfrac{1}{k+1}$. Then $\sum\limits_{i=1}^{n} d_i = 1 - \dfrac{1}{n+1} = \dfrac{n}{n+1}$ (telescoping sum). Substituting

this into our formula for d_{n+1}, we get $d_{n+1} = \dfrac{\left[1 - \dfrac{n}{n+1}\right]^2}{2 - \left(\dfrac{n}{n+1}\right)} = \dfrac{\dfrac{1}{(n+1)^2}}{\dfrac{n+2}{n+1}} = \dfrac{1}{(n+1)(n+2)}$, and the

induction is complete.

Now, we observe that the partial sums $\sum\limits_{i=1}^{n} d_i$ of the diameters of the circles approach 1 as $n \to \infty$; that

is, $\sum\limits_{n=1}^{\infty} a_n = \sum\limits_{n=1}^{\infty} \dfrac{1}{n(n+1)} = 1$, which is what we wanted to prove.

47. The series $1 - 1 + 1 - 1 + 1 - 1 + \cdots$ diverges (geometric series with $r = -1$) so we cannot say
$0 = 1 - 1 + 1 - 1 + 1 - 1 + \cdots$.

49. Suppose on the contrary that $\sum (a_n + b_n)$ converges. Then by Theorem 8(iii), so would
$\sum [(a_n + b_n) - a_n] = \sum b_n$, a contradiction.

51. The partial sums $\{s_n\}$ form an increasing sequence, since $s_n - s_{n-1} = a_n > 0$ for all n. Also, the sequence $\{s_n\}$ is bounded since $s_n \le 1000$ for all n. So by Theorem 10.1.7 , the sequence of partial sums converges, that is, the series $\sum a_n$ is convergent.

53. (a) At the first step, only the interval $\left(\frac{1}{3}, \frac{2}{3}\right)$ (length $\frac{1}{3}$) is removed. At the second step, we remove the intervals $\left(\frac{1}{9}, \frac{2}{9}\right)$ and $\left(\frac{7}{9}, \frac{8}{9}\right)$, which have a total length of $2 \cdot \left(\frac{1}{3}\right)^2$. At the third step, we remove 2^2 intervals, each of length $\left(\frac{1}{3}\right)^3$. In general, at the nth step we remove 2^{n-1} intervals, each of length $\left(\frac{1}{3}\right)^n$, for a length of $2^{n-1} \cdot \left(\frac{1}{3}\right)^n = \frac{1}{3}\left(\frac{2}{3}\right)^{n-1}$. Thus, the total length of all removed intervals is

$$\sum_{n=1}^{\infty} \frac{1}{3}\left(\frac{2}{3}\right)^{n-1} = \frac{1/3}{1-2/3} = 1 \text{(geometric series with } a = \frac{1}{3} \text{ and } r = \frac{2}{3}\text{)}. \text{ Notice that at the } n\text{th step,}$$

the leftmost interval that is removed is $\left(\left(\frac{1}{3}\right)^n, \left(\frac{2}{3}\right)^n\right)$, so we never remove 0, and 0 is in the Cantor set. Also, the rightmost interval removed is $\left(1 - \left(\frac{2}{3}\right)^n, 1 - \left(\frac{1}{3}\right)^n\right)$, so 1 is never removed. Some other numbers in the Cantor set are $\frac{1}{3}, \frac{2}{3}, \frac{1}{9}, \frac{2}{9}, \frac{7}{9}$, and $\frac{8}{9}$.

(b) The area removed at the first step is $\frac{1}{9}$; at the second step, $8 \cdot \left(\frac{1}{9}\right)^2$; at the third step, $(8)^2 \cdot \left(\frac{1}{9}\right)^3$. In general, the area removed at the nth step is $(8)^{n-1}\left(\frac{1}{9}\right)^n = \frac{1}{9}\left(\frac{8}{9}\right)^{n-1}$, so the total area of all removed squares is $\sum_{n=1}^{\infty} \frac{1}{9}\left(\frac{8}{9}\right)^{n-1} = \frac{1/9}{1-8/9} = 1$.

55. (a) $\displaystyle\sum_{n=1}^{\infty} \frac{n}{(n+1)!} \Rightarrow s_1 = \frac{1}{1 \cdot 2} = \frac{1}{2}, s_2 = \frac{1}{2} + \frac{2}{1 \cdot 2 \cdot 3} = \frac{5}{6}, s_3 = \frac{5}{6} + \frac{3}{1 \cdot 2 \cdot 3 \cdot 4} = \frac{23}{24},$

$s_4 = \frac{23}{24} + \frac{4}{1 \cdot 2 \cdot 3 \cdot 4 \cdot 5} = \frac{119}{120}$. The denominators are $(n+1)!$, so a guess would be

$s_n = \dfrac{(n+1)! - 1}{(n+1)!}$.

(b) For $n = 1$, $s_1 = \dfrac{1}{2} = \dfrac{2! - 1}{2!}$, so the formula holds for $n = 1$. Assume $s_k = \dfrac{(k+1)! - 1}{(k+1)!}$. Then

$$s_{k+1} = \frac{(k+1)! - 1}{(k+1)!} + \frac{k+1}{(k+2)!} = \frac{(k+1)! - 1}{(k+1)!} + \frac{k+1}{(k+1)!(k+2)}$$

$$= \frac{(k+2)! - (k+2) + k + 1}{(k+2)!} = \frac{(k+2)! - 1}{(k+2)!}$$

Thus, the formula is true for $n = k + 1$. So by induction, the guess is correct.

(c) $\displaystyle\lim_{n\to\infty} s_n = \lim_{n\to\infty} \frac{(n+1)! - 1}{(n+1)!} = \lim_{n\to\infty}\left[1 - \frac{1}{(n+1)!}\right] = 1$ and so $\displaystyle\sum_{n=0}^{\infty} \frac{n}{(n+1)!} = 1$.

Section 8.3   The Integral and Comparison Tests; Estimating Sums

1.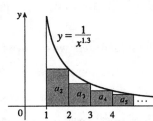
The picture shows that $a_2 = \dfrac{1}{2^{1.3}} < \displaystyle\int_1^2 \dfrac{1}{x^{1.3}}\, dx$,

$a_3 = \dfrac{1}{3^{1.3}} < \displaystyle\int_2^3 \dfrac{1}{x^{1.3}}\, dx$, and so on, so $\displaystyle\sum_{n=2}^{\infty} \dfrac{1}{n^{1.3}} < \displaystyle\int_1^{\infty} \dfrac{1}{x^{1.3}}\, dx$.

The integral converges by (5.9.2) with $p = 1.3 > 1$, so the series converges.

3. (a) We cannot say anything about $\sum a_n$. If $a_n > b_n$ for all n and $\sum b_n$ is convergent, then $\sum a_n$ could be convergent or divergent.

(b) If $a_n < b_n$ for all n, then $\sum a_n$ is convergent. [This is part (a) of the Comparison Test.]

5. $\displaystyle\sum_{n=1}^{\infty} n^b$ is a p-series with $p = -b$. $\displaystyle\sum_{n=1}^{\infty} b^n$ is a geometric series. By (1), the p-series is convergent if $p > 1$. In this case, $\displaystyle\sum_{n=1}^{\infty} n^b = \displaystyle\sum_{n=1}^{\infty} (1/n^{-b})$, so $-b > 1 \ \Leftrightarrow\ b < -1$ are the values for which the series converge. A geometric series $\displaystyle\sum_{n=1}^{\infty} ar^{n-1}$ converges if $|r| < 1$, so $\displaystyle\sum_{n=1}^{\infty} b^n$ converges if $|b| < 1 \ \Leftrightarrow$ $-1 < b < 1$.

7. $f(x) = xe^{-x^2}$ is continuous and positive on $[1, \infty)$, and since $f'(x) = e^{-x^2}(1 - 2x^2) < 0$ for $x > 1$, f is decreasing as well. Thus, we can use the Integral Test.
$\displaystyle\int_1^{\infty} xe^{-x^2}\, dx = \lim_{t \to \infty} \left[-\tfrac{1}{2}e^{-x^2}\right]_1^t = 0 - \left(-\dfrac{e^{-1}}{2}\right) = \dfrac{1}{2e}$. Since the integral converges, the series converges.

9. $f(x) = \dfrac{x}{x^2 + 1}$ is continuous and positive on $[1, \infty)$, and since $f'(x) = \dfrac{1 - x^2}{(x^2 + 1)^2} < 0$ for $x > 1$, f is also decreasing. Using the Integral Test, $\displaystyle\int_1^{\infty} \dfrac{x}{x^2 + 1}\, dx = \lim_{t \to \infty} \left[\dfrac{\ln(x^2 + 1)}{2}\right]_1^t = \infty$, so the series diverges.

11. $f(x) = \dfrac{1}{x \ln x}$ is continuous and positive on $[2, \infty)$, and also decreasing since
$f'(x) = -\dfrac{1 + \ln x}{x^2 (\ln x)^2} < 0$ for $x > 2$, so we can use the Integral Test.
$\displaystyle\int_2^{\infty} \dfrac{1}{x \ln x}\, dx = \lim_{t \to \infty} [\ln(\ln x)]_2^t = \lim_{t \to \infty} [\ln(\ln t) - \ln(\ln 2)] = \infty$, so the series diverges.

13. $\dfrac{1}{n^3 + n^2} < \dfrac{1}{n^3}$ since $n^3 + n^2 > n^3$ for all n, and since $\displaystyle\sum_{n=1}^{\infty} \dfrac{1}{n^3}$ is a convergent p-series ($p = 3 > 1$),

$\displaystyle\sum_{n=1}^{\infty} \dfrac{1}{n^3 + n^2}$ also converges by the Comparison Test [part (a)].

15. $\dfrac{1+5^n}{4^n} > \dfrac{5^n}{4^n} = \left(\dfrac{5}{4}\right)^n$. $\displaystyle\sum_{n=0}^{\infty}\left(\dfrac{5}{4}\right)^n$ is a divergent geometric series ($|r| = \tfrac{5}{4} > 1$), so $\displaystyle\sum_{n=0}^{\infty}\dfrac{1+5^n}{4^n}$ diverges by the Comparison Test.

17. $\dfrac{3}{n\,(n+3)} < \dfrac{3}{n^2}$. $\displaystyle\sum_{n=1}^{\infty}\dfrac{3}{n^2} = 3\sum_{n=1}^{\infty}\dfrac{1}{n^2}$ is a convergent p- series ($p = 2 > 1$), so $\displaystyle\sum_{n=1}^{\infty}\dfrac{3}{n\,(n+3)}$ converges by the Comparison Test.

19. Use the Limit Comparison Test with $a_n = \dfrac{1}{1+\sqrt{n}}$ and $b_n = \dfrac{1}{\sqrt{n}}$: $\displaystyle\lim_{n\to\infty}\dfrac{a_n}{b_n} = \lim_{n\to\infty}\dfrac{\sqrt{n}}{1+\sqrt{n}} = 1 > 0$.

Since $\displaystyle\sum_{n=1}^{\infty}\dfrac{1}{\sqrt{n}}$ is a divergent p- series ($p = \tfrac{1}{2} \le 1$), $\displaystyle\sum_{n=1}^{\infty}\dfrac{1}{1+\sqrt{n}}$ also diverges.

21. Use the Limit Comparison Test with $a_n = \sin(1/n)$ and $b_n = 1/n$:

$\displaystyle\lim_{n\to\infty}\dfrac{a_n}{b_n} = \lim_{n\to\infty}\dfrac{\sin(1/n)}{1/n} = \lim_{\theta\to 0}\dfrac{\sin\theta}{\theta} = 1 > 0$. Since $\displaystyle\sum_{n=1}^{\infty} b_n$ is the divergent harmonic series,

$\displaystyle\sum_{n=1}^{\infty}\sin(1/n)$ also diverges.

23. We have already shown (in Exercise 11) that when $p = 1$ the series $\displaystyle\sum_{n=2}^{\infty}\dfrac{1}{n\,(\ln n)^p}$ diverges, so assume

$p \neq 1$. $f(x) = \dfrac{1}{x\,(\ln x)^p}$ is continuous and positive on $[2,\infty)$, and $f'(x) = -\dfrac{p+\ln x}{x^2\,(\ln x)^{p+1}} < 0$ if

$x > e^{-p}$, so that f is eventually decreasing and we can use the Integral Test.

$\displaystyle\int_2^{\infty}\dfrac{1}{x\,(\ln x)^p}\,dx = \lim_{t\to\infty}\left[\dfrac{(\ln x)^{1-p}}{1-p}\right]_2^t$ (for $p \neq 1$) $= \displaystyle\lim_{t\to\infty}\left[\dfrac{(\ln t)^{1-p}}{1-p}\right] - \dfrac{(\ln 2)^{1-p}}{1-p}$.

This limit exists whenever $1 - p < 0 \iff p > 1$, so the series converges for $p > 1$.

25. (a) $f(x) = \dfrac{1}{x^2}$ is positive and continuous and $f'(x) = \dfrac{-2}{x^3}$ is negative for $x > 1$, and so the Integral

Test applies. $\displaystyle\sum_{n=1}^{\infty}\dfrac{1}{n^2} \approx s_{10} = \dfrac{1}{1^2} + \dfrac{1}{2^2} + \dfrac{1}{3^2} + \cdots + \dfrac{1}{10^2} \approx 1.549768$.

$R_{10} \le \displaystyle\int_{10}^{\infty}\dfrac{1}{x^2}\,dx = \lim_{t\to\infty}\left[\dfrac{-1}{x}\right]_{10}^t = \lim_{t\to\infty}\left(-\dfrac{1}{t} + \dfrac{1}{10}\right) = \dfrac{1}{10}$, so the error is at most 0.1.

(b) $s_{10} + \displaystyle\int_{11}^{\infty}\dfrac{1}{x^2}\,dx \le s \le s_{10} + \int_{10}^{\infty}\dfrac{1}{x^2}\,dx \implies s_{10} + \dfrac{1}{11} \le s \le s_{10} + \dfrac{1}{10} \implies$

$1.549768 + 0.090909 = 1.640677 \le s \le 1.549768 + 0.1 = 1.649768$, so we get $s \approx 1.64522$ (the average of 1.640677 and 1.649768) with error ≤ 0.005 (the maximum of $1.649768 - 1.64522$ and $1.64522 - 1.640677$, rounded up).

(c) $R_n \le \displaystyle\int_n^{\infty}\dfrac{1}{x^2}\,dx = \dfrac{1}{n}$. So $R_n < 0.001$ if $\dfrac{1}{n} < \dfrac{1}{1000} \iff n > 1000$.

27. $f(x) = x^{-3/2}$ is positive and continuous and $f'(x) = -\frac{3}{2}x^{-5/2}$ is negative for $x > 1$, so the Integral Test applies. From the end of Example 7, we see that the error is at most half the length of the interval. From (4), the interval is $\left(s_n + \int_{n+1}^{\infty} f(x)\,dx, s_n + \int_n^{\infty} f(x)\,dx\right)$, so its length is $\int_n^{\infty} f(x)\,dx - \int_{n+1}^{\infty} f(x)\,dx$. Thus, we need n such that

$$0.01 > \tfrac{1}{2}\left(\int_n^{\infty} x^{-3/2}\,dx - \int_{n+1}^{\infty} x^{-3/2}\,dx\right) = \frac{1}{2}\left(\lim_{t\to\infty}\left[\frac{-2}{\sqrt{x}}\right]_n^t - \lim_{t\to\infty}\left[\frac{-2}{\sqrt{x}}\right]_{n+1}^t\right)$$

$$= \frac{1}{\sqrt{n}} - \frac{1}{\sqrt{n+1}} \quad \Leftrightarrow \quad n > 13.08$$

Again from the end of Example 7, we approximate s by the midpoint of this interval. In general, the midpoint is $\tfrac{1}{2}\left[\left(s_n + \int_{n+1}^{\infty} f(x)\,dx\right) + \left(s_n + \int_n^{\infty} f(x)\,dx\right)\right] = s_n + \tfrac{1}{2}\left(\int_{n+1}^{\infty} f(x)\,dx + \int_n^{\infty} f(x)\,dx\right)$. So using $n = 14$, we have $s \approx s_{14} + \tfrac{1}{2}\left(\int_{14}^{\infty} x^{-3/2}\,dx + \int_{15}^{\infty} x^{-3/2}\,dx\right) = 2.0872 + \frac{1}{\sqrt{14}} + \frac{1}{\sqrt{15}} \approx 2.6127$. Any larger value of n will also work. For instance, $s \approx s_{30} + \frac{1}{\sqrt{30}} + \frac{1}{\sqrt{31}} \approx 2.6124$.

29. $\displaystyle\sum_{n=1}^{10} \frac{1}{n^4 + n^2} = \frac{1}{2} + \frac{1}{20} + \frac{1}{90} + \cdots + \frac{1}{10{,}100} \approx 0.567975$. Now $\dfrac{1}{n^4 + n^2} < \dfrac{1}{n^4}$, so using the reasoning and notation of Example 7, the error is

$$R_{10} \le T_{10} = \sum_{n=11}^{\infty} \frac{1}{n^4} \le \int_{10}^{\infty} \frac{dx}{x^4} = \lim_{t\to\infty}\left[-\frac{x^{-3}}{3}\right]_{10}^t = \frac{1}{3000} = 0.000\overline{3}.$$

31. (a) From the figure, $a_2 + a_3 + \cdots + a_n \le \int_1^n f(x)\,dx$, so with

$$f(x) = \frac{1}{x}, \quad \frac{1}{2} + \frac{1}{3} + \frac{1}{4} + \cdots + \frac{1}{n} \le \int_1^n \frac{1}{x}\,dx = \ln n.$$

Thus, $s_n = 1 + \dfrac{1}{2} + \dfrac{1}{3} + \dfrac{1}{4} + \cdots + \dfrac{1}{n} \le 1 + \ln n$.

(b) By part (a), $s_{10^6} \le 1 + \ln 10^6 \approx 14.82 < 15$ and $s_{10^9} \le 1 + \ln 10^9 \approx 21.72 < 22$.

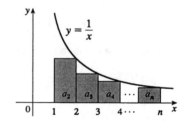

33. Since $\dfrac{d_n}{10^n} \le \dfrac{9}{10^n}$ for each n, and since $\displaystyle\sum_{n=1}^{\infty} \frac{9}{10^n}$ is a convergent geometric series ($|r| = \frac{1}{10} < 1$),

$$0.d_1 d_2 d_3 \ldots = \sum_{n=1}^{\infty} \frac{d_n}{10^n}$$ will always converge by the Comparison Test.

35. Yes. Since $\sum a_n$ converges, its terms approach 0 as $n \to \infty$, so $\displaystyle\lim_{n\to\infty} \frac{\sin a_n}{a_n} = 1$ by Theorem 3.4.2. Thus, $\sum \sin a_n$ converges by the Limit Comparison Test.

Section 8.4 Other Convergence Tests

1. (a) An alternating series is a series whose terms are alternately positive and negative.

(b) An alternating series $\sum_{n=1}^{\infty} (-1)^{n-1} b_n$ converges if $0 < b_{n+1} \leq b_n$ for all n and $\lim_{n\to\infty} b_n = 0$. (This is the Alternating Series Test.)

(c) The error involved in using the partial sum s_n as an approximation to the total sum s is the remainder $R_n = s - s_n$ and the size of the error is smaller than b_{n+1}, that is, $|R_n| \leq b_{n+1}$. (This is the Alternating Series Estimation Theorem.)

3. $\sum_{n=1}^{\infty} (-1)^{n-1} \dfrac{3}{n+4}$. $b_n = \dfrac{3}{n+4} > 0$ and $b_{n+1} \leq b_n$ for all n; $\lim_{n\to\infty} b_n = 0$, so the series converges by the Alternating Series Test.

5. $\sum_{n=1}^{\infty} (-1)^{n+1} \dfrac{n}{5n+1}$. $\lim_{n\to\infty} \dfrac{n}{5n+1} = \dfrac{1}{5}$, so $\lim_{n\to\infty} (-1)^{n+1} \dfrac{n}{5n+1}$ does not exist and the series diverges by the Test for Divergence.

7. $\sum_{n=1}^{\infty} (-1)^n \dfrac{n}{n^2+1}$. $b_n = \dfrac{n}{n^2+1} > 0$ for all n. $b_{n+1} \leq b_n \Leftrightarrow \dfrac{n+1}{(n+1)^2+1} \leq \dfrac{n}{n^2+1} \Leftrightarrow$

$(n+1)(n^2+1) \leq \left[(n+1)^2 + 1\right] n \Leftrightarrow n^3 + n^2 + n + 1 \leq n^3 + 2n^2 + 2n \Leftrightarrow 0 \leq n^2 + n - 1$,

which is true for all $n \geq 1$. Also, $\lim_{n\to\infty} b_n = \lim_{n\to\infty} \dfrac{n}{n^2+1} = \lim_{n\to\infty} \dfrac{1/n}{1 + 1/n^2} = 0$. Therefore, the series converges by the Alternating Series Test.

9. $\sum_{n=1}^{\infty} \dfrac{(-1)^{n-1}}{n} = 1 - \dfrac{1}{2} + \dfrac{1}{3} - \dfrac{1}{4} + \cdots + \dfrac{1}{49} - \dfrac{1}{50} + \dfrac{1}{51} - \dfrac{1}{52} + \cdots$. The 50th partial sum of this series

is an underestimate, since $\sum_{n=1}^{\infty} \dfrac{(-1)^{n-1}}{n} = s_{50} + \left(\dfrac{1}{51} - \dfrac{1}{52} \right) + \left(\dfrac{1}{53} - \dfrac{1}{54} \right) + \cdots$, and the terms in

parentheses are all positive. The result can be seen geometrically in Figure 1.

11. If $p > 0$, $\dfrac{1}{(n+1)^p} \leq \dfrac{1}{n^p}$ and $\lim_{n\to\infty} \dfrac{1}{n^p} = 0$, so the series converges by the Alternating Series Test.

If $p \leq 0$, $\lim_{n\to\infty} \dfrac{(-1)^{n-1}}{n^p}$ does not exist, so the series diverges by the Test for Divergence.

Thus, $\sum_{n=1}^{\infty} \dfrac{(-1)^{n-1}}{n^p}$ converges $\Leftrightarrow p > 0$.

13. $b_7 = 2^7/7! \approx 0.025 > 0.01$ and $b_8 = 2^8/8! \approx 0.006 < 0.01$, so by the Alternating Series Estimation Theorem, $n = 7$. (That is, since the 8th term is less than the desired error, we need to add the first 7 terms to get the sum to the desired accuracy.)

15.

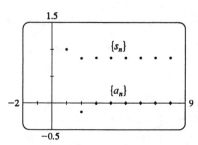

The graph gives us an estimate for the sum of the series $\sum_{n=1}^{\infty} \dfrac{(-1)^{n-1}}{(2n-1)!}$ of 0.84.

$b_5 = \dfrac{1}{(2 \cdot 5 - 1)!} = \dfrac{1}{362{,}880} \approx 0.000003 < 0.00001$, so $\sum_{n=1}^{\infty} \dfrac{(-1)^{n-1}}{(2n-1)!} \approx \sum_{n=1}^{4} \dfrac{(-1)^{n-1}}{(2n-1)!} \approx 0.8415.$

17. $b_6 = \dfrac{1}{2^6 6!} = \dfrac{1}{46{,}080} \approx 0.000022 < 0.00001$, so $\sum_{n=0}^{\infty} \dfrac{(-1)^n}{2^n n!} \approx \sum_{n=0}^{5} \dfrac{(-1)^n}{2^n n!} \approx 0.6065.$

19. Using the Ratio Test, $\lim\limits_{n \to \infty} \left| \dfrac{a_{n+1}}{a_n} \right| = \lim\limits_{n \to \infty} \left| \dfrac{(-3)^{n+1} / (n+1)^3}{(-3)^n / n^3} \right| = 3 \lim\limits_{n \to \infty} \left(\dfrac{n}{n+1} \right)^3 = 3 > 1$, so

the series diverges.

21. $\sum_{n=1}^{\infty} \dfrac{1}{2n+1}$ diverges (use the Integral Test or the Limit Comparison Test with $b_n = 1/n$).

23. $\left| \dfrac{\sin 2n}{n^2} \right| \le \dfrac{1}{n^2}$ and $\sum_{n=1}^{\infty} \dfrac{1}{n^2}$ converges (p-series, $p = 2 > 1$), so $\sum_{n=1}^{\infty} \dfrac{\sin 2n}{n^2}$ converges absolutely by the

Comparison Test.

25. $\lim\limits_{n \to \infty} \left| \dfrac{a_{n+1}}{a_n} \right| = \lim\limits_{n \to \infty} \dfrac{(n+2) \, 5^{n+1} / \left[(n+1) \, 3^{2(n+1)} \right]}{(n+1) \, 5^n / (n 3^{2n})} = \lim\limits_{n \to \infty} \dfrac{5n \, (n+2)}{9 \, (n+1)^2} = \dfrac{5}{9} < 1$, so the series

converges absolutely by the Ratio Test.

27. $\lim\limits_{n \to \infty} \left| \dfrac{a_{n+1}}{a_n} \right| = \lim\limits_{n \to \infty} \dfrac{(n+3)! / \left[(n+1)! 10^{n+1} \right]}{(n+2)! / (n! 10^n)} = \dfrac{1}{10} \lim\limits_{n \to \infty} \dfrac{n+3}{n+1} = \dfrac{1}{10} < 1$, so the series

converges absolutely by the Ratio Test.

29. By the recursive definition, $\lim\limits_{n \to \infty} \left| \dfrac{a_{n+1}}{a_n} \right| = \lim\limits_{n \to \infty} \left| \dfrac{5n+1}{4n+3} \right| = \dfrac{5}{4} > 1$, so the series diverges by the Ratio

Test.

31. (a) $\lim\limits_{n\to\infty} \left| \dfrac{1/(n+1)^3}{1/n^3} \right| = \lim\limits_{n\to\infty} \dfrac{n^3}{(n+1)^3} = \lim\limits_{n\to\infty} \dfrac{1}{(1+1/n)^3} = 1.$ Inconclusive.

(b) $\lim\limits_{n\to\infty} \left| \dfrac{(n+1)}{2^{n+1}} \cdot \dfrac{2^n}{n} \right| = \lim\limits_{n\to\infty} \dfrac{n+1}{2n} = \lim\limits_{n\to\infty} \left(\dfrac{1}{2} + \dfrac{1}{2n} \right) = \dfrac{1}{2}.$ Conclusive (convergent).

(c) $\lim\limits_{n\to\infty} \left| \dfrac{(-3)^n}{\sqrt{n+1}} \cdot \dfrac{\sqrt{n}}{(-3)^{n-1}} \right| = 3 \lim\limits_{n\to\infty} \sqrt{\dfrac{n}{n+1}} = 3 \lim\limits_{n\to\infty} \sqrt{\dfrac{1}{1+1/n}} = 3.$ Conclusive (divergent).

(d) $\lim\limits_{n\to\infty} \left| \dfrac{\sqrt{n+1}}{1+(n+1)^2} \cdot \dfrac{1+n^2}{\sqrt{n}} \right| = \lim\limits_{n\to\infty} \left[\sqrt{1+\dfrac{1}{n}} \cdot \dfrac{1/n^2+1}{1/n^2+(1+1/n)^2} \right] = 1.$ Inconclusive.

33. (a) $\lim\limits_{n\to\infty} \left| \dfrac{a_{n+1}}{a_n} \right| = \lim\limits_{n\to\infty} \dfrac{|x|^{n+1}/(n+1)!}{|x|^n/n!} = |x| \lim\limits_{n\to\infty} \dfrac{1}{n+1} = 0,$ so by the Ratio Test the series converges for all x.

(b) Since the series of part (a) always converges, we must have $\lim\limits_{n\to\infty} \dfrac{x^n}{n!} = 0$ by Theorem 8.2.6.

35. (a) $s_5 = \sum\limits_{n=1}^{5} \dfrac{1}{n2^n} = \dfrac{1}{2} + \dfrac{1}{8} + \dfrac{1}{24} + \dfrac{1}{64} + \dfrac{1}{160} = \dfrac{661}{960} \approx 0.68854.$ Now the ratios

$r_n = \dfrac{a_{n+1}}{a_n} = \dfrac{n2^n}{(n+1)\,2^{n+1}} = \dfrac{n}{2\,(n+1)}$ form an increasing sequence, since

$r_{n+1} - r_n = \dfrac{n+1}{2\,(n+2)} - \dfrac{n}{2\,(n+1)} = \dfrac{(n+1)^2 - n\,(n+2)}{2\,(n+1)\,(n+2)} = \dfrac{1}{2\,(n+1)\,(n+2)} > 0.$ So by

Exercise 34(b), the error is less than $\dfrac{a_6}{1 - \lim\limits_{n\to\infty} r_n} = \dfrac{1/(6 \cdot 2^6)}{1 - 1/2} = \dfrac{1}{192} \approx 0.00521.$

(b) The error in using s_n as an approximation to the sum is $R_n = \dfrac{a_{n+1}}{1 - \frac{1}{2}} = \dfrac{2}{(n+1)\,2^{n+1}}.$ We want

$R_n < 0.00005 \quad\Leftrightarrow\quad \dfrac{1}{(n+1)\,2^n} < 0.00005 \quad\Leftrightarrow\quad (n+1)\,2^n > 20{,}000.$ To find such an n we can

use trial and error or a graph. We calculate $(11+1)\,2^{11} = 24{,}576,$ so $s_{11} = \sum\limits_{n=1}^{11} \dfrac{1}{n2^n} \approx 0.693109$

is within 0.00005 of the actual sum.

Section 8.5 Power Series

Note: "R" stands for "radius of convergence" and "I" stands for "interval of convergence" in this section.

1. A power series is a series of the form $\sum_{n=0}^{\infty} c_n x^n = c_0 + c_1 x + c_2 x^2 + c_3 x^3 + \cdots$, where x is a variable and the c_n's are constants called the coefficients of the series.

More generally, a series of the form $\sum_{n=0}^{\infty} c_n (x - a)^n = c_0 + c_1 (x - a) + c_2 (x - a)^2 + \cdots$ is called a power series in $(x - a)$ or a power series centered at a or a power series about a.

3. (a) We are given that the power series $\sum_{n=0}^{\infty} c_n x^n$ is convergent for $x = 4$. So by Theorem 3 it must converge for at least $-4 < x \leq 4$. In particular, it converges when $x = -2$, that is, $\sum_{n=0}^{\infty} c_n (-2)^n$ is convergent.

(b) It does not follow that $\sum_{n=0}^{\infty} c_n (-4)^n$ is necessarily convergent. [See the comments after Theorem 3 about convergence at the endpoint of an interval. An example is $c_n = (-1)^n / (n4^n)$.]

5. If $a_n = \dfrac{x^n}{n+2}$, then $\lim_{n \to \infty} \left| \dfrac{a_{n+1}}{a_n} \right| = \lim_{n \to \infty} \left| \dfrac{x^{n+1}}{n+3} \cdot \dfrac{n+2}{x^n} \right| = |x| \lim_{n \to \infty} \dfrac{n+2}{n+3} = |x| < 1$ for convergence (by the Ratio Test), and $R = 1$. When $x = 1$, the series is $\sum_{n=0}^{\infty} \dfrac{1}{n+2}$ which diverges (Integral Test or Comparison Test), and when $x = -1$, it is $\sum_{n=0}^{\infty} \dfrac{(-1)^n}{n+2}$ which converges (Alternating Series Test), so $I = [-1, 1)$.

7. If $a_n = \dfrac{x^n}{n!}$, then $\lim_{n \to \infty} \left| \dfrac{a_{n+1}}{a_n} \right| = \lim_{n \to \infty} \left| \dfrac{x^{n+1} / (n+1)!}{x^n / n!} \right| = |x| \lim_{n \to \infty} \dfrac{1}{n+1} = 0 < 1$ for all x. So, by the Ratio Test, $R = \infty$, and $I = (-\infty, \infty)$.

9. If $a_n = \dfrac{(-1)^n x^n}{n2^n}$, then $\lim_{n \to \infty} \left| \dfrac{a_{n+1}}{a_n} \right| = \lim_{n \to \infty} \left| \dfrac{x^{n+1} / [(n+1) 2^{n+1}]}{x^n / (n2^n)} \right| = \left| \dfrac{x}{2} \right| \lim_{n \to \infty} \dfrac{n}{n+1} = \left| \dfrac{x}{2} \right| < 1$ for convergence, so $|x| < 2$ and $R = 2$. When $x = 2$, $\sum_{n=1}^{\infty} \dfrac{(-1)^n x^n}{n2^n} = \sum_{n=1}^{\infty} \dfrac{(-1)^n}{n}$, which converges by the Alternating Series Test. When $x = -2$, $\sum_{n=1}^{\infty} \dfrac{(-1)^n x^n}{n2^n} = \sum_{n=1}^{\infty} \dfrac{1}{n}$, which diverges (harmonic series), so $I = (-2, 2]$.

11. If $a_n = \dfrac{n}{4^n}(2x-1)^n$, then

$$\left|\frac{a_{n+1}}{a_n}\right| = \left|\frac{(n+1)(2x-1)^{n+1}}{4^{n+1}}\cdot\frac{4^n}{n(2x-1)^n}\right| = \left|\frac{2x-1}{4}\left(1+\frac{1}{n}\right)\right| \to \tfrac{1}{2}\left|x-\tfrac{1}{2}\right| \text{ as } n\to\infty. \text{ For}$$

convergence, $\tfrac{1}{2}\left|x-\tfrac{1}{2}\right|<1 \Rightarrow \left|x-\tfrac{1}{2}\right|<2 \Rightarrow R=2$ and $-2<x-\tfrac{1}{2}<2 \Rightarrow$

$-\tfrac{3}{2}<x<\tfrac{5}{2}$. If $x=-\tfrac{3}{2}$, the series becomes $\displaystyle\sum_{n=0}^{\infty}\frac{n}{4^n}(-4)^n = \sum_{n=0}^{\infty}(-1)^n n$, which is divergent by the

Test for Divergence. If $x=\tfrac{5}{2}$, the series is $\displaystyle\sum_{n=0}^{\infty}\frac{n}{4^n}4^n = \sum_{n=0}^{\infty}n$, also divergent by the Test for

Divergence. So $I=\left(-\tfrac{3}{2},\tfrac{5}{2}\right)$.

13. If $a_n = \dfrac{(-1)^n(x-1)^n}{\sqrt{n}}$, then

$$\lim_{n\to\infty}\left|\frac{a_{n+1}}{a_n}\right| = \lim_{n\to\infty}\left|\frac{(x-1)^{n+1}}{\sqrt{n+1}}\cdot\frac{\sqrt{n}}{(x-1)^n}\right| = |x-1|\lim_{n\to\infty}\sqrt{\frac{n}{n+1}}$$

$$= |x-1|<1 \text{ for convergence, or } 0<x<2, \text{ and } R=1.$$

When $x=0$, $\displaystyle\sum_{n=1}^{\infty}\frac{(-1)^n(x-1)^n}{\sqrt{n}} = \sum_{n=1}^{\infty}\frac{1}{\sqrt{n}}$ which is a divergent p-series ($p=\tfrac{1}{2}\le 1$). When $x=2$,

the series is $\displaystyle\sum_{n=1}^{\infty}\frac{(-1)^n}{\sqrt{n}}$, which converges by the Alternating Series Test. So $I=(0,2]$.

15. If $a_n = \dfrac{2^n(x-3)^n}{n+3}$, then

$$\lim_{n\to\infty}\left|\frac{a_{n+1}}{a_n}\right| = \lim_{n\to\infty}\left|\frac{2^{n+1}(x-3)^{n+1}}{n+4}\cdot\frac{n+3}{2^n(x-3)^n}\right| = 2|x-3|\lim_{n\to\infty}\frac{n+3}{n+4}$$

$$= 2|x-3|<1 \text{ for convergence, or}$$

$|x-3|<\tfrac{1}{2} \Leftrightarrow \tfrac{5}{2}<x<\tfrac{7}{2}$, and $R=\tfrac{1}{2}$. When $x=\tfrac{5}{2}$, $\displaystyle\sum_{n=0}^{\infty}\frac{2^n(x-3)^n}{n+3} = \sum_{n=0}^{\infty}\frac{(-1)^n}{n+3}$ which

converges by the Alternating Series Test. When $x=\tfrac{7}{2}$, $\displaystyle\sum_{n=0}^{\infty}\frac{2^n(x-3)^n}{n+3} = \sum_{n=0}^{\infty}\frac{1}{n+3}$, similar to the

harmonic series, which diverges. So $I=\left[\tfrac{5}{2},\tfrac{7}{2}\right)$.

17. If $a_n = n!(2x-1)^n$, then

$$\lim_{n\to\infty}\left|\frac{a_{n+1}}{a_n}\right| = \lim_{n\to\infty}\left|\frac{(n+1)!(2x-1)^{n+1}}{n!(2x-1)^n}\right| = \lim_{n\to\infty}(n+1)|2x-1| \to \infty \text{ as } n\to\infty \text{ for all}$$

$x\ne\tfrac{1}{2}$. Since the series diverges for all $x\ne\tfrac{1}{2}$, $R=0$ and $I=\left\{\tfrac{1}{2}\right\}$.

19. If $a_n = \dfrac{(n!)^k}{(kn)!} x^n$, then

$$\lim_{n\to\infty}\left|\frac{a_{n+1}}{a_n}\right| = \lim_{n\to\infty}\frac{[(n+1)!]^k\,(kn)!}{(n!)^k\,[k\,(n+1)]!}\,|x| = \lim_{n\to\infty}\frac{(n+1)^k}{(kn+k)\,(kn+k-1)\cdots(kn+2)\,(kn+1)}\,|x|$$

$$= \lim_{n\to\infty}\left[\frac{(n+1)}{(kn+1)}\frac{(n+1)}{(kn+2)}\cdots\frac{(n+1)}{(kn+k)}\right]|x|$$

$$= \lim_{n\to\infty}\left[\frac{n+1}{kn+1}\right]\lim_{n\to\infty}\left[\frac{n+1}{kn+2}\right]\cdots\lim_{n\to\infty}\left[\frac{n+1}{kn+k}\right]|x| = \left(\frac{1}{k}\right)^k |x| < 1$$

$\Leftrightarrow$ $|x| < k^k$ for convergence, and the radius of convergence is $R = k^k$.

21. (a) If $a_n = \dfrac{(-1)^n\,x^{2n+1}}{n!\,(n+1)!2^{2n+1}}$, then $\lim\limits_{n\to\infty}\left|\dfrac{a_{n+1}}{a_n}\right| = \left(\dfrac{x}{2}\right)^2\lim\limits_{n\to\infty}\dfrac{1}{(n+1)\,(n+2)} = 0$ for all x. So
$J_1(x)$ converges for all x; the domain is $(-\infty,\infty)$.

(b),(c) The initial terms of $J_1(x)$ up to $n=5$ are

$$a_0 = \frac{x}{2},\ a_1 = -\frac{x^3}{16},\ a_2 = \frac{x^5}{384},$$

$$a_3 = -\frac{x^7}{18,432},\ a_4 = \frac{x^9}{1,474,560},\text{ and}$$

$$a_5 = -\frac{x^{11}}{176,947,200}.$$

The partial sums seem to approximate
$J_1(x)$ well near the origin, but as $|x|$
increases, we need to take a large number
of terms to get a good approximation.

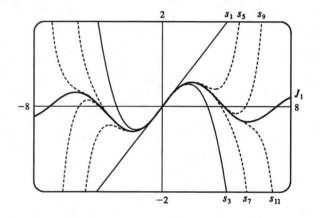

23. $s_{2n-1} = 1 + 2x + x^2 + 2x^3 + \cdots + x^{2n-2} + 2x^{2n-1} = (1+2x)\left(1 + x^2 + x^4 + \cdots + x^{2n-2}\right)$

$$= (1+2x)\frac{1-x^{2n}}{1-x^2}\quad\text{[by (8.2.3) with }r = x^2]\ \to\ \frac{1+2x}{1-x^2}\text{ as }n\to\infty\text{ [by (8.2.4)]},$$

when $|x| < 1$. Also $s_{2n} = s_{2n-1} + x^{2n} \to \dfrac{1+2x}{1-x^2}$ since $x^{2n} \to 0$ for $|x| < 1$. Therefore,

$s_n \to \dfrac{1+2x}{1-x^2}$ since s_{2n} and s_{2n-1} both approach $\dfrac{1+2x}{1-x^2}$ as $n\to\infty$. Thus, the interval of convergence

is $(-1,1)$ and $f(x) = \dfrac{1+2x}{1-x^2}$.

25. $\sum (c_n + d_n)\,x^n = \sum c_n x^n + \sum d_n x^n$ on the interval $(-2,2)$, since both series converge there. So the
radius of convergence must be at least 2. Now since $\sum c_n x^n$ has $R = 2$, it must diverge either at
$x = -2$ or at $x = 2$. So by Exercise 8.2.49, $\sum (c_n + d_n)\,x^n$ diverges either at $x = -2$ or at $x = 2$, and
so its radius of convergence is 2.

Section 8.6   Representations of Functions as Power Series

Note: "R" stands for "radius of convergence" and "I" stands for "interval of convergence" in this section.

1. If $f(x) = \sum\limits_{n=0}^{\infty} c_n x^n$ has radius of convergence 10, then $f'(x) = \sum\limits_{n=1}^{\infty} n c_n x^{n-1}$ also has radius of convergence 10 by Theorem 2.

3. $f(x) = \dfrac{1}{1+x} = \dfrac{1}{1-(-x)} = \sum\limits_{n=0}^{\infty} (-x)^n = \sum\limits_{n=0}^{\infty} (-1)^n x^n$ with $|-x| < 1 \quad\Leftrightarrow\quad |x| < 1$, so $R = 1$

and $I = (-1, 1)$.

5. $f(x) = \dfrac{1}{1+4x^2} = \sum\limits_{n=0}^{\infty} (-1)^n \left(4x^2\right)^n$ (substituting $4x^2$ for x in the series from Exercise 3)

$= \sum\limits_{n=0}^{\infty} (-1)^n 4^n x^{2n}$, with $\left|4x^2\right| < 1 \quad\Leftrightarrow\quad x^2 < \frac{1}{4} \quad\Leftrightarrow\quad |x| < \frac{1}{2}$, so $R = \frac{1}{2}$ and $I = \left(-\frac{1}{2}, \frac{1}{2}\right)$.

7. $f(x) = \dfrac{x}{x-3} = 1 + \dfrac{3}{x-3} = 1 - \dfrac{1}{1-x/3} = 1 - \sum\limits_{n=0}^{\infty} \left(\dfrac{x}{3}\right)^n = -\sum\limits_{n=1}^{\infty} \left(\dfrac{x}{3}\right)^n$ (since the 0th term

of the series is 1). For convergence, $\dfrac{|x|}{3} < 1 \quad\Leftrightarrow\quad |x| < 3$, so $R = 3$ and $I = (-3, 3)$.

Another Method: $\dfrac{x}{x-3} = -\dfrac{x}{3(1-x/3)} = -\dfrac{x}{3} \sum\limits_{n=0}^{\infty} \left(\dfrac{x}{3}\right)^n = -\sum\limits_{n=0}^{\infty} \dfrac{x^{n+1}}{3^{n+1}} = -\sum\limits_{n=1}^{\infty} \dfrac{x^n}{3^n}$

9. $f(x) = \dfrac{1}{(1+x)^2} = -\dfrac{d}{dx}\left(\dfrac{1}{1+x}\right) = -\dfrac{d}{dx}\left(\sum\limits_{n=0}^{\infty} (-1)^n x^n\right)$ (from Exercise 3)

$= \sum\limits_{n=1}^{\infty} (-1)^{n+1} n x^{n-1}$ [from Theorem 2(a)] $= \sum\limits_{n=0}^{\infty} (-1)^n (n+1) x^n$ with $R = 1$.

11. $f(x) = \dfrac{1}{(1+x)^3} = -\dfrac{1}{2}\dfrac{d}{dx}\left[\dfrac{1}{(1+x)^2}\right] = -\dfrac{1}{2}\dfrac{d}{dx}\left[\sum\limits_{n=0}^{\infty} (-1)^n (n+1) x^n\right]$ (from Exercise 9)

$= -\dfrac{1}{2} \sum\limits_{n=1}^{\infty} (-1)^n (n+1) n x^{n-1} = \dfrac{1}{2} \sum\limits_{n=0}^{\infty} (-1)^n (n+2)(n+1) x^n$ with $R = 1$.

13. $f(x) = \ln(5-x) = -\displaystyle\int \dfrac{dx}{5-x} = -\dfrac{1}{5}\displaystyle\int \dfrac{dx}{1-x/5}$

$= -\dfrac{1}{5}\displaystyle\int \left[\sum\limits_{n=0}^{\infty} \left(\dfrac{x}{5}\right)^n\right] dx = C - \dfrac{1}{5} \sum\limits_{n=0}^{\infty} \dfrac{x^{n+1}}{5^n (n+1)} = C - \sum\limits_{n=1}^{\infty} \dfrac{x^n}{n5^n}$

Putting $x = 0$, we get $C = \ln 5$. The series converges for $|x/5| < 1 \quad\Leftrightarrow\quad |x| < 5$, so $R = 5$.

15. $f(x) = \ln(3+x) = \int \dfrac{dx}{3+x} = \dfrac{1}{3}\int \dfrac{dx}{1+x/3} = \dfrac{1}{3}\int \sum_{n=0}^{\infty} (-1)^n \left(\dfrac{x}{3}\right)^n dx$ (from Exercise 3)

$= C + \dfrac{1}{3}\sum_{n=0}^{\infty} \dfrac{(-1/3)^n}{n+1} x^{n+1} = \ln 3 + \dfrac{1}{3}\sum_{n=1}^{\infty} \dfrac{(-1/3)^{n-1}}{n} x^n$ $[C = f(0) = \ln 3]$

$= \ln 3 + \sum_{n=1}^{\infty} \dfrac{(-1)^{n-1}}{n3^n} x^n$ with $R = 3$.

The terms of the series are $a_0 = \ln 3$, $a_1 = \dfrac{x}{3}$, $a_2 = -\dfrac{x^2}{18}$, $a_3 = \dfrac{x^3}{81}$, $a_4 = -\dfrac{x^4}{324}$, $a_5 = \dfrac{x^5}{1215}$, ...

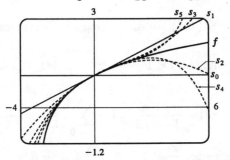

As n increases, $s_n(x)$ approximates f better on the interval of convergence, which is $(-3, 3)$.

17. $f(x) = \ln\left(\dfrac{1+x}{1-x}\right) = \ln(1+x) - \ln(1-x) = \int \dfrac{dx}{1+x} + \int \dfrac{dx}{1-x}$

$= \int \left[\sum_{n=0}^{\infty} (-1)^n x^n + \sum_{n=0}^{\infty} x^n\right] dx = \int \sum_{n=0}^{\infty} 2x^{2n} dx = \sum_{n=0}^{\infty} \dfrac{2x^{2n+1}}{2n+1} + C$

But $f(0) = \ln\frac{1}{1} = 0$, so $C = 0$ and we have $f(x) = \sum_{n=0}^{\infty} \dfrac{2x^{2n+1}}{2n+1}$ with $R = 1$. If $x = \pm 1$, then

$f(x) = \pm 2 \sum_{n=0}^{\infty} \dfrac{1}{2n+1}$, which both diverge by the Limit Comparison Test with $b_n = \dfrac{1}{n}$.

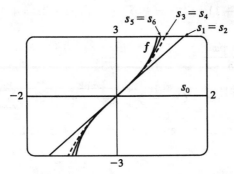

As n increases, $s_n(x)$ approximates f better on the interval of convergence, which is $(-1, 1)$.

19. $\displaystyle\int \frac{dx}{1+x^4} = \int \sum_{n=0}^{\infty} (-1)^n \, x^{4n} \, dx = C + \sum_{n=0}^{\infty} \frac{(-1)^n \, x^{4n+1}}{4n+1}$ with $R=1$.

21. By Example 7, $\displaystyle\arctan x = \sum_{n=0}^{\infty} (-1)^n \frac{x^{2n+1}}{2n+1}$, so

$$\int \frac{\arctan x}{x} \, dx = \int \sum_{n=0}^{\infty} (-1)^n \frac{x^{2n}}{2n+1} \, dx = C + \sum_{n=0}^{\infty} (-1)^n \frac{x^{2n+1}}{(2n+1)^2} \text{ with } R=1.$$

23. We use the representation $\displaystyle\int \frac{dx}{1+x^4} = C + \sum_{n=0}^{\infty} \frac{(-1)^n \, x^{4n+1}}{4n+1}$ from Exercise 19 with $C=0$.

So $\displaystyle\int_0^{0.2} \frac{dx}{1+x^4} = \left[x - \frac{x^5}{5} + \frac{x^9}{9} - \frac{x^{13}}{13} + \cdots \right]_0^{0.2} = 0.2 - \frac{0.2^5}{5} + \frac{0.2^9}{9} - \frac{0.2^{13}}{13} + \cdots$. Since the

series is alternating, the error in the n th-order approximation is less than the first neglected term, by The

Alternating Series Estimation Theorem. If we use only the first two terms of the series, then the error is

at most $0.2^9/9 \approx 5.7 \times 10^{-8}$. So, to six decimal places, $\displaystyle\int_0^{0.2} \frac{dx}{1+x^4} \approx 0.2 - \frac{0.2^5}{5} = 0.199936$.

25. We substitute x^4 for x in Example 7, and find that

$$\int x^2 \tan^{-1}\left(x^4\right) dx = \int x^2 \sum_{n=0}^{\infty} (-1)^n \frac{\left(x^4\right)^{2n+1}}{2n+1} \, dx$$

$$= \int \sum_{n=0}^{\infty} (-1)^n \frac{x^{8n+6}}{2n+1} \, dx = C + \sum_{n=0}^{\infty} (-1)^n \frac{x^{8n+7}}{(2n+1)(8n+7)}$$

So $\displaystyle\int_0^{1/3} x^2 \tan^{-1}\left(x^4\right) dx = \left[\frac{x^7}{7} - \frac{x^{15}}{45} + \cdots \right]_0^{1/3} = \frac{1}{7 \cdot 3^7} - \frac{1}{45 \cdot 3^{15}} + \cdots$. The series is alternating,

so if we use only one term, the error is at most $1/\left(45 \cdot 3^{15}\right) \approx 1.5 \times 10^{-9}$. So

$\displaystyle\int_0^{1/3} x^2 \tan^{-1}\left(x^4\right) dx \approx 1/\left(7 \cdot 3^7\right) \approx 0.000065$ to six decimal places.

27. Using the result of Example 6, $\displaystyle\ln(1-x) = -\sum_{n=1}^{\infty} \frac{x^n}{n}$, with $x = -0.1$, we have

$$\ln 1.1 = \ln[1-(-0.1)] = 0.1 - \frac{0.01}{2} + \frac{0.001}{3} - \frac{0.0001}{4} + \frac{0.00001}{5} - \cdots. \text{ The series is alternating, so}$$

if we use only the first four terms, the error is at most $\dfrac{0.00001}{5} = 0.000002$. So

$$\ln 1.1 \approx 0.1 - \frac{0.01}{2} + \frac{0.001}{3} - \frac{0.0001}{4} \approx 0.09531.$$

29. (a) $J_0(x) = \sum_{n=0}^{\infty} \dfrac{(-1)^n x^{2n}}{2^{2n}(n!)^2}$, $J_0'(x) = \sum_{n=1}^{\infty} \dfrac{(-1)^n 2n x^{2n-1}}{2^{2n}(n!)^2}$, and

$J_0''(x) = \sum_{n=1}^{\infty} \dfrac{(-1)^n 2n(2n-1) x^{2n-2}}{2^{2n}(n!)^2}$, so

$x^2 J_0''(x) + x J_0'(x) + x^2 J_0(x) = \sum_{n=1}^{\infty} \dfrac{(-1)^n 2n(2n-1) x^{2n}}{2^{2n}(n!)^2} + \sum_{n=1}^{\infty} \dfrac{(-1)^n 2n x^{2n}}{2^{2n}(n!)^2} + \sum_{n=0}^{\infty} \dfrac{(-1)^n x^{2n+2}}{2^{2n}(n!)^2}$

$= \sum_{n=1}^{\infty} \dfrac{(-1)^n 2n(2n-1) x^{2n}}{2^{2n}(n!)^2} + \sum_{n=1}^{\infty} \dfrac{(-1)^n 2n x^{2n}}{2^{2n}(n!)^2} + \sum_{n=1}^{\infty} \dfrac{(-1)^{n-1} x^{2n}}{2^{2n-2}[(n-1)!]^2}$

$= \sum_{n=1}^{\infty} (-1)^n \left[\dfrac{2n(2n-1) + 2n - 2^2 n^2}{2^{2n}(n!)^2} \right] x^{2n} = \sum_{n=1}^{\infty} (-1)^n \left[\dfrac{4n^2 - 2n + 2n - 4n^2}{2^{2n}(n!)^2} \right] x^{2n} = 0$

(b) $\displaystyle\int_0^1 J_0(x)\,dx = \int_0^1 \left[\sum_{n=0}^{\infty} \dfrac{(-1)^n x^{2n}}{2^{2n}(n!)^2} \right] dx = \int_0^1 \left(1 - \dfrac{x^2}{4} + \dfrac{x^4}{64} - \dfrac{x^6}{2304} + \cdots \right) dx$

$= \left[x - \dfrac{x^3}{3\cdot 4} + \dfrac{x^5}{5\cdot 64} - \dfrac{x^7}{7\cdot 2304} + \cdots \right]_0^1 = 1 - \dfrac{1}{12} + \dfrac{1}{320} - \dfrac{1}{16{,}128} + \cdots$

Since $\frac{1}{16{,}128} \approx 0.000062$, it follows from The Alternating Series Estimation Theorem that, correct to
three decimal places, $\int_0^1 J_0(x)\,dx \approx 1 - \frac{1}{12} + \frac{1}{320} \approx 0.920$.

31. (a) $f(x) = \sum_{n=0}^{\infty} \dfrac{x^n}{n!} \;\Rightarrow\; f'(x) = \sum_{n=1}^{\infty} \dfrac{n x^{n-1}}{n!} = \sum_{n=1}^{\infty} \dfrac{x^{n-1}}{(n-1)!} = \sum_{n=0}^{\infty} \dfrac{x^n}{n!} = f(x)$.

(b) By Theorem 7.5.2, the only solution to the differential equation $df(x)/dx = f(x)$ is $f(x) = Ke^x$,
but $f(0) = 1$, so $K = 1$ and $f(x) = e^x$.
Or: We could solve the equation $df(x)/dx = f(x)$ as a separable differential equation.

33. If $a_n = \dfrac{x^n}{n^2}$, then by the Ratio Test, $\lim\limits_{n\to\infty} \left| \dfrac{a_{n+1}}{a_n} \right| = |x| \lim\limits_{n\to\infty} \left(\dfrac{n}{n+1} \right)^2 = |x| < 1$ for convergence, so

$R = 1$. When $x = \pm 1$, $\sum_{n=1}^{\infty} \left| \dfrac{x^n}{n^2} \right| = \sum_{n=1}^{\infty} \dfrac{1}{n^2}$ which is a convergent p-series ($p = 2 > 1$), so the interval
of convergence for f is $[-1, 1]$. By Theorem 2, the radii of convergence of f' and f'' are both 1, so we

need only check the endpoints. $f(x) = \sum_{n=1}^{\infty} \dfrac{x^n}{n^2} \;\Rightarrow\; f'(x) = \sum_{n=1}^{\infty} \dfrac{n x^{n-1}}{n^2} = \sum_{n=0}^{\infty} \dfrac{x^n}{n+1}$, and this
series diverges for $x = 1$ (harmonic series) and converges for $x = -1$ (Alternating Series Test), so the

interval of convergence is $[-1, 1)$. $f''(x) = \sum_{n=1}^{\infty} \dfrac{n x^{n-1}}{n+1}$ diverges at both 1 and -1 (Test for

Divergence) since $\lim\limits_{n\to\infty} \dfrac{n}{n+1} = 1 \neq 0$, so its interval of convergence is $(-1, 1)$.

Section 8.7 Taylor and Maclaurin Series

1. Using Theorem 5 with $\sum\limits_{n=0}^{\infty} b_n (x-5)^n$, $b_n = \dfrac{f^{(n)}(a)}{n!}$, so $b_8 = \dfrac{f^{(8)}(5)}{8!}$.

3.

n	$f^{(n)}(x)$	$f^{(n)}(0)$
0	$\cos x$	1
1	$-\sin x$	0
2	$-\cos x$	-1
3	$\sin x$	0
4	$\cos x$	1
$\dots$	$\dots$	$\dots$

$$\cos x = f(0) + f'(0)\,x + \frac{f''(0)}{2!}\,x^2$$
$$+\frac{f^{(3)}(0)}{3!}\,x^3 + \frac{f^{(4)}(0)}{4!}\,x^4 + \cdots$$
$$= 1 - \frac{x^2}{2!} + \frac{x^4}{4!} - \cdots = \sum_{n=0}^{\infty} \frac{(-1)^n x^{2n}}{(2n)!}$$

If $a_n = \dfrac{(-1)^n x^{2n}}{(2n)!}$, then $\lim\limits_{n\to\infty}\left|\dfrac{a_{n+1}}{a_n}\right| = x^2 \lim\limits_{n\to\infty} \dfrac{1}{(2n+2)(2n+1)} = 0 < 1$ for all x. So $R = \infty$ (Ratio Test).

5.

n	$f^{(n)}(x)$	$f^{(n)}(0)$
0	$(1+x)^{-2}$	1
1	$-2(1+x)^{-3}$	-2
2	$2\cdot 3(1+x)^{-4}$	$2\cdot 3$
3	$-2\cdot 3\cdot 4(1+x)^{-5}$	$-2\cdot 3\cdot 4$
4	$2\cdot 3\cdot 4\cdot 5(1+x)^{-6}$	$2\cdot 3\cdot 4\cdot 5$
$\dots$	$\dots$	$\dots$

So $f^{(n)}(0) = (-1)^n (n+1)!$ and

$$\frac{1}{(1+x)^2} = \sum_{n=0}^{\infty} \frac{(-1)^n (n+1)!}{n!}\,x^n$$
$$= \sum_{n=0}^{\infty} (-1)^n (n+1)\,x^n$$

$a_n = (-1)^n (n+1)\,x^n$, then $\lim\limits_{n\to\infty}\left|\dfrac{a_{n+1}}{a_n}\right| = |x| < 1$ for convergence, so $R = 1$.

7. Clearly, $f^{(n)}(x) = e^x$, so $f^{(n)}(3) = e^3$ and $e^x = \sum\limits_{n=0}^{\infty} \dfrac{e^3}{n!}(x-3)^n$. If $a_n = \dfrac{e^3}{n!}(x-3)^n$, then

$$\lim_{n\to\infty}\left|\frac{a_{n+1}}{a_n}\right| = \lim_{n\to\infty}\frac{|x-3|}{n+1} = 0 \text{ for all } x, \text{ so } R = \infty.$$

9.

n	$f^{(n)}(x)$	$f^{(n)}(1)$
0	x^{-1}	1
1	$-x^{-2}$	-1
2	$2x^{-3}$	2
3	$-3\cdot 2x^{-4}$	$-3\cdot 2$
4	$4\cdot 3\cdot 2x^{-5}$	$4\cdot 3\cdot 2$
$\dots$	$\dots$	$\dots$

So $f^{(n)}(1) = (-1)^n n!$, and

$$\frac{1}{x} = \sum_{n=0}^{\infty} \frac{(-1)^n n!}{n!}(x-1)^n = \sum_{n=0}^{\infty} (-1)^n (x-1)^n.$$ If

$a_n = (-1)^n (x-1)^n$ then $\lim\limits_{n\to\infty}\left|\dfrac{a_{n+1}}{a_n}\right| = |x-1| < 1$ for convergence, so $0 < x < 2$ and $R = 1$.

11.

n	$f^{(n)}(x)$	$f^{(n)}\left(\frac{\pi}{4}\right)$
0	$\sin x$	$\sqrt{2}/2$
1	$\cos x$	$\sqrt{2}/2$
2	$-\sin x$	$-\sqrt{2}/2$
3	$-\cos x$	$-\sqrt{2}/2$
4	$\sin x$	$\sqrt{2}/2$
...	...	...

$$\sin x = f\left(\tfrac{\pi}{4}\right) + f'\left(\tfrac{\pi}{4}\right)\left(x - \tfrac{\pi}{4}\right) + \frac{f''\left(\tfrac{\pi}{4}\right)}{2!}\left(x - \tfrac{\pi}{4}\right)^2$$

$$+ \frac{f^{(3)}\left(\tfrac{\pi}{4}\right)}{3!}\left(x - \tfrac{\pi}{4}\right)^3 + \frac{f^{(4)}\left(\tfrac{\pi}{4}\right)}{4!}\left(x - \tfrac{\pi}{4}\right)^4 + \cdots$$

$$= \frac{\sqrt{2}}{2}\left[1 + \left(x - \tfrac{\pi}{4}\right) - \tfrac{1}{2!}\left(x - \tfrac{\pi}{4}\right)^2\right.$$

$$\left. - \tfrac{1}{3!}\left(x - \tfrac{\pi}{4}\right)^3 + \tfrac{1}{4!}\left(x - \tfrac{\pi}{4}\right)^4 + \cdots\right]$$

$$= \frac{\sqrt{2}}{2}\sum_{n=0}^{\infty}\frac{(-1)^{n(n-1)/2}\left(x - \tfrac{\pi}{4}\right)^n}{n!}$$

If $a_n = \dfrac{(-1)^{n(n-1)/2}\left(x - \tfrac{\pi}{4}\right)^n}{n!}$, then $\displaystyle\lim_{n\to\infty}\left|\frac{a_{n+1}}{a_n}\right| = \lim_{n\to\infty}\frac{\left|x - \tfrac{\pi}{4}\right|}{n+1} = 0 < 1$ for all x, so $R = \infty$.

13. If $f(x) = \cos x$, then by Formula 9 with $a = 0$, $|R_n(x)| \leq \dfrac{\left|f^{(n+1)}(x)\right|}{(n+1)!}|x|^{n+1}$. But

$f^{(n+1)}(x) = \pm\sin x$ or $\pm\cos x$. In each case, $\left|f^{(n+1)}(x)\right| \leq 1$, so $|R_n(x)| \leq \dfrac{1}{(n+1)!}|x|^{n+1} \to 0$ as

$n \to \infty$ by Equation 10. So $\displaystyle\lim_{n\to\infty} R_n(x) = 0$ and, by Theorem 8, the series in Exercise 3 represents

$\cos x$ for all x.

15. $e^x = \displaystyle\sum_{n=0}^{\infty}\frac{x^n}{n!} \;\Rightarrow\; e^{3x} = \sum_{n=0}^{\infty}\frac{(3x)^n}{n!} = \sum_{n=0}^{\infty}\frac{3^n x^n}{n!}$, with $R = \infty$.

17. $\cos x = \displaystyle\sum_{n=0}^{\infty}\frac{(-1)^n x^{2n}}{(2n)!} \;\Rightarrow\; x^2\cos x = x^2\sum_{n=0}^{\infty}\frac{(-1)^n x^{2n}}{(2n)!} = \sum_{n=0}^{\infty}\frac{(-1)^n x^{2n+2}}{(2n)!}$, $R = \infty$

19. $\sin x = \displaystyle\sum_{n=0}^{\infty}\frac{(-1)^n x^{2n+1}}{(2n+1)!} \;\Rightarrow\; x\sin\left(\frac{x}{2}\right) = x\sum_{n=0}^{\infty}\frac{(-1)^n (x/2)^{2n+1}}{(2n+1)!} = \sum_{n=0}^{\infty}\frac{(-1)^n x^{2n+2}}{(2n+1)!\,2^{2n+1}}$, with

$R = \infty$.

21. $\sin^2 x = \frac{1}{2}[1 - \cos 2x] = \dfrac{1}{2}\left[1 - \displaystyle\sum_{n=0}^{\infty}\frac{(-1)^n (2x)^{2n}}{(2n)!}\right] = 2^{-1}\left[1 - 1 - \sum_{n=1}^{\infty}\frac{(-1)^n (2x)^{2n}}{(2n)!}\right]$

$$= \sum_{n=1}^{\infty}\frac{(-1)^{n+1} 2^{2n-1} x^{2n}}{(2n)!}, \text{ with } R = \infty.$$

23.

n	$f^{(n)}(x)$	$f^{(n)}(0)$
0	$(1+x)^{1/2}$	1
1	$\frac{1}{2}(1+x)^{-1/2}$	$\frac{1}{2}$
2	$-\frac{1}{4}(1+x)^{-3/2}$	$-\frac{1}{4}$
3	$\frac{3}{8}(1+x)^{-5/2}$	$\frac{3}{8}$
4	$-\frac{15}{16}(1+x)^{-7/2}$	$-\frac{15}{16}$
...	...	...

So $f^{(n)}(0) = \dfrac{(-1)^{n-1}\, 1\cdot 3\cdot 5\cdot\cdots\cdot(2n-3)}{2^n}$ for $n \geq 2$, and

$$\sqrt{1+x} = 1 + \frac{x}{2} + \sum_{n=2}^{\infty} \frac{(-1)^{n-1}\, 1\cdot 3\cdot 5\cdot\cdots\cdot(2n-3)}{2^n n!}\, x^n.$$

If $a_n = \dfrac{(-1)^{n-1}\, 1\cdot 3\cdot 5\cdot\cdots\cdot(2n-3)}{2^n n!}\, x^n$, then

$$\lim_{n\to\infty}\left|\frac{a_{n+1}}{a_n}\right| = \frac{|x|}{2}\lim_{n\to\infty}\frac{2n-1}{n+1} = |x| < 1 \text{ for convergence,}$$

so $R = 1$.

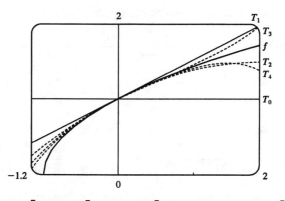

25. $f(x) = (1+x)^{-3} = -\dfrac{1}{2}\dfrac{d}{dx}\left[\dfrac{1}{(1+x)^2}\right] = -\dfrac{1}{2}\dfrac{d}{dx}\left[\displaystyle\sum_{n=0}^{\infty}(-1)^n(n+1)x^n\right]$ (from Exercise 5)

$$= -\frac{1}{2}\sum_{n=1}^{\infty}(-1)^n n(n+1)x^{n-1} = \sum_{n=1}^{\infty}\frac{(-1)^{n+1}n(n+1)x^{n-1}}{2}$$

$$= \sum_{n=0}^{\infty}\frac{(-1)^n(n+1)(n+2)x^n}{2} \text{ with } R = 1 \text{ as in Exercise 5.}$$

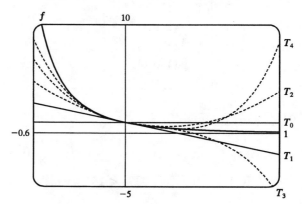

27. $\ln(1+x) = \displaystyle\int \frac{dx}{1+x} = \int \sum_{n=0}^{\infty} (-1)^n\, x^n\, dx = C + \sum_{n=0}^{\infty} (-1)^n \frac{x^{n+1}}{n+1} = \sum_{n=1}^{\infty} \frac{(-1)^{n-1}\, x^n}{n}$ with

$C = 0$ and $R = 1$, so $\ln(1.1) = \displaystyle\sum_{n=1}^{\infty} \frac{(-1)^{n-1}\,(0.1)^n}{n}$. This is an alternating series with

$b_5 = \dfrac{(0.1)^5}{5} = 0.000002$, so to five decimal places, $\ln(1.1) \approx \displaystyle\sum_{n=1}^{4} \frac{(-1)^{n-1}\,(0.1)^n}{n} \approx 0.09531$.

29. $\displaystyle\int \sin(x^2)\, dx = \int \sum_{n=0}^{\infty} (-1)^n \frac{(x^2)^{2n+1}}{(2n+1)!}\, dx = \int \sum_{n=0}^{\infty} \frac{(-1)^n\, x^{4n+2}}{(2n+1)!}\, dx = C + \sum_{n=0}^{\infty} \frac{(-1)^n\, x^{4n+3}}{(4n+3)\,(2n+1)!}$

31. Using the series from Exercise 23 and substituting x^3 for x, we get

$$\int \sqrt{x^3+1}\, dx = \int \left[1 + \frac{x^3}{2} + \sum_{n=2}^{\infty} \frac{(-1)^{n-1}\, 1\cdot 3\cdot 5 \cdots (2n-3)}{2^n n!}\, x^{3n} \right] dx$$

$$= C + x + \frac{x^4}{8} + \sum_{n=2}^{\infty} \frac{(-1)^{n-1}\, 1\cdot 3\cdot 5 \cdots (2n-3)}{2^n n!\,(3n+1)}\, x^{3n+1}$$

33. Using our series from Exercise 29, we get

$$\int_0^1 \sin(x^2)\, dx = \sum_{n=0}^{\infty} \left[\frac{(-1)^n\, x^{4n+3}}{(4n+3)\,(2n+1)!} \right]_0^1 = \sum_{n=0}^{\infty} \frac{(-1)^n}{(4n+3)\,(2n+1)!} \quad \text{and}$$

$|c_3| = \dfrac{1}{75{,}600} < 0.000014$, so by the Alternating Series Estimation Theorem, we have

$$\sum_{n=0}^{2} \frac{(-1)^n}{(4n+3)\,(2n+1)!} \approx \frac{1}{3} - \frac{1}{42} + \frac{1}{1320} \approx 0.310 \text{ (correct to three decimal places)}.$$

35. We first find a series representation for $f(x) = (1+x)^{-1/2}$, and then substitute.

n	$f^{(n)}(x)$	$f^{(n)}(0)$
0	$(1+x)^{-1/2}$	1
1	$-\frac{1}{2}(1+x)^{-3/2}$	$-\frac{1}{2}$
2	$\frac{3}{4}(1+x)^{-5/2}$	$\frac{3}{4}$
3	$-\frac{15}{8}(1+x)^{-7/2}$	$-\frac{15}{8}$
...	...	...

$\dfrac{1}{\sqrt{1+x}} = 1 - \dfrac{x}{2} + \dfrac{3}{4}\left(\dfrac{x^2}{2!}\right) - \dfrac{15}{8}\left(\dfrac{x^3}{3!}\right) + \cdots \quad\Rightarrow\quad \dfrac{1}{\sqrt{1+x^3}} = 1 - \frac{1}{2}x^3 + \frac{3}{8}x^6 - \frac{5}{16}x^9 + \cdots \quad\Rightarrow$

$\displaystyle\int_0^{0.1} \frac{dx}{\sqrt{1+x^3}} = \left[x - \frac{1}{8}x^4 + \frac{3}{56}x^7 - \frac{1}{32}x^{10} + \cdots \right]_0^{0.1} \approx (0.1) - \frac{1}{8}(0.1)^4$, by the Alternating Series

Estimation Theorem, since $\frac{3}{56}(0.1)^7 \approx 0.0000000054 < 10^{-8}$, which is the maximum desired error.

Therefore, $\displaystyle\int_0^{0.1} \frac{dx}{\sqrt{1+x^3}} \approx 0.09998750$.

37. $\lim\limits_{x\to 0}\dfrac{x-\tan^{-1}x}{x^3}=\lim\limits_{x\to 0}\dfrac{x-\left(x-\frac{1}{3}x^3+\frac{1}{5}x^5-\frac{1}{7}x^7+\cdots\right)}{x^3}=\lim\limits_{x\to 0}\dfrac{\frac{1}{3}x^3-\frac{1}{5}x^5+\frac{1}{7}x^7-\cdots}{x^3}$

$\qquad\qquad\quad=\lim\limits_{x\to 0}\left(\frac{1}{3}-\frac{1}{5}x^2+\frac{1}{7}x^4-\cdots\right)=\frac{1}{3}$

since power series are continuous functions.

39. $\lim\limits_{x\to 0}\dfrac{\sin x-x+\frac{1}{6}x^3}{x^5}=\lim\limits_{x\to 0}\dfrac{\left(x-\frac{1}{3!}x^3+\frac{1}{5!}x^5-\frac{1}{7!}x^7+\cdots\right)-x+\frac{1}{6}x^3}{x^5}$

$\qquad\qquad\quad=\lim\limits_{x\to 0}\dfrac{\frac{1}{5!}x^5-\frac{1}{7!}x^7+\cdots}{x^5}=\lim\limits_{x\to 0}\left(\dfrac{1}{5!}-\dfrac{x^2}{7!}+\dfrac{x^4}{9!}-\cdots\right)=\frac{1}{5!}=\frac{1}{120}$

since power series are continuous functions.

41. As in Example 8(a), we have $e^{-x^2}=1-\dfrac{x^2}{1!}+\dfrac{x^4}{2!}-\dfrac{x^6}{3!}+\cdots$ and we know that

$\cos x=1-\dfrac{x^2}{2!}+\dfrac{x^4}{4!}-\cdots$ from Equation 16. Therefore,

$e^{-x^2}\cos x=\left(1-x^2+\frac{1}{2}x^4-\cdots\right)\left(1-\frac{1}{2}x^2+\frac{1}{24}x^4-\cdots\right)$. Writing only the terms with degree ≤ 4,

we get $e^{-x^2}\cos x=1-\frac{1}{2}x^2+\frac{1}{24}x^4-x^2+\frac{1}{2}x^4+\frac{1}{2}x^4+\cdots=1-\frac{3}{2}x^2+\frac{25}{24}x^4+\cdots$.

43.

$$
\begin{array}{r}
-x+\ \tfrac{1}{2}x^2-\ \ \tfrac{1}{3}x^3+\ \ \cdots \\[2pt]
\hline
1+x+\tfrac{1}{2}x^2+\tfrac{1}{6}x^3+\cdots \ \Big|\quad -x-\ \tfrac{1}{2}x^2-\ \ \tfrac{1}{3}x^3-\ \ \cdots \\[4pt]
-x-\ \ x^2-\ \ \tfrac{1}{2}x^3-\ \ \cdots \\[2pt]
\hline
\tfrac{1}{2}x^2+\ \ \tfrac{1}{6}x^3-\ \ \cdots \\[2pt]
\tfrac{1}{2}x^2+\ \ \tfrac{1}{2}x^3+\ \ \cdots \\[2pt]
\hline
-\tfrac{1}{3}x^3+\ \ \cdots \\[2pt]
-\tfrac{1}{3}x^3+\ \ \cdots \\[2pt]
\hline
\cdots
\end{array}
$$

From Example 6 in Section 8.6, we have $\ln(1-x)=-x-\frac{1}{2}x^2-\frac{1}{3}x^3-\cdots$, $|x|<1$. Therefore,

$y=\dfrac{\ln(1-x)}{e^x}=\dfrac{-x-\frac{1}{2}x^2-\frac{1}{3}x^3-\cdots}{1+x+\frac{1}{2}x^2+\frac{1}{6}x^3+\cdots}$. So by the long division above,

$\dfrac{\ln(1-x)}{e^x}=-x+\dfrac{x^2}{2}-\dfrac{x^3}{3}+\cdots$, $|x|<1$.

45. $\displaystyle\sum_{n=0}^{\infty}(-1)^n\dfrac{x^{4n}}{n!}=\sum_{n=0}^{\infty}\dfrac{\left(-x^4\right)^n}{n!}=e^{-x^4}$, by (11).

47. $\displaystyle\sum_{n=0}^{\infty}\dfrac{(-1)^n\,\pi^{2n+1}}{4^{2n+1}(2n+1)!}=\sum_{n=0}^{\infty}\dfrac{(-1)^n\left(\frac{\pi}{4}\right)^{2n+1}}{(2n+1)!}=\sin\frac{\pi}{4}=\frac{1}{\sqrt{2}}$, by (15).

49. $\displaystyle\sum_{n=0}^{\infty}\dfrac{x^{n+1}}{(n+1)!}=\dfrac{x}{1!}+\dfrac{x^2}{2!}+\dfrac{x^3}{3!}+\cdots=\left(1+\dfrac{x}{1!}+\dfrac{x^2}{2!}+\dfrac{x^3}{3!}+\cdots\right)-1=e^x-1$, by (11).

Section 8.8   The Binomial Series

1. The general binomial series in (2) is

$$(1+x)^k = \sum_{n=0}^{\infty} \binom{k}{n} x^n = 1 + kx + \frac{k\,(k-1)}{2!} x^2 + \frac{k\,(k-1)\,(k-2)}{3!} x^3 + \cdots .$$

$$(1+x)^{1/2} = \sum_{n=0}^{\infty} \binom{\frac{1}{2}}{n} x^n = 1 + \left(\frac{1}{2}\right) x + \frac{\left(\frac{1}{2}\right)\left(-\frac{1}{2}\right)}{2!} x^2 + \frac{\left(\frac{1}{2}\right)\left(-\frac{1}{2}\right)\left(-\frac{3}{2}\right)}{3!} x^3 + \cdots$$

$$= 1 + \frac{x}{2} - \frac{x^2}{2^2 \cdot 2!} + \frac{1 \cdot 3 \cdot x^3}{2^3 \cdot 3!} - \frac{1 \cdot 3 \cdot 5 \cdot x^4}{2^4 \cdot 4!} + \cdots$$

$$= 1 + \frac{x}{2} + \sum_{n=2}^{\infty} \frac{(-1)^{n-1}\, 1 \cdot 3 \cdot 5 \cdots (2n-3)\, x^n}{2^n \cdot n!}, \, R = 1$$

3. $[1 + (2x)]^{-4} = 1 + (-4)(2x) + \dfrac{(-4)(-5)}{2!}(2x)^2 + \dfrac{(-4)(-5)(-6)}{3!}(2x)^3 + \cdots$

$$= 1 + \sum_{n=1}^{\infty} \frac{(-1)^n\, 2^n \cdot 4 \cdot 5 \cdot 6 \cdots (n+3)}{n!} x^n$$

$$= 1 + \sum_{n=1}^{\infty} \frac{(-1)^n\, 2^n \cdot 2 \cdot 3 \cdot 4 \cdot 5 \cdot 6 \cdots (n+1)(n+2)(n+3)}{2 \cdot 3 \cdot n!} x^n$$

$$= \sum_{n=0}^{\infty} (-1)^n \frac{2^n\, (n+1)(n+2)(n+3)}{6} x^n, |2x| < 1 \quad \Leftrightarrow \quad |x| < \tfrac{1}{2}, \text{ so } R = \tfrac{1}{2}.$$

5. $\left[1 + \left(-x^4\right)\right]^{1/4} = 1 + \left(\dfrac{1}{4}\right)\left(-x^4\right) + \dfrac{\left(\frac{1}{4}\right)\left(-\frac{3}{4}\right)}{2!}\left(-x^4\right)^2 + \dfrac{\left(\frac{1}{4}\right)\left(-\frac{3}{4}\right)\left(-\frac{7}{4}\right)}{3!}\left(-x^4\right)^3 + \cdots$

$$= 1 - \frac{x^4}{4} - \sum_{n=2}^{\infty} \frac{3 \cdot 7 \cdot 11 \cdots (4n-5)}{4^n \cdot n!} x^{4n}, \text{ with } R = 1.$$

7. $\dfrac{1}{\sqrt[3]{8+x}} = (8+x)^{-1/3} = 8^{-1/3}\left(1+\dfrac{x}{8}\right)^{-1/3} = \dfrac{1}{2}\left(1+\dfrac{x}{8}\right)^{-1/3}$

$\qquad = \dfrac{1}{2}\left[1+\left(-\dfrac{1}{3}\right)\left(\dfrac{x}{8}\right) + \dfrac{\left(-\frac{1}{3}\right)\left(-\frac{4}{3}\right)}{2!}\left(\dfrac{x}{8}\right)^2 + \cdots\right]$

$\qquad = \dfrac{1}{2}\left[1 + \sum_{n=1}^{\infty} \dfrac{(-1)^n\, 1\cdot 4 \cdot 7 \cdots \cdot (3n-2)}{3^n \cdot n!\, 8^n} x^n\right]$ and $\left|\dfrac{x}{8}\right| < 1 \quad \Leftrightarrow \quad |x| < 8,$ so $R = 8.$

The three Taylor polynomials are $T_1(x) = \dfrac{1}{2} - \dfrac{1}{48}x$, $T_2(x) = \dfrac{1}{2} - \dfrac{1}{48}x + \dfrac{1}{576}x^2$, and

$T_3(x) = \dfrac{1}{2} - \dfrac{1}{48}x + \dfrac{1}{576}x^2 - \dfrac{4\cdot 7}{2\cdot 27\cdot 6\cdot 512}x^3 = \dfrac{1}{2} - \dfrac{1}{48}x + \dfrac{1}{576}x^2 - \dfrac{7}{41,472}x^3.$

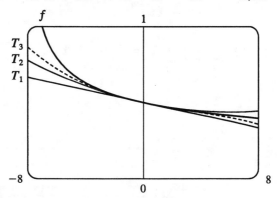

9. (a) $\left[1 + (-x^2)\right]^{-1/2} = 1 + \left(-\dfrac{1}{2}\right)(-x^2) + \dfrac{\left(-\frac{1}{2}\right)\left(-\frac{3}{2}\right)}{2!}(-x^2)^2 + \dfrac{\left(-\frac{1}{2}\right)\left(-\frac{3}{2}\right)\left(-\frac{5}{2}\right)}{3!}(-x^2)^3 + \cdots$

$\qquad = 1 + \sum_{n=1}^{\infty} \dfrac{1\cdot 3 \cdot 5 \cdots \cdot (2n-1)}{2^n \cdot n!} x^{2n}$

(b) $\sin^{-1}x = \displaystyle\int \dfrac{1}{\sqrt{1-x^2}}\, dx = C + x + \sum_{n=1}^{\infty} \dfrac{1\cdot 3 \cdot 5 \cdots \cdot (2n-1)}{(2n+1)\, 2^n \cdot n!} x^{2n+1}$

$\qquad = x + \sum_{n=1}^{\infty} \dfrac{1\cdot 3 \cdot 5 \cdots \cdot (2n-1)}{(2n+1)\, 2^n \cdot n!} x^{2n+1}$ since $0 = \sin^{-1}0 = C.$

11. (a) $\left[1+(-x)\right]^{-2} = 1 + (-2)(-x) + \dfrac{(-2)(-3)}{2!}(-x)^2 + \dfrac{(-2)(-3)(-4)}{3!}(-x)^3 + \cdots$ so

$\qquad = 1 + 2x + 3x^2 + 4x^3 + \cdots = \sum_{n=0}^{\infty} (n+1)x^n,$

$\dfrac{x}{(1-x)^2} = \sum_{n=0}^{\infty} (n+1)x^{n+1} = \sum_{n=1}^{\infty} nx^n.$

(b) With $x = \dfrac{1}{2}$ in part (a), we have $\displaystyle\sum_{n=1}^{\infty} \dfrac{n}{2^n} = \dfrac{\frac{1}{2}}{\left(1-\frac{1}{2}\right)^2} = 2.$

13. (a) $(1 + x^2)^{1/2} = 1 + \left(\frac{1}{2}\right) x^2 + \dfrac{\left(\frac{1}{2}\right)\left(-\frac{1}{2}\right)}{2!} \left(x^2\right)^2 + \dfrac{\left(\frac{1}{2}\right)\left(-\frac{1}{2}\right)\left(-\frac{3}{2}\right)}{3!} \left(x^2\right)^3 + \cdots$

$$= 1 + \frac{x^2}{2} + \sum_{n=2}^{\infty} \frac{(-1)^{n-1} \, 1 \cdot 3 \cdot 5 \cdot \, \cdots \, \cdot (2n-3)}{2^n \cdot n!} x^{2n}$$

(b) The coefficient of x^{10} (corresponding to $n = 5$) in the above Maclaurin series is $\dfrac{f^{(10)}(0)}{10!}$, so

$$\frac{f^{(10)}(0)}{10!} = \frac{(-1)^4 \cdot 1 \cdot 3 \cdot 5 \cdot 7}{2^5 \cdot 5!} \quad \Rightarrow \quad f^{(10)}(0) = 10! \left(\frac{1 \cdot 3 \cdot 5 \cdot 7}{2^5 \cdot 5!} \right) = 99{,}225.$$

15. (a) $g(x) = \displaystyle\sum_{n=0}^{\infty} \binom{k}{n} x^n \quad \Rightarrow \quad g'(x) = \sum_{n=1}^{\infty} \binom{k}{n} n x^{n-1}$, so

$$(1+x) \, g'(x) = (1+x) \sum_{n=1}^{\infty} \binom{k}{n} n x^{n-1} = \sum_{n=1}^{\infty} \binom{k}{n} n x^{n-1} + \sum_{n=1}^{\infty} \binom{k}{n} n x^{n}$$

$$= \sum_{n=0}^{\infty} \binom{k}{n+1} (n+1) x^n + \sum_{n=0}^{\infty} \binom{k}{n} n x^n \qquad \begin{bmatrix} \text{Replace } n \text{ with } n+1 \\ \text{in the first series} \end{bmatrix}$$

$$= \sum_{n=0}^{\infty} (n+1) \frac{k(k-1)(k-2)\cdots(k-n+1)(k-n)}{(n+1)!} x^n$$

$$+ \sum_{n=0}^{\infty} \left[(n) \frac{k(k-1)(k-2)\cdots(k-n+1)}{n!} \right] x^n$$

$$= \sum_{n=0}^{\infty} \frac{(n+1)\, k(k-1)(k-2)\cdots(k-n+1)}{(n+1)!} \, [(k-n)+n] \, x^n$$

$$= k \sum_{n=0}^{\infty} \frac{k(k-1)(k-2)\cdots(k-n+1)}{n!} x^n = k \sum_{n=0}^{\infty} \binom{k}{n} x^n$$

$$= k g(x)$$

Thus, $g'(x) = \dfrac{k g(x)}{1+x}$.

(b) $h(x) = (1+x)^{-k} g(x) \quad \Rightarrow$

$$h'(x) = -k(1+x)^{-k-1} g(x) + (1+x)^{-k} g'(x) \qquad \text{[Product Rule]}$$

$$= -k(1+x)^{-k-1} g(x) + (1+x)^{-k} \frac{k g(x)}{1+x} \qquad \text{[from part (a)]}$$

$$= -k(1+x)^{-k-1} g(x) + k(1+x)^{-k-1} g(x)$$

$$= 0$$

(c) From part (b) we see that $h(x)$ must be constant for $x \in (-1, 1)$, so $h(x) = h(0) = 1$ for $x \in (-1, 1)$. Thus, $h(x) = 1 = (1+x)^{-k} g(x) \quad \Leftrightarrow \quad g(x) = (1+x)^k$ for $x \in (-1, 1)$.

Section 8.9 Applications of Taylor Polynomials

1. (a)

n	$f^{(n)}(x)$	$f^{(n)}(0)$	$T_n(x)$
0	$\cos x$	1	1
1	$-\sin x$	0	1
2	$-\cos x$	-1	$1 - \frac{1}{2}x^2$
3	$\sin x$	0	$1 - \frac{1}{2}x^2$
4	$\cos x$	1	$1 - \frac{1}{2}x^2 + \frac{1}{24}x^4$
5	$-\sin x$	0	$1 - \frac{1}{2}x^2 + \frac{1}{24}x^4$
6	$-\cos x$	-1	$1 - \frac{1}{2}x^2 + \frac{1}{24}x^4 - \frac{1}{720}x^6$

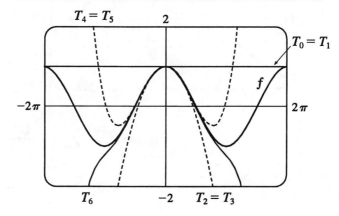

(b)

x	f	$T_0 = T_1$	$T_2 = T_3$	$T_4 = T_5$	T_6
$\frac{\pi}{4}$	0.7071	1	0.6916	0.7074	0.7071
$\frac{\pi}{2}$	0	1	-0.2337	0.0200	-0.0009
π	-1	1	-3.9348	0.1239	-1.2114

(c) As n increases, $T_n(x)$ is a good approximation to $f(x)$ on a larger and larger interval.

3.

n	$f^{(n)}(x)$	$f^{(n)}\left(\frac{\pi}{6}\right)$
0	$\sin x$	$\frac{1}{2}$
1	$\cos x$	$\frac{\sqrt{3}}{2}$
2	$-\sin x$	$-\frac{1}{2}$
3	$-\cos x$	$-\frac{\sqrt{3}}{2}$

$$T_3(x) = \sum_{n=0}^{3} \frac{f^{(n)}\left(\frac{\pi}{6}\right)}{n!}\left(x - \tfrac{\pi}{6}\right)^n$$

$$= \tfrac{1}{2} + \tfrac{\sqrt{3}}{2}\left(x - \tfrac{\pi}{6}\right) - \tfrac{1}{4}\left(x - \tfrac{\pi}{6}\right)^2 - \tfrac{\sqrt{3}}{12}\left(x - \tfrac{\pi}{6}\right)^3$$

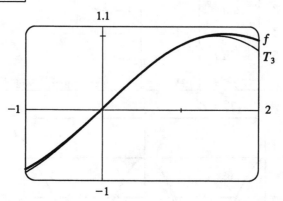

5.

n	$f^{(n)}(x)$	$f^{(n)}(0)$
0	$\tan x$	0
1	$\sec^2 x$	1
2	$2\sec^2 x \tan x$	0
3	$4\sec^2 x \tan^2 x + 2\sec^4 x$	2
4	$8\sec^2 x \tan^3 x + 16\sec^4 x \tan x$	0

$$T_4(x) = \sum_{n=0}^{4} \frac{f^{(n)}(0)}{n!}x^n = x + \frac{2x^3}{3!}$$

$$= x + \frac{x^3}{3}$$

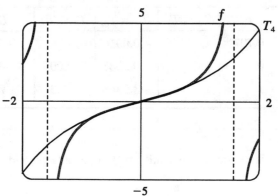

7.

n	$f^{(n)}(x)$	$f^{(n)}\left(\frac{\pi}{3}\right)$
0	$\sec x$	2
1	$\sec x \tan x$	$2\sqrt{3}$
2	$\sec x \tan^2 x + \sec^3 x$	14
3	$\sec x \tan^3 x$ $+5\sec^3 x \tan x$	$46\sqrt{3}$

$$T_3(x) = \sum_{n=0}^{3} \frac{f^{(n)}\left(\frac{\pi}{3}\right)}{n!}\left(x - \frac{\pi}{3}\right)^n$$
$$= 2 + 2\sqrt{3}\left(x - \frac{\pi}{3}\right) + 7\left(x - \frac{\pi}{3}\right)^2$$
$$+ \frac{23\sqrt{3}}{3}\left(x - \frac{\pi}{3}\right)^3$$

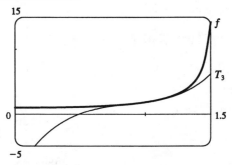

9. In Maple, we can find the Taylor polynomials by the following method: first define
`f:=sec(x);` and then set `T2:=convert(taylor(f,x=0,3),polynom);`,
`T4:=convert(taylor(f,x=0,5),polynom);`, etc. (The third argument in the `taylor`
function is one more than the degree of the desired polynomial). We must `convert` to the type
`polynom` because the output of the `taylor` function contains an error term which we do not want. In
Mathematica, we use `Tn:=Normal[Series[f,{x,0,n}]]`, with n=2, 4, etc. Note that in
Mathematica, the "degree" argument is the same as the degree of the desired polynomial. In Derive,
author $\sec x$, then enter `Calculus, Taylor, 8, 0`; and then simplify the expression. The eighth
Taylor polynomial is $T_8(x) = 1 + \frac{1}{2}x^2 + \frac{5}{24}x^4 + \frac{61}{720}x^6 + \frac{277}{8064}x^8$.

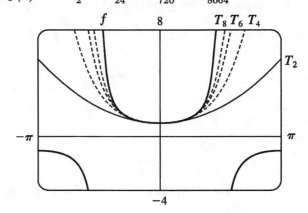

11.

$$f(x) = (1+x)^{1/2} \qquad\qquad f(0) = 1$$
$$f'(x) = \tfrac{1}{2}(1+x)^{-1/2} \qquad\qquad f'(0) = \tfrac{1}{2}$$
$$f''(x) = -\tfrac{1}{4}(1+x)^{-3/2} = -\frac{1}{4(1+x)^{3/2}}$$

(a) $(1+x)^{1/2} \approx T_1(x) = 1 + \tfrac{1}{2}x$

(c)

(b) By Taylor's Inequality, the remainder is

$$|R_1(x)| \le \frac{M}{2!}|x|^2, \text{ where } |f''(x)| \le M.$$

Now $0 \le x \le 0.1 \;\Rightarrow\; 0 \le x^2 \le 0.01$,
and letting $x = 0$ gives $M = 0.25$, so
$$|R_1(x)| \le \tfrac{0.25}{2}(0.01) = 0.00125.$$

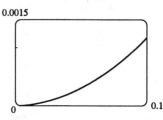

0.0015

0   0.1

From the graph of
$$|R_1(x)| = \left|\sqrt{1+x} - \left(1 + \tfrac{1}{2}x\right)\right|, \text{ it seems}$$
that the error is at most 0.0012 on $[0, 0.1]$.

13.

$$f(x) = \sin x \qquad f\left(\tfrac{\pi}{4}\right) = \tfrac{\sqrt{2}}{2} \qquad\qquad f^{(4)}(x) = \sin x \qquad f^{(4)}\left(\tfrac{\pi}{4}\right) = \tfrac{\sqrt{2}}{2}$$
$$f'(x) = \cos x \qquad f'\left(\tfrac{\pi}{4}\right) = \tfrac{\sqrt{2}}{2} \qquad\qquad f^{(5)}(x) = \cos x \qquad f^{(5)}\left(\tfrac{\pi}{4}\right) = \tfrac{\sqrt{2}}{2}$$
$$f''(x) = -\sin x \qquad f''\left(\tfrac{\pi}{4}\right) = -\tfrac{\sqrt{2}}{2} \qquad\qquad f^{(6)}(x) = -\sin x$$
$$f'''(x) = -\cos x \qquad f'''\left(\tfrac{\pi}{4}\right) = -\tfrac{\sqrt{2}}{2}$$

(a) $\sin x \approx T_5(x)$
$$= \tfrac{\sqrt{2}}{2} + \tfrac{\sqrt{2}}{2}\left(x - \tfrac{\pi}{4}\right) - \tfrac{\sqrt{2}}{4}\left(x - \tfrac{\pi}{4}\right)^2 - \tfrac{\sqrt{2}}{12}\left(x - \tfrac{\pi}{4}\right)^3 + \tfrac{\sqrt{2}}{48}\left(x - \tfrac{\pi}{4}\right)^4 + \tfrac{\sqrt{2}}{240}\left(x - \tfrac{\pi}{4}\right)^5$$

(b) $|R_5(x)| \le \dfrac{M}{6!}\left|x - \tfrac{\pi}{4}\right|^6$, where $\left|f^{(6)}(x)\right| \le M$. Now $0 \le x \le \tfrac{\pi}{2} \;\Rightarrow\; \left(x - \tfrac{\pi}{4}\right)^6 \le \left(\tfrac{\pi}{4}\right)^6$, and

letting $x = \tfrac{\pi}{2}$ gives $M = 1$, so $|R_5(x)| \le \tfrac{1}{6!}\left(\tfrac{\pi}{4}\right)^6 = \tfrac{1}{720}\left(\tfrac{\pi}{4}\right)^6 \approx 0.00033.$

(c)

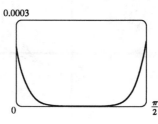

0.0003

0 $\tfrac{\pi}{2}$

From the graph of $|R_5(x)| = |\sin x - T_5(x)|$, it seems that the error is less than 0.00026 on $\left[0, \tfrac{\pi}{2}\right]$.

15.

$$f(x) = e^{x^2} \qquad\qquad f(0) = 1$$
$$f'(x) = e^{x^2}(2x) \qquad\qquad f'(0) = 0$$
$$f''(x) = e^{x^2}(2 + 4x^2) \qquad\qquad f''(0) = 2$$
$$f'''(x) = e^{x^2}(12x + 8x^3) \qquad\qquad f'''(0) = 0$$
$$f^{(4)}(x) = e^{x^2}(12 + 48x^2 + 16x^4)$$

(a) $e^{x^2} \approx T_3(x) = 1 + \frac{2}{2!}x^2 = 1 + x^2$

(b) $|R_3(x)| \le \dfrac{M}{4!}\,|x|^4$, where $\left|f^{(4)}(x)\right| \le M$. Now $0 \le x \le 0.1 \;\Rightarrow\; x^4 \le (0.1)^4$, and letting

$x = 0.1$ gives $|R_3(x)| \le \dfrac{e^{0.01}(12 + 0.48 + 0.0016)}{24}(0.1)^4 < 0.00006.$

(c)

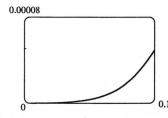

0.00008

0   0.1

From the graph of $|R_3(x)| = \left|e^{x^2} - (1 + x^2)\right|$, it appears that the error is less than 0.000051 on $[0, 0.1]$.

17. From Exercise 3, $\sin x = \frac{1}{2} + \frac{\sqrt{3}}{2}\left(x - \frac{\pi}{6}\right) - \frac{1}{4}\left(x - \frac{\pi}{6}\right)^2 - \frac{\sqrt{3}}{12}\left(x - \frac{\pi}{6}\right)^3 + R_3(x)$, where

$R_3(x) \le \dfrac{M}{4!}\left|x - \frac{\pi}{6}\right|^4$ with $\left|f^{(4)}(x)\right| = |\sin x| \le M = 1$. Now $35° = \left(\frac{\pi}{6} + \frac{\pi}{36}\right)$ radians, so the error

is $\left|R_3\left(\frac{\pi}{36}\right)\right| \le \dfrac{\left(\frac{\pi}{36}\right)^4}{4!} < 0.000003$. Therefore, to five decimal places,

$\sin 35° \approx \frac{1}{2} + \frac{\sqrt{3}}{2}\left(\frac{\pi}{36}\right) - \frac{1}{4}\left(\frac{\pi}{36}\right)^2 - \frac{\sqrt{3}}{12}\left(\frac{\pi}{36}\right)^3 \approx 0.57358.$

19. All derivatives of e^x are e^x, so $|R_n(x)| \le \dfrac{e^x}{(n+1)!}\,|x|^{n+1}$, where $0 < x < 0.1$. Letting $x = 0.1$,

$R_n(0.1) \le \dfrac{e^{0.1}}{(n+1)!}(0.1)^{n+1} < 0.00001$, and by trial and error we find that $n = 3$ satisfies this

inequality since $R_3(0.1) < 0.0000046$. Thus, by adding the three terms of the Maclaurin series for e^x corresponding to $n = 0$, 1, and 2, we can estimate $e^{0.1}$ to within 0.00001.

21. $\sin x = x - \frac{1}{3!}x^3 + \frac{1}{5!}x^5 - \cdots$. By the Alternating
Series Estimation Theorem, the error in the
approximation $\sin x = x - \frac{1}{3!}x^3$ is less than
$\left|\frac{1}{5!}x^5\right| < 0.01 \iff |x^5| < 120\,(0.01) \iff$
$|x| < (1.2)^{1/5} \approx 1.037$.
The curves intersect at $x \approx 1.043$, so the graph
confirms our estimate. Since both the sine function and
the given approximation are odd functions, we only
need to check the estimate for $x > 0$.

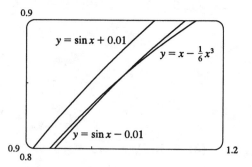

23. Let $s\,(t)$ be the position function of the car, and for convenience set $s\,(0) = 0$. The velocity of the car is
$v\,(t) = s'\,(t)$ and the acceleration is $a\,(t) = s''\,(t)$, so the second degree Taylor polynomial is
$T_2\,(t) = s\,(0) + v\,(0)\,t + \dfrac{a\,(0)}{2}t^2 = 20t + t^2$. We estimate the distance travelled during the next second
to be $s\,(1) \approx T_2\,(1) = 20 + 1 = 21$ m. The function $T_2\,(t)$ would not be accurate over a full minute,
since the car could not possibly maintain an acceleration of 2 m/s^2 for that long (if it did, its final speed
would be 140 m/s ≈ 315 mi/h!)

25. $E = \dfrac{q}{D^2} - \dfrac{q}{(D+d)^2} = \dfrac{q}{D^2} - \dfrac{q}{D^2\,(1+d/D)^2} = \dfrac{q}{D^2}\left[1 - \left(1 + \dfrac{d}{D}\right)^{-2}\right]$.

We use the Binomial Series to expand $(1 + d/D)^{-2}$:

$$E = \frac{q}{D^2}\left[1 - \left(1 - 2\left(\frac{d}{D}\right) + \frac{2\cdot 3}{2!}\left(\frac{d}{D}\right)^2 - \frac{2\cdot 3\cdot 4}{3!}\left(\frac{d}{D}\right)^3 + \cdots\right)\right]$$

$$= \frac{q}{D^2}\left[2\left(\frac{d}{D}\right) - 3\left(\frac{d}{D}\right)^2 + 4\left(\frac{d}{D}\right)^3 - \cdots\right] \approx 2qd \cdot \frac{1}{D^3}$$

when D is much larger than d, that is, when P is far away from the dipole.

27. Using $f\,(x) = T_n\,(x) + R_n\,(x)$ with $n = 1$ and $x = r$, we have $f\,(r) = T_1\,(r) + R_1\,(r)$, where T_1 is the
first-degree Taylor polynomial of f at a. Because $a = x_n$, $f\,(r) = f\,(x_n) + f'\,(x_n)\,(r - x_n) + R_1\,(r)$.
But r is a root of f, so $f\,(r) = 0$ and we have $0 = f\,(x_n) + f'\,(x_n)\,(r - x_n) + R_1\,(r)$. Taking the first
two terms to the left side and dividing by $f'\,(x_n)$, we have $f'\,(x_n)\,(x_n - r) - f\,(x_n) = R_1\,(r) \Rightarrow$
$x_n - r - \dfrac{f\,(x_n)}{f'\,(x_n)} = \dfrac{R_1\,(r)}{f'\,(x_n)}$. By the formula for Newton's method, the left side of the preceding

equation is $x_{n+1} - r$, so $|x_{n+1} - r| = \left|\dfrac{R_1\,(r)}{f'\,(x_n)}\right|$. Taylor's Inequality gives us

$|R_1\,(r)| \le \dfrac{|f''\,(r)|}{2!}\,|r - x_n|^2$. Combining this inequality with the facts $|f''\,(x)| \le M$ and $|f'\,(x)| \ge K$

gives us $|x_{n+1} - r| \le \dfrac{M}{2K}\,|x_n - r|^2$.

Section 8.10   Using Series to Solve Differential Equations

1. Let $y(x) = \sum\limits_{n=0}^{\infty} c_n x^n$. Thus, $y'(x) = \sum\limits_{n=1}^{\infty} n c_n x^{n-1}$, which can be written as

$y' = \sum\limits_{n=0}^{\infty} (n+1) c_{n+1} x^n$ by replacing n with $n+1$. $y' = 6y \;\Rightarrow\; y' - 6y = 0 \;\Rightarrow$

$\sum\limits_{n=0}^{\infty} [(n+1) c_{n+1} - 6c_n] x^n = 0$. Hence, the recurrence relation is $c_{n+1} = \dfrac{6c_n}{n+1}$, $n = 0, 1, 2, \ldots$.

Then $c_1 = 6c_0$, $c_2 = \dfrac{6c_1}{2} = \dfrac{6^2 c_0}{2}$, $c_3 = \dfrac{6c_2}{3} = \dfrac{6^3 c_0}{2 \cdot 3}$, $\ldots$, $c_n = \dfrac{6c_{n-1}}{n} = \dfrac{6^n c_0}{n!}$. Thus, the solution is

$$y(x) = \sum\limits_{n=0}^{\infty} c_0 \frac{6^n}{n!} x^n = \sum\limits_{n=0}^{\infty} \left[c_0 \frac{(6x)^n}{n!} \right] = c_0 e^{6x}.$$

3. Assuming $y(x) = \sum\limits_{n=0}^{\infty} c_n x^n$, we have $y'(x) = \sum\limits_{n=1}^{\infty} n c_n x^{n-1} = \sum\limits_{n=0}^{\infty} (n+1) c_{n+1} x^n$ and

$-x^2 y = -\sum\limits_{n=0}^{\infty} c_n x^{n+2} = -\sum\limits_{n=2}^{\infty} c_{n-2} x^n$. Hence, the equation $y' = x^2 y$ becomes

$\sum\limits_{n=0}^{\infty} (n+1) c_{n+1} x^n - \sum\limits_{n=2}^{\infty} c_{n-2} x^n = 0$ or $c_1 + 2c_2 x + \sum\limits_{n=2}^{\infty} [(n+1) c_{n+1} - c_{n-2}] x^n = 0$. Equating

coefficients gives $c_1 = c_2 = 0$ and $c_{n+1} = \dfrac{c_{n-2}}{n+1}$ for $n = 2, 3, \ldots$. But $c_1 = 0$, so $c_4 = 0$ and $c_7 = 0$

and in general $c_{3n+1} = 0$. Similarly $c_2 = 0$ so $c_{3n+2} = 0$. Finally $c_3 = \dfrac{c_0}{3}$, $c_6 = \dfrac{c_3}{6} = \dfrac{c_0}{6 \cdot 3} = \dfrac{c_0}{3^2 \cdot 2!}$,

$c_9 = \dfrac{c_6}{9} = \dfrac{c_0}{9 \cdot 6 \cdot 3} = \dfrac{c_0}{3^3 \cdot 3!}$, $\ldots$, and $c_{3n} = \dfrac{c_0}{3^n \cdot n!}$. Thus, the solution is

$$y(x) = \sum\limits_{n=0}^{\infty} c_n x^n = \sum\limits_{n=0}^{\infty} c_{3n} x^{3n} = c_0 \sum\limits_{n=0}^{\infty} \frac{\left(x^3 / 3 \right)^n}{n!} = c_0 e^{x^3/3}.$$

5. Assuming $y(x) = \sum\limits_{n=0}^{\infty} c_n x^n$, $y''(x) = \sum\limits_{n=2}^{\infty} n(n-1)c_n x^{n-2} = \sum\limits_{n=0}^{\infty} (n+2)(n+1)c_{n+2}x^n$,

$3xy'(x) = 3x \sum\limits_{n=0}^{\infty} nc_n x^{n-1} = \sum\limits_{n=0}^{\infty} 3nc_n x^n$ and the equation $y'' + 3xy' + 3y = 0$

becomes $\sum\limits_{n=0}^{\infty} (n+2)(n+1)c_{n+2}x^n + \sum\limits_{n=0}^{\infty} 3nc_n x^n + \sum\limits_{n=0}^{\infty} 3c_n x^n = 0 \quad \Leftrightarrow$

$\sum\limits_{n=0}^{\infty} [(n+2)(n+1)c_{n+2} + 3(n+1)c_n] x^n = 0$. Thus, the recurrence relation is $c_{n+2} = -\dfrac{3c_n}{n+2}$ for

$n = 0, 1, 2, \ldots$. Given c_0 and c_1, $c_2 = -\dfrac{3c_0}{2}$, $c_4 = -\dfrac{3c_2}{4} = (-1)^2 \dfrac{3^2 c_0}{2^2 \cdot 2!}$,

$c_6 = -\dfrac{3c_4}{6} = (-1)^3 \dfrac{3^3 c_0}{2^3 \cdot 3!}, \ldots, c_{2n} = (-1)^n \dfrac{3^n c_0}{2^n n!} = (-1)^n \left(\dfrac{3}{2}\right)^n \dfrac{c_0}{n!}$. Also,

$c_3 = -\dfrac{3c_1}{3}$, $c_5 = -\dfrac{3c_3}{5} = (-1)^2 \dfrac{3^2 c_1}{5 \cdot 3}$, $c_7 = -\dfrac{3c_5}{7} = (-1)^3 \dfrac{3^3 c_1}{7 \cdot 5 \cdot 3}, \ldots,$

$c_{2n+1} = (-1)^n \dfrac{3^n c_1}{(2n+1)(2n-1)\cdots\cdot 5\cdot 3} = (-1)^n \dfrac{3^n c_1 2^n n!}{(2n+1)!} = (-1)^n \dfrac{6^n n! c_1}{(2n+1)!}$ because

$(2n+1)(2n-1)\cdots\cdot 5 \cdot 3 = \dfrac{(2n+1)!}{2^n \cdot n!}$. Thus, the solution is

$y(x) = \sum\limits_{n=0}^{\infty} c_{2n}x^{2n} + \sum\limits_{n=0}^{\infty} c_{2n+1}x^{2n+1} = c_0 \sum\limits_{n=0}^{\infty} (-1)^n \dfrac{\left(3x^2/2\right)^n}{n!} + c_1 \sum\limits_{n=0}^{\infty} \left[(-1)^n \dfrac{6^n n! \, x^{2n+1}}{(2n+1)!}\right]$

$= c_0 e^{-3x^2/2} + c_1 \sum\limits_{n=0}^{\infty} \left[(-1)^n \dfrac{6^n n! \, x^{2n+1}}{(2n+1)!}\right]$

7. Let $y(x) = \sum\limits_{n=0}^{\infty} c_n x^n$. Then $y''(x) = \sum\limits_{n=0}^{\infty} (n+2)(n+1)c_{n+2}x^n$, $-xy'(x) = -\sum\limits_{n=0}^{\infty} nc_n x^n$ and the

equation $y'' - xy' - y = 0$ becomes $\sum\limits_{n=0}^{\infty} [(n+2)(n+1)c_{n+2} - (n+1)c_n] x^n = 0$. Thus, the

recurrence relation is $c_{n+2} = \dfrac{c_n}{n+2}$ for $n = 0, 1, 2, \ldots$. But $c_0 = y(0) = 1$, so $c_2 = \dfrac{1}{2}$,

$c_4 = \dfrac{c_2}{4} = \dfrac{1}{2 \cdot 4}$, $c_6 = \dfrac{c_4}{6} = \dfrac{1}{2 \cdot 4 \cdot 6}, \ldots, c_{2n} = \dfrac{1}{2^n n!}$. Also, $c_1 = y'(0) = 0$ and by the recurrence

relation, $c_{2n+1} = 0$ for $n = 0, 1, 2, \ldots$. Thus, the solution to the initial-value problem is

$y(x) = \sum\limits_{n=0}^{\infty} c_n x^n = \sum\limits_{n=0}^{\infty} \dfrac{x^{2n}}{2^n n!} = \sum\limits_{n=0}^{\infty} \dfrac{\left(x^2/2\right)^n}{n!} = e^{x^2/2}$.

9. Assuming that $y(x) = \sum\limits_{n=0}^{\infty} c_n x^n$, we have

$$y''(x)$$

$$= \sum_{n=2}^{\infty} n(n-1) c_n x^{n-2}$$

$$= \sum_{n=-1}^{\infty} (n+3)(n+2) c_{n+3} x^{n+1} \quad \text{[replace } n \text{ with } n+3]$$

$$= 2c_2 + \sum_{n=0}^{\infty} (n+3)(n+2) c_{n+3} x^{n+1}$$

$$x^2 y' = x^2 \sum_{n=1}^{\infty} n c_n x^{n-1} = \sum_{n=0}^{\infty} n c_n x^{n+1}, \ xy = x \sum_{n=0}^{\infty} c_n x^n = \sum_{n=0}^{\infty} c_n x^{n+1}, \text{ and the equation}$$

$y'' + x^2 y' + xy = 0$ becomes $2c_2 + \sum\limits_{n=0}^{\infty} [(n+3)(n+2) c_{n+3} + (n+1) c_n] x^{n+1} = 0$. So $c_2 = 0$ and

the recurrence relation is $c_{n+3} = -\dfrac{(n+1) c_n}{(n+3)(n+2)}$, $n = 0, 1, 2, \ldots$. But $c_0 = y(0) = 0 = c_2$ and by

the recurrence relation, $c_{3n} = c_{3n+2} = 0$ for $n = 0, 1, 2, \ldots$. Also, $c_1 = y'(0) = 1$ so

$$c_4 = -\frac{2}{4 \cdot 3}, \ c_7 = -\frac{5 c_4}{7 \cdot 6} = (-1)^2 \frac{2 \cdot 5}{7 \cdot 6 \cdot 4 \cdot 3} = (-1)^2 \frac{2^2 5^2}{7!}, \ \ldots,$$

$$c_{3n+1} = (-1)^n \frac{2^2 5^2 \cdots (3n-1)^2}{(3n+1)!}. \text{ Thus, the solution is}$$

$$y(x) = \sum_{n=0}^{\infty} c_n x^n$$

$$= x + \sum_{n=1}^{\infty} \left[(-1)^n \frac{2^2 5^2 \cdots (3n-1)^2 x^{3n+1}}{(3n+1)!}\right]$$

Chapter 8 Review

Concept Check

1. (a) See Definition 8.1.1.

   (b) See Definition 8.2.2.

   (c) The terms of the sequence $\{a_n\}$ approach 3 as n becomes large.

   (d) By adding sufficiently many terms of the series, we can make the partial sums as close to 3 as we like.

2. (a) See the definition on page 566.

   (b) A sequence is monotonic if it is either increasing or decreasing.

   (c) By Theorem 8.1.7, every bounded, monotonic sequence is convergent.

3. (a) See (4) in Section 8.2.

   (b) See (1) in Section 8.3.

4. If $\sum a_n = 3$, then $\lim\limits_{n \to \infty} a_n = 0$ and $\lim\limits_{n \to \infty} s_n = 3$.

5. (a) See the Test for Divergence on page 575.

   (b) See the Integral Test on page 581.

   (c) See the Comparison Test on page 583.

   (d) See the Limit Comparison Test on page 584.

   (e) See the Alternating Series Test on page 589.

   (f) See the Ratio Test on page 594.

6. (a) See the definition on page 592.

   (b) By (8.4.1), it is convergent.

7. (a) Use either (3) or (4) in Section 8.3.

   (b) See Example 8 in Section 8.3.

   (c) By adding terms until you reach the desired accuracy given by the Alternating Series Estimation Theorem on page 591.

8. (a) $\sum\limits_{n=0}^{\infty} c_n (x - a)^n$

   (b) Given the power series $\sum\limits_{n=0}^{\infty} c_n (x - a)^n$, the radius of convergence is:

   (i) 0 if the series converges only when $x = a$,

   (ii) ∞ if the series converges for all x, or

   (iii) a positive number R such that the series converges if $|x - a| < R$ and diverges if $|x - a| > R$.

(c) The interval of convergence of a power series is the interval that consists of all values of x for which the series converges. Corresponding to the cases in part (b), the interval of convergence is: (i) the single point $\{a\}$, (ii) all real numbers, that is, the real number line $(-\infty, \infty)$, or (iii) an interval with endpoints $a - R$ and $a + R$ which can contain neither, either, or both of the endpoints. In this case, we must test the series for convergence at each endpoint to determine the interval of convergence.

9. **(a), (b)** See Theorem 2 on page 604.

10. **(a)** $T_n(x) = \sum_{i=0}^{n} \frac{f^{(i)}(a)}{i!} (x - a)^i$ **(b)** $\sum_{n=0}^{\infty} \frac{f^{(n)}(a)}{n!} (x - a)^n$

    (c) $\sum_{n=0}^{\infty} \frac{f^{(n)}(0)}{n!} x^n$ $[a = 0$ in part (b)$]$ **(d)** See Theorem 8 on page 611.

    (e) See (9) on page 611.

11. **(a)–(e)** See page 616.

12. See (2) on page 621 for the expansion. The radius of convergence for the binomial series is 1.

True-False Quiz

1. False. See Note 2 after Theorem 8.2.6.

3. False. For example, take $c_n = (-1)^n / (n6^n)$.

5. False, since $\lim_{n \to \infty} \left| \frac{a_{n+1}}{a_n} \right| = \lim_{n \to \infty} \left| \frac{n^3}{(n+1)^3} \right| = \lim_{n \to \infty} \frac{1}{(1 + 1/n)^3} = 1.$

7. False. See the note after Example 4 in Section 8.3.

9. True. See (6) in Section 8.1.

11. True. By Theorem 8.7.5 the coefficient of x^3 is $\dfrac{f'''(0)}{3!} = \dfrac{1}{3} \implies f'''(0) = 2.$

    Or: Use Theorem 8.6.2 to differentiate f three times.

13. False. For example, let $a_n = b_n = (-1)^n$. Then $\{a_n\}$ and $\{b_n\}$ are divergent, but $a_n b_n = 1$, so $\{a_n b_n\}$ is convergent.

15. True by Theorem 8.4.1. $\left[\sum (-1)^n a_n \text{ is absolutely convergent and hence convergent.}\right]$

Exercises

1. $\lim\limits_{n\to\infty} a_n = \lim\limits_{n\to\infty}\dfrac{n}{2n+5} = \lim\limits_{n\to\infty}\dfrac{1}{2+5/n} = \dfrac{1}{2}$. Since the limit exists, the sequence is convergent.

3. $\{2n+5\}$ is divergent since $2n+5 \to \infty$ as $n \to \infty$.

5. $\{\sin n\}$ is divergent since $\lim\limits_{n\to\infty} \sin n$ does not exist.

7. $\left\{\left(1+\dfrac{3}{n}\right)^{4n}\right\}$ is convergent. Let $y = \left(1+\dfrac{3}{x}\right)^{4x}$. Then

$$\lim_{x\to\infty}\ln y = \lim_{x\to\infty} 4x\ln(1+3/x) = \lim_{x\to\infty}\frac{\ln(1+3/x)}{1/(4x)} \overset{H}{=} \lim_{x\to\infty}\frac{\dfrac{1}{1+3/x}\left(-\dfrac{3}{x^2}\right)}{-1/(4x^2)}$$

$$= \lim_{x\to\infty}\frac{12}{1+3/x} = 12, \text{ so } \lim_{x\to\infty} y = \lim_{n\to\infty}\left(1+\frac{3}{n}\right)^{4n} = e^{12}.$$

9. Use the Limit Comparison Test with $a_n = \dfrac{n^2}{n^3+1}$ and $b_n = \dfrac{1}{n}$. Then

$$\lim_{n\to\infty}\frac{a_n}{b_n} = \lim_{n\to\infty}\frac{n^2/(n^3+1)}{1/n} = \lim_{n\to\infty}\frac{1}{1+1/n^3} = 1 > 0. \text{ Since } \sum_{n=1}^{\infty}\frac{1}{n} \text{ (the harmonic series)}$$

diverges, $\displaystyle\sum_{n=1}^{\infty}\frac{n^2}{n^3+1}$ also diverges.

11. This is an alternating series with $a_n = \dfrac{1}{n^{1/4}}$, $a_n > 0$ for all n, and $a_n > a_{n+1}$.

$\lim\limits_{n\to\infty} a_n = \lim\limits_{n\to\infty}\dfrac{1}{n^{1/4}} = 0$, so the series converges by the Alternating Series Test.

13. $\lim\limits_{n\to\infty}\left|\dfrac{a_{n+1}}{a_n}\right| = \lim\limits_{n\to\infty}\dfrac{4^{n+1}}{(n+1)3^{n+1}}\cdot\dfrac{n3^n}{4^n} = \dfrac{4}{3}\lim\limits_{n\to\infty}\dfrac{n}{n+1} = \dfrac{4}{3} > 1$ so the series diverges by the Ratio Test.

15. $\left|\dfrac{\sin n}{1+n^2}\right| \le \dfrac{1}{1+n^2} < \dfrac{1}{n^2}$ and since $\displaystyle\sum_{n=1}^{\infty}\dfrac{1}{n^2}$ converges (p-series with $p = 2 > 1$), so does

$\displaystyle\sum_{n=1}^{\infty}\left|\dfrac{\sin n}{1+n^2}\right|$ by the Comparison Test, and so does $\displaystyle\sum_{n=1}^{\infty}\dfrac{\sin n}{1+n^2}$ by Theorem 8.4.1.

17. $\lim\limits_{n\to\infty}\left|\dfrac{a_{n+1}}{a_n}\right| = \lim\limits_{n\to\infty}\dfrac{1\cdot3\cdot5\cdots(2n-1)(2n+1)}{5^{n+1}(n+1)!}\cdot\dfrac{5^n n!}{1\cdot3\cdot5\cdots(2n-1)} = \lim\limits_{n\to\infty}\dfrac{2n+1}{5(n+1)}$

$= \dfrac{2}{5} < 1$, so the series converges by the Ratio Test.

19. Convergent geometric series. $\displaystyle\sum_{n=1}^{\infty}\frac{2^{2n+1}}{5^n} = \sum_{n=1}^{\infty}\frac{(2^2)^n\cdot2^1}{5^n} = 2\sum_{n=1}^{\infty}\frac{4^n}{5^n} = 2\left(\frac{\frac{4}{5}}{1-\frac{4}{5}}\right) = 8.$

21. $\displaystyle\sum_{n=1}^{\infty} \left[\tan^{-1}(n+1) - \tan^{-1} n\right] = \lim_{n\to\infty} \left[(\tan^{-1} 2 - \tan^{-1} 1) + (\tan^{-1} 3 - \tan^{-1} 2) + \cdots\right.$

$$+ \left.(\tan^{-1}(n+1) - \tan^{-1} n)\right]$$

$$= \lim_{n\to\infty} \left[\tan^{-1}(n+1) - \tan^{-1} 1\right] = \tfrac{\pi}{2} - \tfrac{\pi}{4} = \tfrac{\pi}{4}$$

23. $1.2 + 0.0\overline{345} = \dfrac{12}{10} + \dfrac{345/10{,}000}{1 - 1/1000} = \dfrac{12}{10} + \dfrac{345}{9990} = \dfrac{4111}{3330}$

25. $\displaystyle\sum_{n=1}^{\infty} \dfrac{(-1)^{n+1}}{n^5} = 1 - \dfrac{1}{32} + \dfrac{1}{243} - \dfrac{1}{1024} + \dfrac{1}{3125} - \dfrac{1}{7776} + \dfrac{1}{16{,}807} - \dfrac{1}{32{,}768} + \cdots.$ Since

$\dfrac{1}{32{,}768} < 0.000031, \displaystyle\sum_{n=1}^{\infty} \dfrac{(-1)^{n+1}}{n^5} \approx \sum_{n=1}^{7} \dfrac{(-1)^{n+1}}{n^5} \approx 0.9721.$

27. $\displaystyle\sum_{n=1}^{\infty} \dfrac{1}{2 + 5^n} \approx \sum_{n=1}^{8} \dfrac{1}{2 + 5^n} \approx 0.18976224.$ To estimate the error, note that $\dfrac{1}{2 + 5^n} < \dfrac{1}{5^n}$, so the

remainder term is $R_8 = \displaystyle\sum_{n=9}^{\infty} \dfrac{1}{2 + 5^n} < \sum_{n=9}^{\infty} \dfrac{1}{5^n} = \dfrac{1/5^9}{1 - 1/5} = 6.4 \times 10^{-7}$ (geometric series with

$a = \tfrac{1}{5^9}$ and $r = \tfrac{1}{5}$).

29. Use the Limit Comparison Test. $\displaystyle\lim_{n\to\infty} \left| \dfrac{\left(\dfrac{n+1}{n}\right) a_n}{a_n} \right| = \lim_{n\to\infty} \dfrac{n+1}{n} = \lim_{n\to\infty} \left(1 + \dfrac{1}{n}\right) = 1 > 0.$ Since

$\sum |a_n|$ is convergent, so is $\displaystyle\sum \left| \left(\dfrac{n+1}{n}\right) a_n \right|$ by the Limit Comparison Test.

31. $\displaystyle\lim_{n\to\infty} \left| \dfrac{a_{n+1}}{a_n} \right| = \lim_{n\to\infty} \left| \dfrac{x^{n+1}}{3^{n+1}(n+1)^3} \cdot \dfrac{3^n n^3}{x^n} \right| = \dfrac{|x|}{3} \lim_{n\to\infty} \left(\dfrac{n}{n+1} \right)^3 = \dfrac{|x|}{3} < 1$ for convergence

(Ratio Test) $\Rightarrow$ $|x| < 3$ and the radius of convergence is 3. When $x = \pm 3$, $\displaystyle\sum_{n=1}^{\infty} |a_n| = \sum_{n=1}^{\infty} \dfrac{1}{n^3}$,

which is a convergent p-series ($p = 3 > 1$), so the interval of convergence is $[-3, 3]$.

33. $\displaystyle\lim_{n\to\infty} \left| \dfrac{a_{n+1}}{a_n} \right| = \lim_{n\to\infty} \left| \dfrac{2^{n+1}(x-3)^{n+1}}{\sqrt{n+4}} \cdot \dfrac{\sqrt{n+3}}{2^n (x-3)^n} \right| = 2|x-3| \lim_{n\to\infty} \sqrt{\dfrac{n+3}{n+4}} = 2|x-3| <$

$1 \Longleftrightarrow |x - 3| < \tfrac{1}{2}$, so the radius of convergence is $\tfrac{1}{2}$. For $x = \tfrac{7}{2}$, the series becomes

$\displaystyle\sum_{n=0}^{\infty} \dfrac{1}{\sqrt{n+3}} = \sum_{n=3}^{\infty} \dfrac{1}{n^{1/2}}$, which diverges ($p = \tfrac{1}{2} \le 1$), but for $x = \tfrac{5}{2}$, we get $\displaystyle\sum_{n=0}^{\infty} \dfrac{(-1)^n}{\sqrt{n+3}}$, which is a

convergent alternating series, so the interval of convergence is $\left[\tfrac{5}{2}, \tfrac{7}{2}\right)$.

35.

$$
\begin{array}{ll}
f\left(x\right)=\sin x & f\left(\frac{\pi}{6}\right)=\frac{1}{2}\\[4pt]
f'\left(x\right)=\cos x & f'\left(\frac{\pi}{6}\right)=\frac{\sqrt{3}}{2}\\[4pt]
f''\left(x\right)=-\sin x & f''\left(\frac{\pi}{6}\right)=-\frac{1}{2}\\[4pt]
f'''\left(x\right)=-\cos x & f'''\left(\frac{\pi}{6}\right)=-\frac{\sqrt{3}}{2}\\[4pt]
f^{(4)}\left(x\right)=\sin x & f^{(4)}\left(\frac{\pi}{6}\right)=\frac{1}{2}
\end{array}
$$

$$\cdots \qquad\qquad \cdots$$

$$f^{(2n)}\left(\tfrac{\pi}{6}\right)=(-1)^n\cdot\tfrac{1}{2} \text{ and } f^{(2n+1)}\left(\tfrac{\pi}{6}\right)=(-1)^n\cdot\tfrac{\sqrt{3}}{2}.$$

$$\sin x = \sum_{n=0}^{\infty}\frac{f^{(n)}\left(\frac{\pi}{6}\right)}{n!}\left(x-\tfrac{\pi}{6}\right)^n$$

$$= \sum_{n=0}^{\infty}\frac{(-1)^n}{2\,(2n)!}\left(x-\tfrac{\pi}{6}\right)^{2n}+\sum_{n=0}^{\infty}\frac{(-1)^n\sqrt{3}}{2\,(2n+1)!}\left(x-\tfrac{\pi}{6}\right)^{2n+1}$$

37. $\dfrac{1}{1+x}=\dfrac{1}{1-(-x)}=\displaystyle\sum_{n=0}^{\infty}(-1)^n\,x^n$ for $|x|<1$ $\Rightarrow$ $\dfrac{x^2}{1+x}=\displaystyle\sum_{n=0}^{\infty}(-1)^n\,x^{n+2}$ with $R=1$.

39. $\dfrac{1}{1-x}=\displaystyle\sum_{n=0}^{\infty}x^n$ for $|x|<1$ $\Rightarrow$ $\ln\left(1-x\right)=-\displaystyle\int\frac{dx}{1-x}=-\int\sum_{n=0}^{\infty}x^n\,dx=C-\sum_{n=0}^{\infty}\frac{x^{n+1}}{n+1}.$

$\ln\left(1-0\right)=C-0$ $\Rightarrow$ $C=0$ $\Rightarrow$ $\ln\left(1-x\right)=-\displaystyle\sum_{n=0}^{\infty}\frac{x^{n+1}}{n+1}=\sum_{n=1}^{\infty}\frac{-x^n}{n}$ with $R=1$.

41. $\sin x=\displaystyle\sum_{n=0}^{\infty}\frac{(-1)^n\,x^{2n+1}}{(2n+1)!}$ $\Rightarrow$ $\sin\left(x^4\right)=\displaystyle\sum_{n=0}^{\infty}\frac{(-1)^n\left(x^4\right)^{2n+1}}{(2n+1)!}=\sum_{n=0}^{\infty}\frac{(-1)^n\,x^{8n+4}}{(2n+1)!}$ for all x, so the radius of convergence is ∞.

43. $f\left(x\right)=1/\sqrt[4]{16-x}=(16-x)^{-1/4}=\tfrac{1}{2}\left(1-\tfrac{1}{16}x\right)^{-1/4}$

$$=\tfrac{1}{2}\left[1+\left(-\tfrac{1}{4}\right)\left(-\tfrac{x}{16}\right)+\frac{\left(-\tfrac{1}{4}\right)\left(-\tfrac{5}{4}\right)}{2!}\left(-\tfrac{x}{16}\right)^2+\cdots\right]$$

$$=\frac{1}{2}+\sum_{n=1}^{\infty}\frac{1\cdot5\cdot9\cdots\cdots(4n-3)}{2\cdot4^n\cdot n!\cdot16^n}x^n=\frac{1}{2}+\sum_{n=1}^{\infty}\frac{1\cdot5\cdot9\cdots\cdots(4n-3)}{2^{6n+1}\,n!}x^n$$

for $\left|-\dfrac{x}{16}\right|<1$ $\Rightarrow$ $R=16$.

45. $e^x=\displaystyle\sum_{n=0}^{\infty}\frac{x^n}{n!}$ so $\dfrac{e^x}{x}=\dfrac{1}{x}+\displaystyle\sum_{n=1}^{\infty}\frac{x^{n-1}}{n!}$ and $\displaystyle\int\frac{e^x}{x}\,dx=C+\ln|x|+\sum_{n=1}^{\infty}\frac{x^n}{n\cdot n!}$

47. (a)

$$f(x) = x^{1/2} \qquad\qquad f(1) = 1$$
$$f'(x) = \tfrac{1}{2}x^{-1/2} \qquad\qquad f'(1) = \tfrac{1}{2}$$
$$f''(x) = -\tfrac{1}{4}x^{-3/2} \qquad\qquad f''(1) = -\tfrac{1}{4}$$
$$f'''(x) = \tfrac{3}{8}x^{-5/2} \qquad\qquad f'''(1) = \tfrac{3}{8}$$
$$f^{(4)}(x) = -\tfrac{15}{16}x^{-7/2}$$

$$\sqrt{x} \approx T_3(x) = 1 + \frac{1/2}{1!}(x-1) - \frac{1/4}{2!}(x-1)^2 + \frac{3/8}{3!}(x-1)^3$$

$$= 1 + \frac{1}{2}(x-1) - \frac{1}{8}(x-1)^2 + \frac{1}{16}(x-1)^3$$

(b)

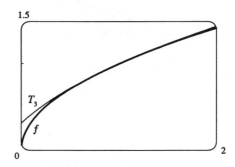

(c) $|R_3(x)| \le \dfrac{M}{4!}|x-1|^4$, where $\left|f^{(4)}(x)\right| \le M$ with $f^{(4)}(x) = -\tfrac{15}{16}x^{-7/2}$. Now $0.9 \le x \le 1.1$

$\Rightarrow \quad (x-1)^4 \le (0.1)^4$, and letting $x = 0.9$ gives $M = \dfrac{15}{16(0.9)^{7/2}}$, so

$$|R_3(x)| \le \frac{15}{16(0.9)^{7/2}\,4!}(0.1)^4 \approx 0.000005648.$$

(d)

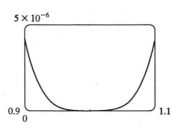

From the graph of $|R_3(x)| = |\sqrt{x} - T_3(x)|$, it appears that the error is less than 5×10^{-6} on $[0.9, 1.1]$.

49. $e^x = \sum\limits_{n=0}^{\infty} \dfrac{x^n}{n!} \;\Rightarrow\; e^{-1/x^2} = \sum\limits_{n=0}^{\infty} \dfrac{\left(-1/x^2\right)^n}{n!} = 1 - \dfrac{1}{x^2} + \dfrac{1}{2x^4} - \cdots \;\Rightarrow\;$

$x^2 \left(1 - e^{-1/x^2}\right) = x^2 \left(\dfrac{1}{x^2} - \dfrac{1}{2x^4} + \cdots\right) = 1 - \dfrac{1}{2x^2} + \cdots \to 1$ as $x \to \infty$.

51. Let $y(x) = \sum\limits_{n=0}^{\infty} c_n x^n$. Then $y''(x) = \sum\limits_{n=2}^{\infty} n(n-1) c_n x^{n-2} = \sum\limits_{n=0}^{\infty} (n+2)(n+1) c_{n+2} x^n$,

$xy' = x \sum\limits_{n=1}^{\infty} n c_n x^{n-1} = \sum\limits_{n=0}^{\infty} n c_n x^n$, and the equation $y'' + xy' + y = 0$ becomes

$\sum\limits_{n=0}^{\infty} \left[(n+2)(n+1) c_{n+2} + (n+1) c_n\right] x^n = 0$. Thus, the recurrence relation is

$c_{n+2} = -c_n / (n+2)$ for $n = 0, 1, 2, \ldots$. But $c_0 = y(0) = 0$, so $c_{2n} = 0$ for $n = 0, 1, 2, \ldots$. Also,

$c_1 = y'(0) = 1$, so $c_3 = -\dfrac{1}{3}$, $c_5 = \dfrac{(-1)^2}{3 \cdot 5} = \dfrac{(-1)^2 \, 2^2 2!}{5!}$, $c_7 = \dfrac{(-1)^3}{3 \cdot 5 \cdot 7} = \dfrac{(-1)^3 \, 2^3 3!}{7!}, \ldots,$

$c_{2n+1} = \dfrac{(-1)^n \, 2^n n!}{(2n+1)!}$ for $n = 0, 1, 2, \ldots$. Note that $2^n n! = (2 \cdot 1) \cdot (2 \cdot 2) \cdot (2 \cdot 3) \cdots \cdots (2 \cdot n)$. Thus,

the solution to the initial-value problem is $y(x) = \sum\limits_{n=0}^{\infty} c_n x^n = \sum\limits_{n=0}^{\infty} \dfrac{(-1)^n \, 2^n n! x^{2n+1}}{(2n+1)!}$.

Focus on Problem Solving

1. It would be far too much work to compute 15 derivatives of f. The key idea is to remember that $f^{(n)}(0)$ occurs in the coefficient of x^n in the Maclaurin series of f. We start with the Maclaurin series for sin:

$$\sin x = x - \frac{x^3}{3!} + \frac{x^5}{5!} - \cdots. \text{ Then } \sin\left(x^3\right) = x^3 - \frac{x^9}{3!} + \frac{x^{15}}{5!} - \cdots \text{ and so the coefficient of } x^{15} \text{ is}$$

$$\frac{f^{(15)}(0)}{15!} = \frac{1}{5!}. \text{ Therefore, } f^{(15)}(0) = \frac{15!}{5!} = 6 \cdot 7 \cdot 8 \cdot 9 \cdot 10 \cdot 11 \cdot 12 \cdot 13 \cdot 14 \cdot 15 = 10{,}897{,}286{,}400.$$

3. (a) From the addition formula for tan in the front endpapers, with $x = y = \theta$, we get

$$\tan 2\theta = \frac{2\tan\theta}{1 - \tan^2\theta}, \text{ so } \cot 2\theta = \frac{1 - \tan^2\theta}{2\tan\theta} \quad \Rightarrow \quad 2\cot 2\theta = \frac{1 - \tan^2\theta}{\tan\theta} = \cot\theta - \tan\theta.$$

Replacing θ by $\frac{1}{2}x$, we get $2\cot x = \cot\frac{1}{2}x - \tan\frac{1}{2}x$, or $\tan\frac{1}{2}x = \cot\frac{1}{2}x - 2\cot x$.

(b) From part (a), $\tan\dfrac{x}{2^n} = \cot\dfrac{x}{2^n} - 2\cot\dfrac{x}{2^{n-1}}$, so the nth partial sum of $\displaystyle\sum_{n=1}^{\infty} \frac{1}{2^n}\tan\frac{x}{2^n}$ is

$$s_n = \frac{\tan(x/2)}{2} + \frac{\tan(x/4)}{4} + \frac{\tan(x/8)}{8} + \cdots + \frac{\tan(x/2^n)}{2^n}$$

$$= \left[\frac{\cot(x/2)}{2} - \cot x\right] + \left[\frac{\cot(x/4)}{4} - \frac{\cot(x/2)}{2}\right] + \left[\frac{\cot(x/8)}{8} - \frac{\cot(x/4)}{4}\right] + \cdots$$

$$+ \left[\frac{\cot(x/2^n)}{2^n} - \frac{\cot(x/2^{n-1})}{2^{n-1}}\right] = -\cot x + \frac{\cot(x/2^n)}{2^n} \quad \text{(telescoping sum)}$$

Now $\dfrac{\cot(x/2^n)}{2^n} = \dfrac{\cos(x/2^n)}{2^n \sin(x/2^n)} = \dfrac{\cos(x/2^n)}{x} \cdot \dfrac{x/2^n}{\sin(x/2^n)} \to \dfrac{1}{x} \cdot 1 = \dfrac{1}{x}$ as $n \to \infty$

since $x/2^n \to 0$ for $x \ne 0$. Therefore, if $x \ne 0$ and $x \ne n\pi$, then

$$\sum_{n=1}^{\infty} \frac{1}{2^n}\tan\frac{x}{2^n} = \lim_{n\to\infty}\left(-\cot x + \frac{1}{2^n}\cot\frac{x}{2^n}\right) = -\cot x + \frac{1}{x}. \text{ If } x = 0, \text{ then all terms in the}$$

series are 0, so the sum is 0.

5. (a) At each stage, each side is replaced by four shorter sides, each of length $\frac{1}{3}$ of the side length at the preceding stage. Writing s_0 and ℓ_0 for the number of sides and the length of the side of the initial triangle, we generate the table at right.

In general, we have $s_n = 3 \cdot 4^n$ and $\ell_n = \left(\frac{1}{3}\right)^n$, so the length of the perimeter at the nth stage of construction is

$s_0 = 3$	$\ell_0 = 1$
$s_1 = 3 \cdot 4$	$\ell_1 = 1/3$
$s_2 = 3 \cdot 4^2$	$\ell_2 = 1/3^2$
$s_3 = 3 \cdot 4^3$	$\ell_3 = 1/3^3$
$\cdots$	$\cdots$

$$p_n = s_n\ell_n = 3 \cdot 4^n \cdot \left(\tfrac{1}{3}\right)^n = 3 \cdot \left(\tfrac{4}{3}\right)^n.$$

(b) $p_n = \dfrac{4^n}{3^{n-1}} = 4\left(\dfrac{4}{3}\right)^{n-1}$. Since $\frac{4}{3} > 1$, $p_n \to \infty$ as $n \to \infty$.

(c) The area of each of the small triangles added at a given stage is one-ninth of the area of the triangle added at the preceding stage. Let a be the area of the original triangle. Then the area a_n of each of the small triangles added at stage n is $a_n = a \cdot \dfrac{1}{9^n} = \dfrac{a}{9^n}$. Since a small triangle is added to each side at every stage, it follows that the total area A_n added to the figure at the nth stage is

$$A_n = s_{n-1} \cdot a_n = 3 \cdot 4^{n-1} \cdot \frac{a}{9^n} = a \cdot \frac{4^{n-1}}{3^{2n-1}}.$$ Then the total area enclosed by the snowflake curve

is $A = a + A_1 + A_2 + A_3 + \cdots = a + a \cdot \dfrac{1}{3} + a \cdot \dfrac{4}{3^3} + a \cdot \dfrac{4^2}{3^5} + a \cdot \dfrac{4^3}{3^7} + \cdots$. After the first term,

this is a geometric series with common ratio $\frac{4}{9}$, so $A = a + \dfrac{a/3}{1 - \frac{4}{9}} = a + \dfrac{a}{3} \cdot \dfrac{9}{5} = \dfrac{8a}{5}$. But the area

of the original equilateral triangle with side 1 is $a = \frac{1}{2} \cdot 1 \cdot \sin \frac{\pi}{3} = \frac{\sqrt{3}}{4}$. So the area enclosed by the

snowflake curve is $\frac{8}{5} \cdot \frac{\sqrt{3}}{4} = \frac{2\sqrt{3}}{5}$.

7. (a) Let $a = \arctan x$ and $b = \arctan y$. Then, from Formula 14b in Appendix C,

$$\tan(a - b) = \frac{\tan a - \tan b}{1 + \tan a \tan b} = \frac{\tan(\arctan x) - \tan(\arctan y)}{1 + \tan(\arctan x)\tan(\arctan y)} \quad \Rightarrow \quad \tan(a - b) = \frac{x - y}{1 + xy}$$

$$\Rightarrow \quad \arctan x - \arctan y = a - b = \arctan \frac{x - y}{1 + xy} \quad \text{since } -\frac{\pi}{2} < \arctan x - \arctan y < \frac{\pi}{2}.$$

(b) From part (a) we have

$$\arctan \tfrac{120}{119} - \arctan \tfrac{1}{239} = \arctan \frac{\frac{120}{119} - \frac{1}{239}}{1 + \frac{120}{119} \cdot \frac{1}{239}} = \arctan \frac{\frac{28,561}{28,441}}{\frac{28,561}{28,441}} = \arctan 1 = \tfrac{\pi}{4}.$$

(c) Replacing y by $-y$ in the formula of part (a), we get $\arctan x + \arctan y = \arctan \dfrac{x + y}{1 - xy}$. So

$$4 \arctan \tfrac{1}{5} = 2\left(\arctan \tfrac{1}{5} + \arctan \tfrac{1}{5}\right) = 2\arctan \frac{\frac{1}{5} + \frac{1}{5}}{1 - \frac{1}{5} \cdot \frac{1}{5}} = 2\arctan \tfrac{5}{12}$$

$$= \arctan \tfrac{5}{12} + \arctan \tfrac{5}{12}$$

$$= \arctan \frac{\frac{5}{12} + \frac{5}{12}}{1 - \frac{5}{12} \cdot \frac{5}{12}} = \arctan \tfrac{120}{119}$$

Thus, from part (b), we have $4 \arctan \tfrac{1}{5} - \arctan \tfrac{1}{239} = \arctan \tfrac{120}{119} - \arctan \tfrac{1}{239} = \tfrac{\pi}{4}$.

(d) From Example 7 in Section 8.6 we have $\arctan x = x - \dfrac{x^3}{3} + \dfrac{x^5}{5} - \dfrac{x^7}{7} + \dfrac{x^9}{9} - \dfrac{x^{11}}{11} + \cdots$, so

$\arctan \dfrac{1}{5} = \dfrac{1}{5} - \dfrac{1}{3 \cdot 5^3} + \dfrac{1}{5 \cdot 5^5} - \dfrac{1}{7 \cdot 5^7} + \dfrac{1}{9 \cdot 5^9} - \dfrac{1}{11 \cdot 5^{11}} + \cdots$. This is an alternating series and

the size of the terms decreases to 0, so by the Alternating Series Estimation Theorem, the sum lies

between s_5 and s_6, that is, $0.197395560 < \arctan \frac{1}{5} < 0.197395562$.

(e) From the series in part (d) we get $\arctan \dfrac{1}{239} = \dfrac{1}{239} - \dfrac{1}{3 \cdot 239^3} + \dfrac{1}{5 \cdot 239^5} - \cdots$. The third term is less than 2.6×10^{-13}, so by the Alternating Series Estimation Theorem, we have, to nine decimal places, $\arctan \frac{1}{239} \approx s_2 \approx 0.004184076$. Thus, $0.004184075 < \arctan \frac{1}{239} < 0.004184077$.

(f) From part (c) we have $\pi = 16 \arctan \frac{1}{5} - 4 \arctan \frac{1}{239}$, so from parts (d) and (e) we have
$$16 \,(0.197395560) - 4 \,(0.004184077) < \pi < 16\,(0.197395562) - 4\,(0.004184075) \quad \Rightarrow$$
$3.141592652 < \pi < 3.141592692$. So, to 7 decimal places, $\pi \approx 3.1415927$.

9. We start with the geometric series $\displaystyle\sum_{n=0}^{\infty} x^n = \dfrac{1}{1-x}$, $|x| < 1$, and differentiate:

$$\sum_{n=1}^{\infty} n x^{n-1} = \frac{d}{dx}\left(\sum_{n=0}^{\infty} x^n\right) = \frac{d}{dx}\left(\frac{1}{1-x}\right) = \frac{1}{(1-x)^2} \text{ for } |x| < 1 \quad \Rightarrow$$

$$\sum_{n=1}^{\infty} n x^n = x \sum_{n=1}^{\infty} n x^{n-1} = \frac{x}{(1-x)^2} \text{ for } |x| < 1. \text{ Differentiate again:}$$

$$\sum_{n=1}^{\infty} n^2 x^{n-1} = \frac{d}{dx}\frac{x}{(1-x)^2} = \frac{(1-x)^2 - x \cdot 2(1-x)(-1)}{(1-x)^4} = \frac{x+1}{(1-x)^3} \quad \Rightarrow$$

$$\sum_{n=1}^{\infty} n^2 x^n = \frac{x^2 + x}{(1-x)^3} \quad \Rightarrow$$

$$\sum_{n=1}^{\infty} n^3 x^{n-1} = \frac{d}{dx}\frac{x^2 + x}{(1-x)^3} = \frac{(1-x)^3 (2x+1) - (x^2+x)\,3(1-x)^2(-1)}{(1-x)^6} = \frac{x^2 + 4x + 1}{(1-x)^4} \quad \Rightarrow$$

$$\sum_{n=1}^{\infty} n^3 x^n = \frac{x^3 + 4x^2 + x}{(1-x)^4}, \ |x| < 1. \text{ The radius of convergence is 1 because that is the radius of}$$

convergence for the geometric series we started with. If $x = \pm 1$, the series is $\sum n^3 (\pm 1)^n$, which diverges by the Test For Divergence, so the interval of convergence is $(-1, 1)$.

11. $u = 1 + \dfrac{x^3}{3!} + \dfrac{x^6}{6!} + \dfrac{x^9}{9!} + \cdots$, $v = x + \dfrac{x^4}{4!} + \dfrac{x^7}{7!} + \dfrac{x^{10}}{10!} + \cdots$, $w = \dfrac{x^2}{2!} + \dfrac{x^5}{5!} + \dfrac{x^8}{8!} + \cdots$. The key idea is to differentiate: $\dfrac{du}{dx} = \dfrac{3x^2}{3!} + \dfrac{6x^5}{6!} + \dfrac{9x^8}{9!} + \cdots = \dfrac{x^2}{2!} + \dfrac{x^5}{5!} + \dfrac{x^8}{8!} + \cdots = w$. Similarly, $\dfrac{dv}{dx} = 1 + \dfrac{x^3}{3!} + \dfrac{x^6}{6!} + \dfrac{x^9}{9!} + \cdots = u$, and $\dfrac{dw}{dx} = x + \dfrac{x^4}{4!} + \dfrac{x^7}{7!} + \dfrac{x^{10}}{10!} + \cdots = v$. So $u' = w$, $v' = u$, and $w' = v$. Now differentiate the left hand side of the desired equation:

$$\frac{d}{dx}\left(u^3 + v^3 + w^3 - 3uvw\right) = 3u^2 u' + 3v^2 v' + 3w^2 w' - 3\,(u'vw + uv'w + uvw')$$

$$= 3u^2 w + 3v^2 u + 3w^2 v - 3\,(vw^2 + u^2 w + uv^2) = 0 \quad \Rightarrow$$

$u^3 + v^3 + w^3 - 3uvw = C$. To find the value of the constant C, we put $x = 0$ in the last equation and get $1^3 + 0^3 + 0^3 - 3\,(1 \cdot 0 \cdot 0) = C \quad \Rightarrow \quad C = 1$, so $u^3 + v^3 + w^3 - 3uvw = 1$.

13. If L is the length of a side of the equilateral triangle, then the area is $A = \frac{1}{2}L \cdot \frac{\sqrt{3}}{2}L = \frac{\sqrt{3}}{4}L^2$ and so $L^2 = \frac{4}{\sqrt{3}}A$. Let r be the radius of one of the circles. When there are n rows of circles, the figure shows

that $L = \sqrt{3}r + r + (n-2)(2r) + r + \sqrt{3}r = r\left(2n - 2 + 2\sqrt{3}\right)$, so $r = \dfrac{L}{2\left(n + \sqrt{3} - 1\right)}$. The

number of circles is $1 + 2 + \cdots + n = \dfrac{n(n+1)}{2}$ and so the total area of the circles is

$$
\begin{aligned}
A_n &= \frac{n(n+1)}{2}\pi r^2 = \frac{n(n+1)}{2}\pi \frac{L^2}{4\left(n + \sqrt{3} - 1\right)^2} \\
&= \frac{n(n+1)}{2}\pi \frac{4A/\sqrt{3}}{4\left(n + \sqrt{3} - 1\right)^2} = \frac{n(n+1)}{\left(n + \sqrt{3} - 1\right)^2}\frac{\pi A}{2\sqrt{3}} \quad \Rightarrow
\end{aligned}
$$

$$
\frac{A_n}{A} = \frac{n(n+1)}{\left(n + \sqrt{3} - 1\right)^2}\frac{\pi}{2\sqrt{3}} = \frac{1 + 1/n}{\left[1 + \left(\sqrt{3} - 1\right)/n\right]^2}\frac{\pi}{2\sqrt{3}} \to \frac{\pi}{2\sqrt{3}} \text{ as } n \to \infty.
$$

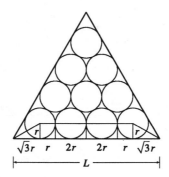

15. Call the series S. We group the terms according to the number of digits in their denominators:

$$
S = \underbrace{\left(1 + \frac{1}{2} + \cdots + \frac{1}{8} + \frac{1}{9}\right)}_{g_1} + \underbrace{\left(\frac{1}{11} + \cdots + \frac{1}{99}\right)}_{g_2} + \underbrace{\left(\frac{1}{111} + \cdots + \frac{1}{999}\right)}_{g_3} + \cdots
$$

Now in the group g_n, there are 9^n terms, since we have 9 choices for each of the n digits in the

denominator. Furthermore, each term in g_n is less than $\dfrac{1}{10^{n-1}}$. So $g_n < 9^n \cdot \dfrac{1}{10^{n-1}} = 9\left(\dfrac{9}{10}\right)^{n-1}$.

Now $\displaystyle\sum_{n=1}^{\infty} 9\left(\frac{9}{10}\right)^{n-1}$ is a geometric series with $a = 9$ and $r = \frac{9}{10} < 1$. Therefore, by the

Comparison Test, $S = \displaystyle\sum_{n=1}^{\infty} g_n < \sum_{n=1}^{\infty} 9\left(\frac{9}{10}\right)^{n-1} = \dfrac{9}{1 - \frac{9}{10}} = 90.$

Chapter 9   Vectors and the Geometry of Space

Section 9.1   Three-Dimensional Coordinate Systems

1. We start at the origin, which has coordinates $(0, 0, 0)$. First we move 4 units along the positive x-axis, affecting only the x-coordinate, bringing us to the point $(4, 0, 0)$. We then move 3 units straight downward, in the negative z-direction. Thus only the z-coordinate is affected, and we arrive at $(4, 0, -3)$.

3. The distance from a point to the xz-plane is the absolute value of the y-coordinate of the point. $Q\,(-5, -1, 4)$ has the y-coordinate with the smallest absolute value, so Q is the point closest to the xz-plane. $R\,(0, 3, 8)$ must lie in the yz-plane since the distance from R to the yz-plane, given by the x-coordinate of R, is 0.

5. The equation $x + y = 2$ represents the set of all points in $\mathbb{R}^3$ whose x- and y-coordinates have a sum of 2, or equivalently where $y = 2 - x$. This is the set $\{(x, 2 - x, z) \mid x \in \mathbb{R}, z \in \mathbb{R}\}$ which is a vertical plane that intersects the xy-plane in the line $y = 2 - x$, $z = 0$.

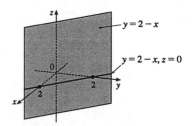

7. We can find the lengths of the sides of the triangle by using the distance formula between pairs of vertices:

$$|AB| = \sqrt{(5 - 3)^2 + [-3 - (-4)]^2 + (0 - 1)^2} = \sqrt{4 + 1 + 1} = \sqrt{6}$$

$$|BC| = \sqrt{(6 - 5)^2 + [-7 - (-3)]^2 + (4 - 0)^2} = \sqrt{1 + 16 + 16} = \sqrt{33}$$

$$|CA| = \sqrt{(3 - 6)^2 + [-4 - (-7)]^2 + (1 - 4)^2} = \sqrt{9 + 9 + 9} = \sqrt{27} = 3\sqrt{3}$$

Since the Pythagorean Theorem is satisfied by $|AB|^2 + |CA|^2 = |BC|^2$, ABC is a right triangle. ABC is not isosceles, as no two sides have the same length.

9. An equation of the sphere with center $(0, 1, -1)$ and radius 4 is $(x - 0)^2 + (y - 1)^2 + [z - (-1)]^2 = 4^2$ or $x^2 + (y - 1)^2 + (z + 1)^2 = 16$. The intersection of this sphere with the yz-plane is the set of points on the sphere whose x-coordinate is 0. Putting $x = 0$ in the equation, we have $(y - 1)^2 + (z + 1)^2 = 16$, $x = 0$, which represents a circle in the yz-plane with center $(0, 1, -1)$ and radius 4. (This intersection is a *great circle* of the sphere.)

11. Completing the squares in the equation gives
$$\left(x^2 + 2x + 1\right) + \left(y^2 + 8y + 16\right) + \left(z^2 - 4z + 4\right) = 28 + 1 + 16 + 4 \quad \Rightarrow$$
$$(x + 1)^2 + (y + 4)^2 + (z - 2)^2 = 49 \text{ which we recognize as an equation of a sphere with center }$$
$(-1, -4, 2)$ and radius 7.

13. (a) If the midpoint of the line segment from $P_1\,(x_1, y_1, z_1)$ to $P_2\,(x_2, y_2, z_2)$ is

$Q = \left(\dfrac{x_1 + x_2}{2}, \dfrac{y_1 + y_2}{2}, \dfrac{z_1 + z_2}{2}\right)$, then the distances $|P_1Q|$ and $|QP_2|$ are equal, and each is half

of $|P_1P_2|$. We verify that this is the case:

$$|P_1P_2| = \sqrt{(x_2 - x_1)^2 + (y_2 - y_1)^2 + (z_2 - z_1)^2}$$

$$|P_1Q| = \sqrt{\left[\tfrac{1}{2}(x_1 + x_2) - x_1\right]^2 + \left[\tfrac{1}{2}(y_1 + y_2) - y_1\right]^2 + \left[\tfrac{1}{2}(z_1 + z_2) - z_1\right]^2}$$

$$= \sqrt{\left(\tfrac{1}{2}x_2 - \tfrac{1}{2}x_1\right)^2 + \left(\tfrac{1}{2}y_2 - \tfrac{1}{2}y_1\right)^2 + \left(\tfrac{1}{2}z_2 - \tfrac{1}{2}z_1\right)^2}$$

$$= \sqrt{\left(\tfrac{1}{2}\right)^2 \left[(x_2 - x_1)^2 + (y_2 - y_1)^2 + (z_2 - z_1)^2\right]}$$

$$= \tfrac{1}{2}\sqrt{(x_2 - x_1)^2 + (y_2 - y_1)^2 + (z_2 - z_1)^2} = \tfrac{1}{2}|P_1P_2|$$

$$|QP_2| = \sqrt{\left[x_2 - \tfrac{1}{2}(x_1 + x_2)\right]^2 + \left[y_2 - \tfrac{1}{2}(y_1 + y_2)\right]^2 + \left[z_2 - \tfrac{1}{2}(z_1 + z_2)\right]^2}$$

$$= \sqrt{\left(\tfrac{1}{2}x_2 - \tfrac{1}{2}x_1\right)^2 + \left(\tfrac{1}{2}y_2 - \tfrac{1}{2}y_1\right)^2 + \left(\tfrac{1}{2}z_2 - \tfrac{1}{2}z_1\right)^2}$$

$$= \sqrt{\left(\tfrac{1}{2}\right)^2 \left[(x_2 - x_1)^2 + (y_2 - y_1)^2 + (z_2 - z_1)^2\right]}$$

$$= \tfrac{1}{2}\sqrt{(x_2 - x_1)^2 + (y_2 - y_1)^2 + (z_2 - z_1)^2} = \tfrac{1}{2}|P_1P_2|$$

So Q is indeed the midpoint of P_1P_2.

(b) By part (a), the midpoints of sides AB, BC and CA are $P_1\left(-\tfrac{1}{2}, 1, 4\right)$, $P_2\left(1, \tfrac{1}{2}, 5\right)$ and $P_3\left(\tfrac{5}{2}, \tfrac{3}{2}, 4\right)$.
Then the lengths of the medians are:

$$|AP_2| = \sqrt{0^2 + \left(\tfrac{1}{2} - 2\right)^2 + (5 - 3)^2} = \sqrt{\tfrac{9}{4} + 4} = \sqrt{\tfrac{25}{4}} = \tfrac{5}{2}$$

$$|BP_3| = \sqrt{\left(\tfrac{5}{2} + 2\right)^2 + \left(\tfrac{3}{2}\right)^2 + (4 - 5)^2} = \sqrt{\tfrac{81}{4} + \tfrac{9}{4} + 1} = \sqrt{\tfrac{94}{4}} = \tfrac{1}{2}\sqrt{94}$$

$$|CP_1| = \sqrt{\left(-\tfrac{1}{2} - 4\right)^2 + (1 - 1)^2 + (4 - 5)^2} = \sqrt{\tfrac{81}{4} + 1} = \tfrac{1}{2}\sqrt{85}$$

15. (a) Since the sphere touches the xy-plane, its radius is the distance from its center, $(2, -3, 6)$, to the
xy-plane, namely 6. Therefore $r = 6$ and an equation of the sphere is
$(x - 2)^2 + (y + 3)^2 + (z - 6)^2 = 6^2 = 36$.

(b) The radius of this sphere is the distance from its center $(2, -3, 6)$ to the yz-plane, which is 2.
Therefore, an equation is $(x - 2)^2 + (y + 3)^2 + (z - 6)^2 = 4$.

(c) Here the radius is the distance from the center $(2, -3, 6)$ to the xz-plane, which is 3. Therefore, an
equation is $(x - 2)^2 + (y + 3)^2 + (z - 6)^2 = 9$.

17. The equation $x = 9$ represents a plane parallel to the yz-plane and 9 units in front of it.

19. The inequality $y > 2$ represents a half-space containing all points to the right of the plane $y = 2$.

21. The inequality $|z| \leq 2$ is equivalent to $-2 \leq z \leq 2$, so it represents all points on and between the two horizontal planes $z = 2$ and $z = -2$.

23. The inequality $x^2 + y^2 + z^2 > 1$ is equivalent to $\sqrt{x^2 + y^2 + z^2} > 1$, so the region consists of those points whose distance from the origin is greater than 1. This is the set of all points outside the sphere with radius 1 and center $(0,0,0)$.

25. Completing the square in z gives $x^2 + y^2 + \left(z^2 - 2z + 1\right) < 3 + 1$ or $x^2 + y^2 + (z-1)^2 < 4$, which is equivalent to $\sqrt{x^2 + y^2 + (z-1)^2} < 2$. Thus the region consists of those points whose distance from the point $(0, 0, 1)$ is less than 2. This is the set of all points inside the sphere with radius 2 and center $(0, 0, 1)$.

27. This describes all points with negative y-coordinates, that is, $y < 0$.

29. This describes a region all of whose points have a distance to the origin which is greater than r, but smaller than R. So inequalities describing the region are $r < \sqrt{x^2 + y^2 + z^2} < R$, or $r^2 < x^2 + y^2 + z^2 < R^2$.

31. (a) To find the x- and y-coordinates of the point P, we project it onto L_2 and project the resulting point Q onto the x- and y-axes. To find the z-coordinate, we project P onto either the xz- plane or the yz-plane (using our knowledge of its x- or y- coordinate) and then project the resulting point onto the z-axis. (Or, we could draw a line parallel to QO from P to the z-axis.) The coordinates of P are $(2, 1, 4)$.

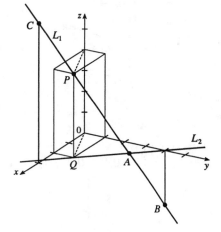

(b) A is the intersection of L_1 and L_2, B is directly below the y-intercept of L_2, and C is directly above the x-intercept of L_2.

33. We need to find a set of points $\{P(x, y, z) \mid |AP| = |BP|\}$.

$$\sqrt{(x+1)^2 + (y-5)^2 + (z-3)^2} = \sqrt{(x-6)^2 + (y-2)^2 + (z+2)^2} \Rightarrow$$
$$(x+1)^2 + (y-5) + (z-3)^2 = (x-6)^2 + (y-2)^2 + (z+2)^2 \Rightarrow$$
$$x^2 + 2x + 1 + y^2 - 10y + 25 + z^2 - 6z + 9 = x^2 - 12x + 36 + y^2 - 4y + 4 + z^2 + 4z + 4 \Rightarrow$$
$14x - 6y - 10z = 9$. Thus the set of points is a plane perpendicular to the line segment joining A and B (since this plane must contain the perpendicular bisector of the line segment AB).

Section 9.2 Vectors

1. (a) The cost of a theater ticket is a scalar, because it has only magnitude.

(b) The current in a river is a vector, because it has both magnitude (the speed of the current) and direction at any given location.

(c) If we assume that the initial path is linear, the initial flight path from Houston to Dallas is a vector, because it has both magnitude (distance) and direction.

(d) The population of the world is a scalar, because it has only magnitude.

3. (a) By the Triangle Law, $\overrightarrow{AB} + \overrightarrow{BC}$ is the vector with initial point A and terminal point C, namely $\overrightarrow{AC}$.

(b) By the Triangle Law, $\overrightarrow{CD} + \overrightarrow{DA}$ is the vector with initial point C and terminal point A, namely $\overrightarrow{CA}$.

(c) First we consider $\overrightarrow{BC} - \overrightarrow{DC}$ as $\overrightarrow{BC} + \left(-\overrightarrow{DC}\right)$. Then since $-\overrightarrow{DC}$ has the same length as $\overrightarrow{CD}$ but points in the opposite direction, we have $-\overrightarrow{DC} = \overrightarrow{CD}$ and so $\overrightarrow{BC} - \overrightarrow{DC} = \overrightarrow{BC} + \overrightarrow{CD} = \overrightarrow{BD}$.

(d) We use the Triangle Law twice: $\overrightarrow{BC} + \overrightarrow{CD} + \overrightarrow{DA} = \left(\overrightarrow{BC} + \overrightarrow{CD}\right) + \overrightarrow{DA} = \overrightarrow{BD} + \overrightarrow{DA} = \overrightarrow{BA}$.

5. $\mathbf{a} = \langle 4 - 1, 4 - 3 \rangle = \langle 3, 1 \rangle$

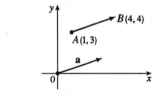

7. $\mathbf{a} = \langle 2 - 0, 3 - 3, -1 - 1 \rangle = \langle 2, 0, -2 \rangle$

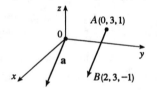

9. $\langle 1, 0, 1 \rangle + \langle 0, 0, 1 \rangle$
$= \langle 1 + 0, 0 + 0, 1 + 1 \rangle = \langle 1, 0, 2 \rangle$

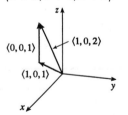

11. $|\mathbf{a}| = \sqrt{5^2 + (-12)^2} = \sqrt{169} = 13$

$\mathbf{a} + \mathbf{b} = \langle 5 - 2, -12 + 8 \rangle = \langle 3, -4 \rangle$

$\mathbf{a} - \mathbf{b} = \langle 5 - (-2), -12 - 8 \rangle = \langle 7, -20 \rangle$

$2\mathbf{a} = \langle 2\,(5), 2\,(-12) \rangle = \langle 10, -24 \rangle$

$3\mathbf{a} + 4\mathbf{b} = \langle 15, -36 \rangle + \langle -8, 32 \rangle = \langle 7, -4 \rangle$

13. $|\mathbf{a}| = \sqrt{1^2 + (-1)^2} = \sqrt{2}$

$\mathbf{a} + \mathbf{b} = (\mathbf{i} - \mathbf{j}) + (\mathbf{i} + \mathbf{j}) = 2\mathbf{i}$

$\mathbf{a} - \mathbf{b} = (\mathbf{i} - \mathbf{j}) - (\mathbf{i} + \mathbf{j}) = -2\mathbf{j}$

$2\mathbf{a} = 2(\mathbf{i} - \mathbf{j}) = 2\mathbf{i} - 2\mathbf{j}$

$3\mathbf{a} + 4\mathbf{b} = 3(\mathbf{i} - \mathbf{j}) + 4(\mathbf{i} + \mathbf{j})$
$= 3\mathbf{i} - 3\mathbf{j} + 4\mathbf{i} + 4\mathbf{j} = 7\mathbf{i} + \mathbf{j}$

15. $|\langle 1, 2 \rangle| = \sqrt{1^2 + 2^2} = \sqrt{5}$. Thus

$\mathbf{u} = \frac{1}{\sqrt{5}} \langle 1, 2 \rangle = \left\langle \frac{1}{\sqrt{5}}, \frac{2}{\sqrt{5}} \right\rangle$.

17. By the Triangle Law, $\overrightarrow{AB} + \overrightarrow{BC} = \overrightarrow{AC}$. Then $\overrightarrow{AB} + \overrightarrow{BC} + \overrightarrow{CA} = \overrightarrow{AC} + \overrightarrow{CA}$, but

$\overrightarrow{AC} + \overrightarrow{CA} = \overrightarrow{AC} + \left(-\overrightarrow{AC} \right) = \mathbf{0}$. So $\overrightarrow{AB} + \overrightarrow{BC} + \overrightarrow{CA} = \mathbf{0}$.

19. $|\mathbf{F}_1| = 10$ lb and $|\mathbf{F}_2| = 12$ lb.

$\mathbf{F}_1 = -|\mathbf{F}_1| \cos 45^\circ \mathbf{i} + |\mathbf{F}_1| \sin 45^\circ \mathbf{j} = -10 \cos 45^\circ \mathbf{i} + 10 \sin 45^\circ \mathbf{j}$
$= -5\sqrt{2}\mathbf{i} + 5\sqrt{2}\mathbf{j}$

$\mathbf{F}_2 = |\mathbf{F}_2| \cos 30^\circ \mathbf{i} + |\mathbf{F}_2| \sin 30^\circ \mathbf{j} = 12 \cos 30^\circ \mathbf{i} + 12 \sin 30^\circ \mathbf{j} = 6\sqrt{3}\mathbf{i} + 6\mathbf{j}$

$\mathbf{F} = \mathbf{F}_1 + \mathbf{F}_2 = \left(6\sqrt{3} - 5\sqrt{2} \right) \mathbf{i} + \left(6 + 5\sqrt{2} \right) \mathbf{j} \approx 3.32\mathbf{i} + 13.07\mathbf{j}$

$|\mathbf{F}| \approx \sqrt{(3.32)^2 + (13.07)^2} \approx 13.5$ lb. $\tan \theta = \dfrac{6 + 5\sqrt{2}}{6\sqrt{3} - 5\sqrt{2}} \quad \Rightarrow \quad \theta = \tan^{-1} \dfrac{6 + 5\sqrt{2}}{6\sqrt{3} - 5\sqrt{2}} \approx 76^\circ$.

21. With respect to the water's surface, the woman's velocity is the vector sum of the velocity of the ship with respect to the water, and the woman's velocity with respect to the ship. If we let north be the positive y-direction, then $\mathbf{v} = \langle 0, 22 \rangle + \langle -3, 0 \rangle = \langle -3, 22 \rangle$. The woman's speed is $|\mathbf{v}| = \sqrt{9 + 484} \approx 22.2$ mi/h. The vector $\mathbf{v}$ makes an angle θ with the east, where $\theta = \tan^{-1} \frac{22}{-3} \approx 98^\circ$. Therefore, the woman's direction is about $\text{N}\,(98 - 90)^\circ\,\text{W} = \text{N}\,8^\circ\,\text{W}$.

23. Let $\mathbf{T}_1$ and $\mathbf{T}_2$ represent the tension vectors in each side of the clothesline as shown in the figure. $\mathbf{T}_1$ and $\mathbf{T}_2$ have equal vertical components and opposite horizontal components, so we can write

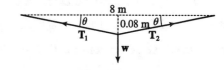

$\mathbf{T}_1 = -a\mathbf{i} + b\mathbf{j}$ and $\mathbf{T}_2 = a\mathbf{i} + b\mathbf{j}$ $(a, b > 0)$.

By similar triangles, $\dfrac{b}{a} = \dfrac{0.08}{4} \quad \Rightarrow \quad a = 50b$. The force due to gravity acting on the shirt has magnitude $0.8g \approx (0.8)(9.8) = 7.84$ N, hence we have $\mathbf{w} = -7.84\mathbf{j}$. The resultant $\mathbf{T}_1 + \mathbf{T}_2$ of the tensile forces counterbalances $\mathbf{w}$, so $\mathbf{T}_1 + \mathbf{T}_2 = -\mathbf{w} \quad \Rightarrow \quad (-a\mathbf{i} + b\mathbf{j}) + (a\mathbf{i} + b\mathbf{j}) = 7.84\mathbf{j} \quad \Rightarrow$
$(-50b\mathbf{i} + b\mathbf{j}) + (50b\mathbf{i} + b\mathbf{j}) = 2b\mathbf{j} = 7.84\mathbf{j} \quad \Rightarrow \quad b = \frac{7.84}{2} = 3.92$ and $a = 50b = 196$. Thus the tensions are $\mathbf{T}_1 = -a\mathbf{i} + b\mathbf{j} = -196\mathbf{i} + 3.92\mathbf{j}$ and $\mathbf{T}_2 = a\mathbf{i} + b\mathbf{j} = 196\mathbf{i} + 3.92\mathbf{j}$.
Alternatively, we can find the value of θ and proceed as in Example 5.

25.

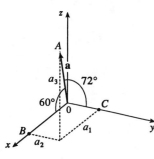

Let $\mathbf{a} = \langle a_1, a_2, a_3 \rangle$, as shown in the figure. Since $|\mathbf{a}| = 1$ and triangle ABO is a right triangle, we have $\cos 60° = \dfrac{a_1}{1} \Rightarrow a_1 = \cos 60°$.

Similarly, triangle ACO is a right triangle, so $a_2 = \cos 72°$. Finally, since $|\mathbf{a}| = 1$ we have $\sqrt{(\cos 60°)^2 + (\cos 72°)^2 + a_3^2} = 1 \Rightarrow$

$a_3^2 = 1 - (\cos 60°)^2 - (\cos 72°)^2 \Rightarrow$

$a_3 = \sqrt{1 - (\cos 60°)^2 - (\cos 72°)^2}$. Thus

$$\mathbf{a} = \left\langle \cos 60°, \cos 72°, \sqrt{1 - (\cos 60°)^2 - (\cos 72°)^2} \right\rangle$$

$$\approx \langle 0.50, 0.31, 0.81 \rangle$$

27. $|\mathbf{r} - \mathbf{r}_0|$ is the distance between the points (x, y, z) and (x_0, y_0, z_0), so the set of points is a sphere with radius 1 and center (x_0, y_0, z_0).

Alternate Method: $|\mathbf{r} - \mathbf{r}_0| = 1 \Leftrightarrow \sqrt{(x - x_0)^2 + (y - y_0)^2 + (z - z_0)^2} = 1 \Leftrightarrow$
$(x - x_0)^2 + (y - y_0)^2 + (z - z_0)^2 = 1$, which is the equation of a sphere with radius 1 and center (x_0, y_0, z_0).

29. $\mathbf{a} + (\mathbf{b} + \mathbf{c}) = \langle a_1, a_2 \rangle + (\langle b_1, b_2 \rangle + \langle c_1, c_2 \rangle) = \langle a_1, a_2 \rangle + \langle b_1 + c_1, b_2 + c_2 \rangle$
$\qquad = \langle a_1 + b_1 + c_1, a_2 + b_2 + c_2 \rangle = \langle (a_1 + b_1) + c_1, (a_2 + b_2) + c_2 \rangle$
$\qquad = \langle a_1 + b_1, a_2 + b_2 \rangle + \langle c_1, c_2 \rangle = (\langle a_1, a_2 \rangle + \langle b_1, b_2 \rangle) + \langle c_1, c_2 \rangle$
$\qquad = (\mathbf{a} + \mathbf{b}) + \mathbf{c}$

31. Consider triangle ABC, where D and E are the midpoints of AB and BC. We know that $\overrightarrow{AB} + \overrightarrow{BC} = \overrightarrow{AC}$ (1) and $\overrightarrow{DB} + \overrightarrow{BE} = \overrightarrow{DE}$ (2). However, $\overrightarrow{DB} = \frac{1}{2}\overrightarrow{AB}$, and $\overrightarrow{BE} = \frac{1}{2}\overrightarrow{BC}$. Substituting these expressions for $\overrightarrow{DB}$ and $\overrightarrow{BE}$ into (2) gives $\frac{1}{2}\overrightarrow{AB} + \frac{1}{2}\overrightarrow{BC} = \overrightarrow{DE}$. Comparing this with (1) gives $\overrightarrow{DE} = \frac{1}{2}\overrightarrow{AC}$. Therefore $\overrightarrow{AC}$ and $\overrightarrow{DE}$ are parallel and $\left|\overrightarrow{DE}\right| = \frac{1}{2}\left|\overrightarrow{AC}\right|$.

Section 9.3   The Dot Product

1. (a) $\mathbf{a} \cdot \mathbf{b}$ is a scalar, and the dot product is defined only for vectors, so $(\mathbf{a} \cdot \mathbf{b}) \cdot \mathbf{c}$ has no meaning.

(b) $(\mathbf{a} \cdot \mathbf{b})\,\mathbf{c}$ is a scalar multiple of a vector, so it does have meaning.

(c) Both $|\mathbf{a}|$ and $\mathbf{b} \cdot \mathbf{c}$ are scalars, so $|\mathbf{a}|\,(\mathbf{b} \cdot \mathbf{c})$ is an ordinary product of real numbers, and has meaning.

(d) Both $\mathbf{a}$ and $\mathbf{b} + \mathbf{c}$ are vectors, so the dot product $\mathbf{a} \cdot (\mathbf{b} + \mathbf{c})$ has meaning.

(e) $\mathbf{a} \cdot \mathbf{b}$ is a scalar, but $\mathbf{c}$ is a vector, and so the two quantities cannot be added and this expression has no meaning.

(f) $|\mathbf{a}|$ is a scalar, and the dot product is defined only for vectors, so $|\mathbf{a}| \cdot (\mathbf{b} + \mathbf{c})$ has no meaning.

3. $\mathbf{a} \cdot \mathbf{b} = (2)\,(3) \cos \frac{\pi}{3} = 6 \cdot \frac{1}{2} = 3$

5. $\mathbf{a} \cdot \mathbf{b} = \langle 4, 7, -1 \rangle \cdot \langle -2, 1, 4 \rangle = (4)\,(-2) + (7)\,(1) + (-1)\,(4) = -5$

7. $\mathbf{u}$, $\mathbf{v}$, and $\mathbf{w}$ are all unit vectors, so the triangle is an equilateral triangle. Thus the angle between $\mathbf{u}$ and $\mathbf{v}$ is $60\,°$ and $\mathbf{u} \cdot \mathbf{v} = |\mathbf{u}|\,|\mathbf{v}| \cos 60\,° = (1)\,(1)\left(\frac{1}{2}\right) = \frac{1}{2}$. If $\mathbf{w}$ is moved so it has the same initial point as $\mathbf{u}$, we can see that the angle between them is $120\,°$ and we have
$\mathbf{u} \cdot \mathbf{w} = |\mathbf{u}|\,|\mathbf{w}| \cos 120\,° = (1)\,(1)\left(-\frac{1}{2}\right) = -\frac{1}{2}$.

9. (a) $\mathbf{i} \cdot \mathbf{j} = \langle 1, 0, 0 \rangle \cdot \langle 0, 1, 0 \rangle = (1)\,(0) + (0)\,(1) + (0)\,(0) = 0$. Similarly
$\mathbf{j} \cdot \mathbf{k} = (0)\,(0) + (1)\,(0) + (0)\,(1) = 0$ and $\mathbf{k} \cdot \mathbf{i} = (0)\,(1) + (0)\,(0) + (1)\,(0) = 0$.
Another Method: Because $\mathbf{i}$, $\mathbf{j}$, and $\mathbf{k}$ are mutually perpendicular, the cosine factor in each dot product is $\cos \frac{\pi}{2} = 0$.

(b) $\mathbf{i} \cdot \mathbf{i} = |\mathbf{i}|^2 = 1^2 = 1$ since $\mathbf{i}$ is a unit vector. Similarly $\mathbf{j} \cdot \mathbf{j} = |\mathbf{j}|^2 = 1$ and $\mathbf{k} \cdot \mathbf{k} = |\mathbf{k}|^2 = 1$.

11. $|\mathbf{a}| = \sqrt{1^2 + 2^2 + 2^2} = 3$, $|\mathbf{b}| = \sqrt{3^2 + 4^2 + 0^2} = 5$, $\mathbf{a} \cdot \mathbf{b} = 3 + 8 + 0 = 11$. $\mathbf{a} \cdot \mathbf{b} = |\mathbf{a}|\,|\mathbf{b}| \cos \theta \;\Rightarrow$
$\cos \theta = \dfrac{\mathbf{a} \cdot \mathbf{b}}{|\mathbf{a}|\,|\mathbf{b}|} = \dfrac{11}{3 \cdot 5}$, so $\theta = \cos^{-1} \frac{11}{15} \approx 43\,°$.

13. $|\mathbf{a}| = \sqrt{36 + 4 + 9} = 7$, $|\mathbf{b}| = \sqrt{3}$, $\mathbf{a} \cdot \mathbf{b} = 6 - 2 - 3 = 1$, so $\cos \theta = \frac{1}{7\sqrt{3}}$ and $\theta = \cos^{-1} \frac{1}{7\sqrt{3}} \approx 85\,°$.

15. (a) Since $\mathbf{a} = -2\mathbf{b}$, $\mathbf{a}$ and $\mathbf{b}$ are parallel vectors (and thus not orthogonal).

(b) $\mathbf{a} \cdot \mathbf{b} = 8 + (-8) = 0$, so $\mathbf{a}$ and $\mathbf{b}$ are orthogonal (and not parallel).

(c) $\mathbf{a} \cdot \mathbf{b} = -2 + 16 + (-15) \neq 0$, so $\mathbf{a}$ and $\mathbf{b}$ are not orthogonal. Also, since $\mathbf{a}$ is not a scalar multiple of $\mathbf{b}$, $\mathbf{a}$ and $\mathbf{b}$ are not parallel.

(d) $\mathbf{a} \cdot \mathbf{b} = 3 + (-1) + (-2) = 0$, so $\mathbf{a}$ and $\mathbf{b}$ are orthogonal.

17. Let $\mathbf{a} = a_1\mathbf{i} + a_2\mathbf{j} + a_3\mathbf{k}$ be a vector orthogonal to both $\mathbf{i} + \mathbf{j}$ and $\mathbf{i} + \mathbf{k}$. Then, by definition, $a_1 + a_2 = 0$ and $a_1 + a_3 = 0$, so $a_1 = -a_2 = -a_3$. Furthermore $\mathbf{a}$ is to be a unit vector, so $1 = a_1^2 + a_2^2 + a_3^2 = 3a_1^2$ implies $a_1 = \pm\frac{1}{\sqrt{3}}$. Thus $\mathbf{a} = \frac{1}{\sqrt{3}}\mathbf{i} - \frac{1}{\sqrt{3}}\mathbf{j} - \frac{1}{\sqrt{3}}\mathbf{k}$ and $\mathbf{a} = -\frac{1}{\sqrt{3}}\mathbf{i} + \frac{1}{\sqrt{3}}\mathbf{j} + \frac{1}{\sqrt{3}}\mathbf{k}$ are two such unit vectors.

19. $|\mathbf{a}| = \sqrt{4+9} = \sqrt{13}$. The scalar projection of $\mathbf{b}$ onto $\mathbf{a}$ is $\text{comp}_\mathbf{a}\,\mathbf{b} = \dfrac{\mathbf{a}\cdot\mathbf{b}}{|\mathbf{a}|} = \dfrac{2\cdot4+3\cdot1}{\sqrt{13}} = \dfrac{11}{\sqrt{13}}$.

The vector projection of $\mathbf{b}$ onto $\mathbf{a}$ is $\text{proj}_\mathbf{a}\,\mathbf{b} = \dfrac{\mathbf{a}\cdot\mathbf{b}}{|\mathbf{a}|^2}\mathbf{a} = \dfrac{11}{\sqrt{13}}\cdot\dfrac{1}{\sqrt{13}}\langle 2,3\rangle = \dfrac{11}{13}\langle 2,3\rangle = \left\langle \dfrac{22}{13}, \dfrac{33}{13}\right\rangle$.

21. $|\mathbf{a}| = \sqrt{1+0+1} = \sqrt{2}$ so the scalar projection of $\mathbf{b}$ onto $\mathbf{a}$ is

$\text{comp}_\mathbf{a}\,\mathbf{b} = \dfrac{\mathbf{a}\cdot\mathbf{b}}{|\mathbf{a}|} = \dfrac{1}{\sqrt{2}}(1+0+0) = \dfrac{1}{\sqrt{2}}$ while the vector projection of $\mathbf{b}$ onto $\mathbf{a}$ is

$\text{proj}_\mathbf{a}\,\mathbf{b} = \dfrac{\mathbf{a}\cdot\mathbf{b}}{|\mathbf{a}|^2}\mathbf{a} = \dfrac{1}{\sqrt{2}}\cdot\dfrac{1}{\sqrt{2}}(\mathbf{i}+\mathbf{k}) = \dfrac{1}{2}(\mathbf{i}+\mathbf{k})$.

23. $(\text{orth}_\mathbf{a}\,\mathbf{b})\cdot\mathbf{a} = (\mathbf{b} - \text{proj}_\mathbf{a}\,\mathbf{b})\cdot\mathbf{a} = \mathbf{b}\cdot\mathbf{a} - (\text{proj}_\mathbf{a}\,\mathbf{b})\cdot\mathbf{a} = \mathbf{b}\cdot\mathbf{a} - \dfrac{\mathbf{a}\cdot\mathbf{b}}{|\mathbf{a}|^2}\mathbf{a}\cdot\mathbf{a}$

$= \mathbf{b}\cdot\mathbf{a} - \dfrac{\mathbf{a}\cdot\mathbf{b}}{|\mathbf{a}|^2}|\mathbf{a}|^2 = \mathbf{b}\cdot\mathbf{a} - \mathbf{a}\cdot\mathbf{b} = 0$

So they are orthogonal by (2).

25. $\text{comp}_\mathbf{a}\,\mathbf{b} = \dfrac{\mathbf{a}\cdot\mathbf{b}}{|\mathbf{a}|} = 2 \;\Leftrightarrow\; \mathbf{a}\cdot\mathbf{b} = 2|\mathbf{a}| = 2\sqrt{10}$. If $\mathbf{b} = \langle b_1, b_2, b_3\rangle$, then we need

$3b_1 + 0b_2 - 1b_3 = 2\sqrt{10}$. One possible solution is obtained by taking $b_1 = 0$, $b_2 = 0$, $b_3 = -2\sqrt{10}$.
In general, $\mathbf{b} = \langle s, t, 3s - 2\sqrt{10}\rangle$, $s, t \in \mathbb{R}$.

27. Here $\mathbf{D} = (4-2)\mathbf{i} + (9-3)\mathbf{j} + (15-0)\mathbf{k} = 2\mathbf{i} + 6\mathbf{j} + 15\mathbf{k}$ so
$W = \mathbf{F}\cdot\mathbf{D} = 20 + 108 - 90 = 38$ joules.

29. $W = |\mathbf{F}|\,|\mathbf{D}|\cos\theta$
$= (25)(10)\cos 20°$
≈ 235 ft-lb

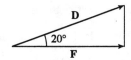

31. First note that $\mathbf{n} = \langle a, b\rangle$ is perpendicular to the line, because if $Q_1 = (a_1, b_1)$ and $Q_2 = (a_2, b_2)$ lie on the line, then $\mathbf{n}\cdot\overrightarrow{Q_1Q_2} = aa_2 - aa_1 + bb_2 - bb_1 = 0$, since $aa_2 + bb_2 = -c = aa_1 + bb_1$ from the equation of the line. Let $P_2 = (x_2, y_2)$ lie on the line. Then the distance from P_1

to the line is the absolute value of the scalar projection of $\overrightarrow{P_1P_2}$ onto $\mathbf{n}$.

$\text{comp}_\mathbf{n}\left(\overrightarrow{P_1P_2}\right) = \dfrac{|\mathbf{n}\cdot\langle x_2 - x_1, y_2 - y_1\rangle|}{|\mathbf{n}|} = \dfrac{|ax_2 - ax_1 + by_2 - by_1|}{\sqrt{a^2+b^2}} = \dfrac{|ax_1 + by_1 + c|}{\sqrt{a^2+b^2}}$ since

$ax_2 + by_2 = -c$. The required distance is $\dfrac{|3\cdot-2 + -4\cdot3 + 5|}{\sqrt{3^2+4^2}} = \dfrac{13}{5}$.

33. For convenience, consider the unit cube positioned so that its back left corner is at the origin, and its edges lie along the coordinate axes. The diagonal of the cube that begins at the origin and ends at $(1, 1, 1)$ has vector representation $\langle 1, 1, 1 \rangle$. The angle θ between this vector and the vector of the edge which also begins at the origin and runs along the x-axis [that is, $\langle 1, 0, 0 \rangle$] is given by

$$\cos \theta = \frac{\langle 1, 1, 1 \rangle \cdot \langle 1, 0, 0 \rangle}{|\langle 1, 1, 1 \rangle| |\langle 1, 0, 0 \rangle|} = \frac{1}{\sqrt{3}} \Rightarrow \quad \theta = \cos^{-1} \frac{1}{\sqrt{3}} \approx 55^\circ.$$

35. Consider the H-C-H combination consisting of the sole carbon atom and the two hydrogen atoms that are at $(1, 0, 0)$ and $(0, 1, 0)$ (or any H-C-H combination, for that matter). Vector representations of the line segments emanating from the carbon atom and extending to these two hydrogen atoms are $\langle 1 - \frac{1}{2}, 0 - \frac{1}{2}, 0 - \frac{1}{2} \rangle = \langle \frac{1}{2}, -\frac{1}{2}, -\frac{1}{2} \rangle$ and $\langle 0 - \frac{1}{2}, 1 - \frac{1}{2}, 0 - \frac{1}{2} \rangle = \langle -\frac{1}{2}, \frac{1}{2}, -\frac{1}{2} \rangle$. The bond angle, θ, is therefore given by $\cos \theta = \frac{\langle \frac{1}{2}, -\frac{1}{2}, -\frac{1}{2} \rangle \cdot \langle -\frac{1}{2}, \frac{1}{2}, -\frac{1}{2} \rangle}{|\langle \frac{1}{2}, -\frac{1}{2}, -\frac{1}{2} \rangle| |\langle -\frac{1}{2}, \frac{1}{2}, -\frac{1}{2} \rangle|} = \frac{-\frac{1}{4} - \frac{1}{4} + \frac{1}{4}}{\sqrt{\frac{3}{4}}\sqrt{\frac{3}{4}}} = -\frac{1}{3} \Rightarrow$

$\theta = \cos^{-1}\left(-\frac{1}{3}\right) \approx 109.5^\circ.$

37. If $c = 0$ then $c\mathbf{a} = \mathbf{0}$, so $(c\mathbf{a}) \cdot \mathbf{b} = \mathbf{0} \cdot \mathbf{b} = 0$ by Property 5. Similarly, $\mathbf{a} \cdot (c\mathbf{b}) = \mathbf{a} \cdot \mathbf{0} = 0$, and $c(\mathbf{a} \cdot \mathbf{b}) = 0 \, (|\mathbf{a}| |\mathbf{b}| \cos \theta) = 0$, thus $(c\mathbf{a}) \cdot \mathbf{b} = c(\mathbf{a} \cdot \mathbf{b}) = \mathbf{a} \cdot (c\mathbf{b})$. If $c > 0$, the angle θ between $\mathbf{a}$ and $\mathbf{b}$ coincides with the angle between $c\mathbf{a}$ and $\mathbf{b}$, so by definition of the dot product, $(c\mathbf{a}) \cdot \mathbf{b} = |c\mathbf{a}| |\mathbf{b}| \cos \theta = |c| |\mathbf{a}| |\mathbf{b}| \cos \theta = c |\mathbf{a}| |\mathbf{b}| \cos \theta$. Similarly, $\mathbf{a} \cdot (c\mathbf{b}) = |\mathbf{a}| |c\mathbf{b}| \cos \theta = |\mathbf{a}| |c| |\mathbf{b}| \cos \theta = c |\mathbf{a}| |\mathbf{b}| \cos \theta$, and $c(\mathbf{a} \cdot \mathbf{b}) = c |\mathbf{a}| |\mathbf{b}| \cos \theta$. Thus, $(c\mathbf{a}) \cdot \mathbf{b} = c(\mathbf{a} \cdot \mathbf{b}) = \mathbf{a} \cdot (c\mathbf{b})$. The case for $c < 0$ is similar. Using components, let $\mathbf{a} = \langle a_1, a_2, a_3 \rangle$ and $\mathbf{b} = \langle b_1, b_2, b_3 \rangle$. Then

$$
\begin{aligned}
(c\mathbf{a}) \cdot \mathbf{b} &= \langle ca_1, ca_2, ca_3 \rangle \cdot \langle b_1, b_2, b_3 \rangle = (ca_1) b_1 + (ca_2) b_2 + (ca_3) b_3 \\
&= c(a_1 b_1 + a_2 b_2 + a_3 b_3) = c(\mathbf{a} \cdot \mathbf{b}) \\
&= a_1 (cb_1) + a_2 (cb_2) + a_3 (cb_3) = \langle a_1, a_2, a_3 \rangle \cdot \langle cb_1, cb_2, cb_3 \rangle = \mathbf{a} \cdot (c\mathbf{b})
\end{aligned}
$$

39. $|\mathbf{a} \cdot \mathbf{b}| = ||\mathbf{a}| |\mathbf{b}| \cos \theta| = |\mathbf{a}| |\mathbf{b}| |\cos \theta|$. Since $|\cos \theta| \leq 1$, $|\mathbf{a} \cdot \mathbf{b}| = |\mathbf{a}| |\mathbf{b}| |\cos \theta| \leq |\mathbf{a}| |\mathbf{b}|$.
Note: We have equality in the case of $\cos \theta = \pm 1$, so $\theta = 0$ or $\theta = \pi$, thus equality when $\mathbf{a}$ and $\mathbf{b}$ are parallel.

41. (a)

The Parallelogram Law states that the sum of the squares of the lengths of the diagonals of a parallelogram equals the sum of the squares of its (four) sides.

(b) $|\mathbf{a} + \mathbf{b}|^2 = (\mathbf{a} + \mathbf{b}) \cdot (\mathbf{a} + \mathbf{b}) = |\mathbf{a}|^2 + 2(\mathbf{a} \cdot \mathbf{b}) + |\mathbf{b}|^2$ and $|\mathbf{a} - \mathbf{b}|^2 = (\mathbf{a} - \mathbf{b}) \cdot (\mathbf{a} - \mathbf{b}) = |\mathbf{a}|^2 - 2(\mathbf{a} \cdot \mathbf{b}) + |\mathbf{b}|^2$. Adding these two equations gives $|\mathbf{a} + \mathbf{b}|^2 + |\mathbf{a} - \mathbf{b}|^2 = 2|\mathbf{a}|^2 + 2|\mathbf{b}|^2$.

Section 9.4 The Cross Product

1. (a) Since $\mathbf{b} \times \mathbf{c}$ is a vector, the dot product $\mathbf{a} \cdot (\mathbf{b} \times \mathbf{c})$ is meaningful and is a scalar.

(b) $\mathbf{b} \cdot \mathbf{c}$ is a scalar, so a $\times$ $(\mathbf{b} \cdot \mathbf{c})$ is meaningless, as the cross product is defined only for two *vectors*.

(c) Since $\mathbf{b} \times \mathbf{c}$ is a vector, the cross product $\mathbf{a} \times (\mathbf{b} \times \mathbf{c})$ is meaningful and results in another vector.

(d) $\mathbf{a} \cdot \mathbf{b}$ is a scalar, so the cross product $(\mathbf{a} \cdot \mathbf{b}) \times \mathbf{c}$ is meaningless.

(e) Since $(\mathbf{a} \cdot \mathbf{b})$ and $(\mathbf{c} \cdot \mathbf{d})$ are both scalars, the cross product $(\mathbf{a} \cdot \mathbf{b}) \times (\mathbf{c} \cdot \mathbf{d})$ is meaningless.

(f) $\mathbf{a} \times \mathbf{b}$ and $\mathbf{c} \times \mathbf{d}$ are both vectors, so the dot product $(\mathbf{a} \times \mathbf{b}) \cdot (\mathbf{c} \times \mathbf{d})$ is meaningful and is a scalar.

3. 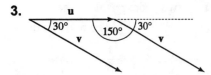 If we sketch $\mathbf{u}$ and $\mathbf{v}$ starting from the same initial point, we see that the angle between them is $30°$, so $|\mathbf{u} \times \mathbf{v}| = |\mathbf{u}|\,|\mathbf{v}| \sin 30° = (6)\,(8)\,\left(\frac{1}{2}\right) = 24$. By the right-hand rule, $\mathbf{u} \times \mathbf{v}$ is directed into the page.

5. The magnitude of the torque is

$$|\mathbf{\tau}| = |\mathbf{r} \times \mathbf{F}| = |\mathbf{r}|\,|\mathbf{F}| \sin\theta = (0.18\text{ m})\,(60\text{ N}) \sin(180 - (70 + 10))° = 10.8 \sin 100° \approx 10.6 \text{ J}.$$

7. $\mathbf{a} \times \mathbf{b} = \begin{vmatrix} \mathbf{i} & \mathbf{j} & \mathbf{k} \\ -2 & 3 & 4 \\ 3 & 0 & 1 \end{vmatrix} = \begin{vmatrix} 3 & 4 \\ 0 & 1 \end{vmatrix} \mathbf{i} - \begin{vmatrix} -2 & 4 \\ 3 & 1 \end{vmatrix} \mathbf{j} + \begin{vmatrix} -2 & 3 \\ 3 & 0 \end{vmatrix} \mathbf{k} = 3\mathbf{i} + 14\mathbf{j} - 9\mathbf{k}$

9. $\mathbf{a} \times \mathbf{b} = \begin{vmatrix} \mathbf{i} & \mathbf{j} & \mathbf{k} \\ 1 & 1 & 1 \\ 1 & 1 & -1 \end{vmatrix} = \begin{vmatrix} 1 & 1 \\ 1 & -1 \end{vmatrix} \mathbf{i} - \begin{vmatrix} 1 & 1 \\ 1 & -1 \end{vmatrix} \mathbf{j} + \begin{vmatrix} 1 & 1 \\ 1 & 1 \end{vmatrix} \mathbf{k} = -2\mathbf{i} + 2\mathbf{j}$

11. We know that the cross product of two vectors is orthogonal to both. So we calculate

$$\langle 1, -1, 1 \rangle \times \langle 0, 4, 4 \rangle = \begin{vmatrix} \mathbf{i} & \mathbf{j} & \mathbf{k} \\ 1 & -1 & 1 \\ 0 & 4 & 4 \end{vmatrix} = \begin{vmatrix} -1 & 1 \\ 4 & 4 \end{vmatrix} \mathbf{i} - \begin{vmatrix} 1 & 1 \\ 0 & 4 \end{vmatrix} \mathbf{j} + \begin{vmatrix} 1 & -1 \\ 0 & 4 \end{vmatrix} \mathbf{k} = -8\mathbf{i} - 4\mathbf{j} + 4\mathbf{k}. \text{ So two unit}$$

vectors orthogonal to both are $\pm \dfrac{\langle -8, -4, 4 \rangle}{\sqrt{64 + 16 + 16}} = \pm \dfrac{\langle -8, -4, 4 \rangle}{4\sqrt{6}}$, that is, $\left\langle -\dfrac{2}{\sqrt{6}}, -\dfrac{1}{\sqrt{6}}, \dfrac{1}{\sqrt{6}} \right\rangle$ and $\left\langle \dfrac{2}{\sqrt{6}}, \dfrac{1}{\sqrt{6}}, -\dfrac{1}{\sqrt{6}} \right\rangle$.

13. We know that the area of the parallelogram determined by two vectors is equal to the length of the cross product of these vectors. The vectors corresponding to $\overrightarrow{AB}$ and $\overrightarrow{AD}$ are $\mathbf{a} = \langle 3, -1, 0 \rangle$ and $\mathbf{b} = \langle 2, -2, 0 \rangle$, so the area of parallelogram $ABCD$ is

$$|\mathbf{a} \times \mathbf{b}| = \left\| \begin{vmatrix} \mathbf{i} & \mathbf{j} & \mathbf{k} \\ 3 & -1 & 0 \\ 2 & -2 & 0 \end{vmatrix} \right\| = |(0)\,\mathbf{i} - (0)\,\mathbf{j} + (-6 + 2)\,\mathbf{k}| = |-4\mathbf{k}| = 4.$$

15. (a) Because the plane through P, Q, and R contains the vectors $\overrightarrow{PQ}$ and $\overrightarrow{PR}$, a vector orthogonal to both of these vectors (such as their cross product) is also orthogonal to the plane. Here $\overrightarrow{PQ} = \langle -1, 2, 0 \rangle$ and $\overrightarrow{PR} = \langle -1, 0, 3 \rangle$, so

$\overrightarrow{PQ} \times \overrightarrow{PR} = \langle (2)(3) - (0)(0), (0)(-1) - (-1)(3), (-1)(0) - (2)(-1) \rangle = \langle 6, 3, 2 \rangle$.

Therefore, $\langle 6, 3, 2 \rangle$ (or any scalar multiple thereof) is orthogonal to the plane through P, Q, and R.

(b) Note that the area of the triangle determined by P, Q, and R is equal to half of the area of the parallelogram determined by the three points. From part (a), the area of the parallelogram is

$\left| \overrightarrow{PQ} \times \overrightarrow{PR} \right| = |\langle 6, 3, 2 \rangle| = \sqrt{36 + 9 + 4} = 7$, so the area of the triangle is $\frac{1}{2}(7) = \frac{7}{2}$.

17. Using the notation of (1), $\mathbf{r} = \langle 0, 0.3, 0 \rangle$ and $\mathbf{F}$ has direction $\langle 0, 3, -4 \rangle$. The angle θ between them can be determined by $\langle 0, 0.3, 0 \rangle \cdot \langle 0, 3, -4 \rangle = |\langle 0, 0.3, 0 \rangle| \, |\langle 0, 3, -4 \rangle| \cos \theta \Rightarrow 0.9 = (0.3)(5) \cos \theta \Rightarrow \cos \theta = 0.6 \Rightarrow \theta \approx 53.1°$. Then $|\tau| = |\mathbf{r}| \, |\mathbf{F}| \sin \theta \Rightarrow 100 = 0.3 \, |\mathbf{F}| \sin 53.1° \Rightarrow |\mathbf{F}| \approx 417$ N.

19. We know that the volume of the parallelepiped determined by $\mathbf{a}$, $\mathbf{b}$ and $\mathbf{c}$ is the magnitude of their scalar triple product, which is

$$\mathbf{a} \cdot (\mathbf{b} \times \mathbf{c}) = \begin{vmatrix} 1 & 0 & 6 \\ 2 & 3 & -8 \\ 8 & -5 & 6 \end{vmatrix} = 1 \begin{vmatrix} 3 & -8 \\ -5 & 6 \end{vmatrix} - 0 + 6 \begin{vmatrix} 2 & 3 \\ 8 & -5 \end{vmatrix}$$

$$= (18 - 40) + 6(-10 - 24)$$

$$= -226$$

Thus the volume of the parallelepiped is $|-226| = 226$ cubic units.

21. $\mathbf{a} = \overrightarrow{PQ} = \langle 1, -1, 2 \rangle$, $\mathbf{b} = \overrightarrow{PR} = \langle 3, 0, 6 \rangle$ and $\mathbf{c} = \overrightarrow{PS} = \langle 2, -2, -3 \rangle$.

$$\mathbf{a} \cdot (\mathbf{b} \times \mathbf{c}) = \begin{vmatrix} 1 & -1 & 2 \\ 3 & 0 & 6 \\ 2 & -2 & -3 \end{vmatrix} = 1 \begin{vmatrix} 0 & 6 \\ -2 & -3 \end{vmatrix} - (-1) \begin{vmatrix} 3 & 6 \\ 2 & -3 \end{vmatrix} + 2 \begin{vmatrix} 3 & 0 \\ 2 & -2 \end{vmatrix} = 12 - 21 - 12 = -21$$, so the volume of the parallelepiped is 21 cubic units.

23. $\mathbf{a} \cdot (\mathbf{b} \times \mathbf{c}) = \begin{vmatrix} 2 & 3 & 1 \\ 1 & -1 & 0 \\ 7 & 3 & 2 \end{vmatrix} = 2 \begin{vmatrix} -1 & 0 \\ 3 & 2 \end{vmatrix} - 3 \begin{vmatrix} 1 & 0 \\ 7 & 2 \end{vmatrix} + 1 \begin{vmatrix} 1 & -1 \\ 7 & 3 \end{vmatrix} = -4 - 6 + 10 = 0$, which says that the volume of the parallelepiped determined by $\mathbf{a}$, $\mathbf{b}$ and $\mathbf{c}$ is 0, and thus these three vectors are coplanar.

25. (a)

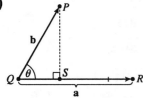

The distance between a point and a line is the length of the perpendicular from the point to the line, here $\left|\overrightarrow{PS}\right| = d$.

But referring to triangle PQS,

$d = \left|\overrightarrow{PS}\right| = \left|\overrightarrow{QP}\right| \sin\theta = |\mathbf{b}| \sin\theta$. But θ is the angle

between $\overrightarrow{QP} = \mathbf{b}$ and $\overrightarrow{QR} = \mathbf{a}$. Thus by the definition

of cross product, $\sin\theta = \dfrac{|\mathbf{a} \times \mathbf{b}|}{|\mathbf{a}|\,|\mathbf{b}|}$ and so $d = |\mathbf{b}| \sin\theta = \dfrac{|\mathbf{b}|\,|\mathbf{a} \times \mathbf{b}|}{|\mathbf{a}|\,|\mathbf{b}|} = \dfrac{|\mathbf{a} \times \mathbf{b}|}{|\mathbf{a}|}$.

(b) $\mathbf{a} = \overrightarrow{QR} = \langle -1, -2, -1 \rangle$ and $\mathbf{b} = \overrightarrow{QP} = \langle 1, -5, -7 \rangle$. Then

$\mathbf{a} \times \mathbf{b} = \langle (-2)\,(-7) - (-1)\,(-5), (-1)\,(1) - (-1)\,(-7), (-1)\,(-5) - (-2)\,(1) \rangle = \langle 9, -8, 7 \rangle$.

Thus the distance is $d = \dfrac{|\mathbf{a} \times \mathbf{b}|}{|\mathbf{a}|} = \dfrac{1}{\sqrt{6}} \sqrt{81 + 64 + 49} = \sqrt{\dfrac{194}{6}} = \sqrt{\dfrac{97}{3}}$.

27. $(\mathbf{a} - \mathbf{b}) \times (\mathbf{a} + \mathbf{b}) = (\mathbf{a} - \mathbf{b}) \times \mathbf{a} + (\mathbf{a} - \mathbf{b}) \times \mathbf{b}$ by Property 3 of the Cross Product

$\phantom{(\mathbf{a} - \mathbf{b}) \times (\mathbf{a} + \mathbf{b})} = \mathbf{a} \times \mathbf{a} + (-\mathbf{b}) \times \mathbf{a} + \mathbf{a} \times \mathbf{b} + (-\mathbf{b}) \times \mathbf{b}$ by Property 4

$\phantom{(\mathbf{a} - \mathbf{b}) \times (\mathbf{a} + \mathbf{b})} = (\mathbf{a} \times \mathbf{a}) - (\mathbf{b} \times \mathbf{a}) + (\mathbf{a} \times \mathbf{b}) - (\mathbf{b} \times \mathbf{b})$ by Property 2 (with $c = -1$)

$\phantom{(\mathbf{a} - \mathbf{b}) \times (\mathbf{a} + \mathbf{b})} = \mathbf{0} - (\mathbf{b} \times \mathbf{a}) + (\mathbf{a} \times \mathbf{b}) - \mathbf{0}$

$\phantom{(\mathbf{a} - \mathbf{b}) \times (\mathbf{a} + \mathbf{b})} = (\mathbf{a} \times \mathbf{b}) + (\mathbf{a} \times \mathbf{b})$ by Property 1

$\phantom{(\mathbf{a} - \mathbf{b}) \times (\mathbf{a} + \mathbf{b})} = 2\,(\mathbf{a} \times \mathbf{b})$

29. $\mathbf{a} \times (\mathbf{b} \times \mathbf{c}) + \mathbf{b} \times (\mathbf{c} \times \mathbf{a}) + \mathbf{c} \times (\mathbf{a} \times \mathbf{b})$

$ = [(\mathbf{a} \cdot \mathbf{c})\,\mathbf{b} - (\mathbf{a} \cdot \mathbf{b})\,\mathbf{c}] + [(\mathbf{b} \cdot \mathbf{a})\,\mathbf{c} - (\mathbf{b} \cdot \mathbf{c})\,\mathbf{a}] + [(\mathbf{c} \cdot \mathbf{b})\,\mathbf{a} - (\mathbf{c} \cdot \mathbf{a})\,\mathbf{b}]$ by Exercise 28

$ = (\mathbf{a} \cdot \mathbf{c})\,\mathbf{b} - (\mathbf{a} \cdot \mathbf{b})\,\mathbf{c} + (\mathbf{a} \cdot \mathbf{b})\,\mathbf{c} - (\mathbf{b} \cdot \mathbf{c})\,\mathbf{a} + (\mathbf{b} \cdot \mathbf{c})\,\mathbf{a} - (\mathbf{a} \cdot \mathbf{c})\,\mathbf{b} = \mathbf{0}$

31. (a) No. If $\mathbf{a} \cdot \mathbf{b} = \mathbf{a} \cdot \mathbf{c}$, then $\mathbf{a} \cdot (\mathbf{b} - \mathbf{c}) = 0$, so $\mathbf{a}$ is perpendicular to $\mathbf{b} - \mathbf{c}$, which can happen if $\mathbf{b} \neq \mathbf{c}$. For example, let $\mathbf{a} = \langle 1, 1, 1 \rangle$, $\mathbf{b} = \langle 1, 0, 0 \rangle$ and $\mathbf{c} = \langle 0, 1, 0 \rangle$.

(b) No. If $\mathbf{a} \times \mathbf{b} = \mathbf{a} \times \mathbf{c}$ then $\mathbf{a} \times (\mathbf{b} - \mathbf{c}) = \mathbf{0}$, which implies that $\mathbf{a}$ is parallel to $\mathbf{b} - \mathbf{c}$, which of course can happen if $\mathbf{b} \neq \mathbf{c}$.

(c) Yes. Since $\mathbf{a} \cdot \mathbf{c} = \mathbf{a} \cdot \mathbf{b}$, $\mathbf{a}$ is perpendicular to $\mathbf{b} - \mathbf{c}$, by part (a). From part (b), $\mathbf{a}$ is also parallel to $\mathbf{b} - \mathbf{c}$. Thus since $\mathbf{a} \neq \mathbf{0}$ but is both parallel and perpendicular to $\mathbf{b} - \mathbf{c}$, we have $\mathbf{b} - \mathbf{c} = \mathbf{0}$, so $\mathbf{b} = \mathbf{c}$.

Section 9.5   Equations of Lines and Planes

1. (a) True; each of the first two lines has a direction vector parallel to the direction vector of the third line, so these vectors are each scalar multiples of the third direction vector. Then the first two direction vectors are also scalar multiples of each other, so these vectors, and hence the two lines, are parallel.

(b) False; for example, the x- and y-axes are both perpendicular to the z-axis, yet the x- and y-axes are not parallel.

(c) True; each of the first two planes has a normal vector parallel to the normal vector of the third plane, so these two normal vectors are parallel to each other and the planes are parallel.

(d) False; for example, the xy- and yz-planes are not parallel, yet they are both perpendicular to the xz-plane.

(e) False; the x- and y-axes are not parallel, yet they are both parallel to the plane $z = 1$.

(f) True; if each line is perpendicular to a plane, then the lines' direction vectors are both parallel to a normal vector for the plane. Thus, the direction vectors are parallel to each other and the lines are parallel.

(g) False; the planes $y = 1$ and $z = 1$ are not parallel, yet they are both parallel to the x-axis.

(h) True; if each plane is perpendicular to a line, then any normal vector for each plane is parallel to a direction vector for the line. Thus, the normal vectors are parallel to each other and the planes are parallel.

(i) True; see Figure 9 and the preceeding discussion on page 679.

(j) False; they can be skew, as in Example 3.

(k) True. Consider any normal vector for the plane and any direction vector for the line. If the normal vector is perpendicular to the direction vector, the line and plane are parallel. Otherwise, the vectors meet at an angle θ, $0° \leq \theta < 90°$, and the line will intersect the plane at an angle $90° - \theta$.

3. For this line, we have $\mathbf{r}_0 = \mathbf{j} + 2\mathbf{k}$ and $\mathbf{v} = 6\mathbf{i} + 3\mathbf{j} + 2\mathbf{k}$, so a vector equation is
$\mathbf{r} = (\mathbf{j} + 2\mathbf{k}) + t(6\mathbf{i} + 3\mathbf{j} + 2\mathbf{k}) = 6t\,\mathbf{i} + (1 + 3t)\,\mathbf{j} + (2 + 2t)\,\mathbf{k}$ and parametric equations are $x = 6t$, $y = 1 + 3t$, $z = 2 + 2t$.

5. A line perpendicular to the given plane has the same direction as a normal vector to the plane, such as $\mathbf{n} = \langle 1, 3, 1 \rangle$. So $\mathbf{r}_0 = \mathbf{i} + 6\mathbf{k}$, and we can take $\mathbf{v} = \mathbf{i} + 3\mathbf{j} + \mathbf{k}$. Then a vector equation is
$\mathbf{r} = (\mathbf{i} + 6\mathbf{k}) + t(\mathbf{i} + 3\mathbf{j} + \mathbf{k}) = (1 + t)\,\mathbf{i} + 3t\mathbf{j} + (6 + t)\,\mathbf{k}$, and parametric equations are
$x = 1 + t, y = 3t, z = 6 + t$.

7. $\mathbf{v} = \langle 3 - 3, 2 - 1, -6 - (-1) \rangle = \langle 0, 1, -5 \rangle$, and letting $P_0 = (3, 1, -1)$, parametric equations are $x = 3$, $y = 1 + t$, $z = -1 - 5t$, while symmetric equations are $x = 3, y - 1 = (z + 1)/(-5)$. Notice here that the direction number $a = 0$, so rather than writing $(x - 3)/0$ in the symmetric equation we must write the equation $x = 3$ separately.

9. Direction vectors of the lines are $v_1 = \langle 6, 9, 12 \rangle$ and $v_2 = \langle 4, 6, 8 \rangle$, and since $v_1 = \frac{3}{2} v_2$, the direction vectors and thus the lines are parallel.

11. (a) A direction vector of the line with parametric equations $x = 1 + 2t$, $y = 3t$, $z = 5 - 7t$ is $v = \langle 2, 3, -7 \rangle$ and the desired parallel line must also have v as a direction vector. Here $P_0 = (0, 2, -1)$, so symmetric equations for the line are $\dfrac{x}{2} = \dfrac{y - 2}{3} = \dfrac{z + 1}{-7}$.

(b) The line intersects the xy-plane when $z = 0$, so we need $\dfrac{x}{2} = \dfrac{y - 2}{3} = \dfrac{1}{-7}$ or $x = -\frac{2}{7}$, $y = \frac{11}{7}$. Thus the point of intersection with the xy- plane is $\left(-\frac{2}{7}, \frac{11}{7}, 0 \right)$. Similarly for the yz-plane, we need $x = 0 \iff 0 = \dfrac{y - 2}{3} = \dfrac{z + 1}{-7} \iff y = 2$, $z = -1$. Thus the line intersects the yz-plane at $(0, 2, -1)$. For the xz-plane, we need $y = 0 \iff \dfrac{x}{2} = -\dfrac{2}{3} = \dfrac{z + 1}{-7} \iff x = -\frac{4}{3}$, $z = \frac{11}{3}$. So the line intersects the xz-plane at $\left(-\frac{4}{3}, 0, \frac{11}{3} \right)$.

13. The lines aren't parallel since the direction vectors $\langle 2, 4, -3 \rangle$ and $\langle 1, 3, 2 \rangle$ aren't parallel, so we check to see if the lines intersect. The parametric equations of the lines are L_1: $x = 4 + 2t$, $y = -5 + 4t$, $z = 1 - 3t$ and L_2: $x = 2 + s$, $y = -1 + 3s$, $z = 2s$. For the lines to intersect we must be able to find one value of t and one value of s satisfying the following three equations: $4 + 2t = 2 + s$, $-5 + 4t = -1 + 3s$, $1 - 3t = 2s$. Solving the first two equations we get $t = -5$, $s = -8$ and checking, we see that these values don't satisfy the third equation. Thus L_1 and L_2 aren't parallel and don't intersect, so they must be skew lines.

15. Since the direction vectors are $v_1 = \langle -6, 9, -3 \rangle$ and $v_2 = \langle 2, -3, 1 \rangle$, we have $v_1 = -3v_2$ so the lines are parallel.

17. Since the plane is perpendicular to the vector $\langle 7, 1, 4 \rangle$, we can take $\langle 7, 1, 4 \rangle$ as a normal vector to the plane. Setting $a = 7$, $b = 1$, $c = 4$, $x_0 = 1$, $y_0 = 4$, $z_0 = 5$ in Equation 6 gives $7(x - 1) + 1(y - 4) + 4(z - 5) = 0$ or $7x + y + 4z = 31$ to be an equation of the plane.

19. Since the two planes are parallel, they will have the same normal vectors. Thus $n = \langle 1, 1, -1 \rangle$ and an equation of the plane is $1(x - 6) + 1(y - 5) + 1(z - 2) = 0$ or $x + y - z = 13$.

21. Here the vectors $a = \langle 1, 1, 1 \rangle$ and $b = \langle 1, 2, 3 \rangle$ lie in the plane, so $a \times b$ is a normal vector to the plane. Thus $n = a \times b = \langle 3 - 2, 1 - 3, 2 - 1 \rangle = \langle 1, -2, 1 \rangle$ and an equation of the plane is $(x - 0) - 2(y - 0) + (z - 0) = 0$ or $x - 2y + z = 0$.

23. If we first find two nonparallel vectors in the plane, their cross product will be a normal vector to the plane. Since the given line lies in the plane, its direction vector $\mathbf{a} = \langle 2, -3, -1 \rangle$ is one vector in the plane. We can verify that the given point $(1, 6, -4)$ does not lie on this line, so to find another nonparallel vector $\mathbf{b}$ which lies in the plane, we can pick any point on the line and find a vector connecting the points. If we put $t = 0$ we see that $(1, 2, 3)$ is on the line, so $\mathbf{b} = \langle 1 - 1, 6 - 2, -4 - 3 \rangle = \langle 0, 4, -7 \rangle$ and $\mathbf{n} = \mathbf{a} \times \mathbf{b} = \langle 21 + 4, 0 + 14, 8 - 0 \rangle = \langle 25, 14, 8 \rangle$. Thus, an equation of the plane is $25(x - 1) + 14(y - 6) + 8(z + 4) = 0$ or $25x + 14y + 8z = 77$.

25. Substituting the parametric equations of the line into the equation of the plane gives $x + y + z = 1 + t + 2t + 3t = 1 \;\Rightarrow\; t = 0$. This value of t corresponds to the point of intersection $(1, 0, 0)$, obtained by substitution of $t = 0$ into the equations of the line.

27. The normal vectors to the planes are $\mathbf{n}_1 = \langle 1, 0, 1 \rangle$ and $\mathbf{n}_2 = \langle 0, 1, 1 \rangle$. Thus the normal vectors (and consequently the planes) aren't parallel. Furthermore, $\mathbf{n}_1 \cdot \mathbf{n}_2 = 1 \neq 0$ so the planes aren't perpendicular. Letting θ be the angle between the two planes, we have $\cos\theta = \dfrac{\mathbf{n}_1 \cdot \mathbf{n}_2}{|\mathbf{n}_1| \, |\mathbf{n}_2|} = \dfrac{1}{\sqrt{2}\sqrt{2}} = \dfrac{1}{2}$ and $\theta = \cos^{-1}\frac{1}{2} = 60°$.

29. The normals are $\mathbf{n}_1 = \langle 1, 4, -3 \rangle$ and $\mathbf{n}_2 = \langle -3, 6, 7 \rangle$, so the normals (and thus the planes) aren't parallel. But $\mathbf{n}_1 \cdot \mathbf{n}_2 = -3 + 24 - 21 = 0$, so the normals (and thus the planes) are perpendicular.

31. (a) To find a point on the line of intersection, set one of the variables equal to a constant, say $z = 0$. (This will only work if the line of intersection crosses the xy-plane; otherwise, try setting x or y equal to 0.) Then the equations of the planes reduce to $x + y = 2$ and $3x - 4y = 6$. Solving these two equations gives $x = 2$, $y = 0$. So a point on the line of intersection is $(2, 0, 0)$. The direction of the line is $\mathbf{v} = \mathbf{n}_1 \times \mathbf{n}_2 = \langle 5 - 4, -3 - 5, -4 - 3 \rangle = \langle 1, -8, -7 \rangle$, and symmetric equations for the line are $x - 2 = \dfrac{y}{-8} = \dfrac{z}{-7}$.

(b) The angle between the planes satisfies $\cos\theta = \dfrac{\mathbf{n}_1 \cdot \mathbf{n}_2}{|\mathbf{n}_1| \, |\mathbf{n}_2|} = \dfrac{3 - 4 - 5}{\sqrt{3}\sqrt{50}} = -\dfrac{\sqrt{6}}{5}$. Therefore $\theta = \cos^{-1}\left(-\frac{\sqrt{6}}{5}\right) \approx 119°$ (or $61°$).

33. The plane contains the points $(a, 0, 0)$, $(0, b, 0)$ and $(0, 0, c)$. Thus the vectors $\mathbf{a} = \langle -a, b, 0 \rangle$ and $\mathbf{b} = \langle -a, 0, c \rangle$ lie in the plane, and $\mathbf{n} = \mathbf{a} \times \mathbf{b} = \langle bc - 0, 0 + ac, 0 + ab \rangle = \langle bc, ac, ab \rangle$ is a normal vector to the plane. The equation of the plane is therefore $bcx + acy + abz = abc + 0 + 0$ or $bcx + acy + abz = abc$. Notice that if $a \neq 0$, $b \neq 0$ and $c \neq 0$ then we can rewrite the equation as $\dfrac{x}{a} + \dfrac{y}{b} + \dfrac{z}{c} = 1$. This is a good equation to remember!

35. Two vectors which are perpendicular to the required line are the normal of the given plane, $\langle 1, 1, 1 \rangle$, and a direction vector for the given line, $\langle 1, -1, 2 \rangle$. So a direction vector for the required line is $\langle 1, 1, 1 \rangle \times \langle 1, -1, 2 \rangle = \langle 3, -1, -2 \rangle$. Thus L is given by $\langle x, y, z \rangle = \langle 0, 1, 2 \rangle + t \langle 3, -1, -2 \rangle$, or in parametric form, $x = 3t$, $y = 1 - t$, $z = 2 - 2t$.

37. Let P_i have normal vector $\mathbf{n}_i$. Then $\mathbf{n}_1 = \langle 4, -2, 6 \rangle$, $\mathbf{n}_2 = \langle 4, -2, -2 \rangle$, $\mathbf{n}_3 = \langle -6, 3, -9 \rangle$, $\mathbf{n}_4 = \langle 2, -1, -1 \rangle$. Now $\mathbf{n}_1 = -\frac{2}{3} \mathbf{n}_3$, so $\mathbf{n}_1$ and $\mathbf{n}_3$ are parallel, and hence P_1 and P_3 are parallel; similarly P_2 and P_4 are parallel because $\mathbf{n}_2 = 2\mathbf{n}_4$. However, $\mathbf{n}_1$ and $\mathbf{n}_2$ are not parallel. $\left(0, 0, \frac{1}{2} \right)$ lies on P_1, but not on P_3, so they are not the same plane, but both P_2 and P_4 contain the point $(0, 0, -3)$, so these two planes are identical.

39. Let $Q = (2, 2, 0)$ and $R = (3, -1, 5)$, points on the line corresponding to $t = 0$ and $t = 1$. Let $P = (1, 2, 3)$. Then $\mathbf{a} = \overrightarrow{QR} = \langle 1, -3, 5 \rangle$, $\mathbf{b} = \overrightarrow{QP} = \langle -1, 0, 3 \rangle$. The distance is

$$d = \frac{|\mathbf{a} \times \mathbf{b}|}{|\mathbf{a}|} = \frac{|\langle 1, -3, 5 \rangle \times \langle -1, 0, 3 \rangle|}{|\langle 1, -3, 5 \rangle|} = \frac{|\langle -9, -8, -3 \rangle|}{|\langle 1, -3, 5 \rangle|} = \frac{\sqrt{9^2 + 8^2 + 3^2}}{\sqrt{1^2 + 3^2 + 5^2}} = \frac{\sqrt{154}}{\sqrt{35}} = \sqrt{\frac{22}{5}}.$$

41. By Equation 8, the distance is $D = \dfrac{1}{\sqrt{1 + 4 + 4}} \left[(1)(2) + (-2)(8) + (-2)(5) - 1 \right] = \dfrac{25}{3}$.

43. Put $y = z = 0$ in the equation of the first plane, to get the point $(-1, 0, 0)$ on the plane. Because the planes are parallel, the distance D between them is the distance from $(-1, 0, 0)$ to the second plane. By Equation 8, $D = \dfrac{|3(-1) + 6(0) - 3(0) - 4|}{\sqrt{3^2 + 6^2 + (-3)^2}} = \dfrac{7}{3\sqrt{6}}$.

45. The distance between two parallel planes is the same as the distance between a point on one of the planes and the other plane. Let $P_0 = (x_0, y_0, z_0)$ be a point on the plane given by $ax + by + cz = d_1$. Then $ax_0 + by_0 + cz_0 = d_1$ and the distance between P_0 and the plane given by $ax + by + cz = d_2$ is

$$D = \frac{1}{\sqrt{a^2 + b^2 + c^2}} |ax_0 + by_0 + cz_0 - d_2| = \frac{|d_1 - d_2|}{\sqrt{a^2 + b^2 + c^2}}.$$

47. L_1: $x = y = z$ $\Rightarrow$ $x = y$ (1). L_2: $x + 1 = y/2 = z/3$ $\Rightarrow$ $x + 1 = y/2$ (2). The solution of (1)

and (2) is $x = y = -2$. However, when $x = -2$, $x = z$ $\Rightarrow$ $z = -2$, but $x + 1 = z/3$ $\Rightarrow$ $z = -3$,

a contradiction. Hence the lines do not intersect. For L_1, $\mathbf{v}_1 = \langle 1, 1, 1 \rangle$, and for L_2, $\mathbf{v}_2 = \langle 1, 2, 3 \rangle$, so

the lines are not parallel. Thus the lines are skew lines. If two lines are skew, they can be viewed as

lying in two parallel planes and so the distance between the skew lines would be the same as the distance

between these parallel planes. The common normal vector to the planes must be perpendicular to

both $\langle 1, 1, 1 \rangle$ and $\langle 1, 2, 3 \rangle$, the direction vectors of the two lines. So set

$\mathbf{n} = \langle 1, 1, 1 \rangle \times \langle 1, 2, 3 \rangle = \langle 3 - 2, -3 + 1, 2 - 1 \rangle = \langle 1, -2, 1 \rangle$. From above, we know that

$(-2, -2, -2)$ and $(-2, -2, -3)$ are points of L_1 and L_2 respectively. So in the notation of Equation 7,

$d_1 = 1(-2) - 2(-2) + 1(-2) = 0$ and $d_2 = 1(-2) - 2(-2) + 1(-3) = -1$. By Exercise 45, the

distance between these two skew lines is $D = \dfrac{|0 - (-1)|}{\sqrt{1 + 4 + 1}} = \dfrac{1}{\sqrt{6}}$.

Alternate solution (without reference to planes): A vector which is perpendicular to both of the lines is

$\mathbf{n} = \langle 1, 1, 1 \rangle \times \langle 1, 2, 3 \rangle = \langle 1, -2, 1 \rangle$. Pick any point on each of the lines, say $(-2, -2, -2)$ and

$(-2, -2, -3)$, and form the vector $\mathbf{b} = \langle 0, 0, 1 \rangle$ connecting the two points. The distance between the

two skew lines is the absolute value of the scalar projection of $\mathbf{b}$ along $\mathbf{n}$, that is,

$$D = \frac{|\mathbf{n} \cdot \mathbf{b}|}{|\mathbf{n}|} = \frac{|1 \cdot 0 - 2 \cdot 0 + 1 \cdot 1|}{\sqrt{1 + 4 + 1}} = \frac{1}{\sqrt{6}}.$$

49. If $a \neq 0$, then $ax + by + cz = d$ $\Rightarrow$ $ax - d + by + cz = 0$ $\Rightarrow$

$a(x - d/a) + b(y - 0) + c(z - 0) = 0$ which by (6) is the scalar equation of the plane through the

point $(d/a, 0, 0)$ with normal vector $\langle a, b, c \rangle$. Similarly, if $b \neq 0$ (or if $c \neq 0$) the equation of the

plane can be rewritten as $a(x - 0) + b(y - d/b) + c(z - 0) = 0$ [or as

$a(x - 0) + b(y - 0) + c(z - d/c) = 0$] which by (6) is the scalar equation of a plane through the point

$(0, d/b, 0)$ [or the point $(0, 0, d/c)$] with normal vector $\langle a, b, c \rangle$.

Section 9.6 Functions and Surfaces

1. (a) According to Table 1, $f(40, 15) = 25$, which means that if a 40-knot wind has been blowing in the open sea for 15 hours, it will create waves with estimated heights of 25 feet.

(b) $h = f(30, t)$ means we fix v at 30 and allow t to vary, resulting in a function of one variable. Thus here, $h = f(30, t)$ gives the wave heights produced by 30-knot winds blowing for t hours. From the table (look at the row corresponding to $v = 30$), the function increases but at a declining rate as t increases. In fact, the function values appear to be approaching a limiting value of approximately 19, which suggests that 30-knot winds cannot produce waves higher than about 19 feet.

(c) $h = f(v, 30)$ means we fix t at 30, again giving a function of one variable. So, $h = f(v, 30)$ gives the wave heights produced by winds of speed v blowing for 30 hours. From the table (look at the column corresponding to $t = 30$), the function appears to increase at an increasing rate, with no apparent limiting value. This suggests that faster winds (lasting 30 hours) always create higher waves.

3. (a) $f(2, 4) = e^{2^2 - 4} = e^0 = 1$.

(b) The exponential function is defined everywhere, so no matter what values of x and y we use, $e^{x^2 - y}$ is defined. So the domain of f is $\mathbb{R}^2$.

(c) Because the range of $g(x, y) = x^2 - y$ is $\mathbb{R}$, and the range of e^x is $(0, \infty)$, the range of $e^{g(x,y)} = e^{x^2 - y}$ is $\{z \mid z > 0\}$.

5. $y - 2x \geq 0$ so $D = \{(x, y) \mid y \geq 2x\}$.

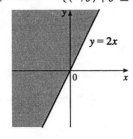

7. $D = \{(x, y) \mid x^2 + y \geq 0\}$
$= \{(x, y) \mid y \geq -x^2\}$

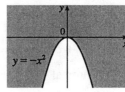

9. $z = 3$, a horizontal plane through the point $(0, 0, 3)$.

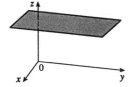

11. $z = 1 - x - y$ or $x + y + z = 1$, a plane with intercepts 1, 1, and 1.

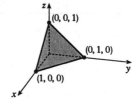

13. $z = 1 - x^2$, a parabolic cylinder.

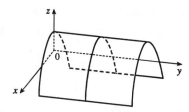

15. All six graphs have different traces in the planes $x = 0$ and $y = 0$, so we investigate these for each function.

(a) $f(x, y) = |x| + |y|$. The trace in $x = 0$ is $z = |y|$, and in $y = 0$ is $z = |x|$, so it must be graph VI.

(b) $f(x, y) = |xy|$. The trace in $x = 0$ is $z = 0$, and in $y = 0$ is $z = 0$, so it must be graph V.

(c) $f(x, y) = \dfrac{1}{1 + x^2 + y^2}$. The trace in $x = 0$ is $z = \dfrac{1}{1 + y^2}$, and in $y = 0$ is $z = \dfrac{1}{1 + x^2}$. In addition, we can see that f is close to 0 for large values of x and y, so this is graph I.

(d) $f(x, y) = \left(x^2 - y^2\right)^2$. The trace in $x = 0$ is $z = y^4$, and in $y = 0$ is $z = x^4$. Both graph II and graph IV seem plausible; notice the trace in $z = 0$ is $0 = \left(x^2 - y^2\right)^2 \Rightarrow y = \pm x$, so it must be graph IV.

(e) $f(x, y) = (x - y)^2$. The trace in $x = 0$ is $z = y^2$, and in $y = 0$ is $z = x^2$. Both graph II and graph IV seem plausible; notice the trace in $z = 0$ is $0 = (x - y)^2 \Rightarrow y = x$, so it must be graph II.

(f) $f(x, y) = \sin(|x| + |y|)$. The trace in $x = 0$ is $z = \sin|y|$, and in $y = 0$ is $z = \sin|x|$. In addition, notice that the oscillating nature of the graph is characteristic of trigonometric functions. So this is graph III.

17. The equation of the graph is $z = x^2 + 9y^2$. The traces in $x = k$ are

$z = 9y^2 + k$, a family of parabolas opening upward. In $y = k$, we have

$z = x^2 + 9k^2$, again a family of parabolas opening upward. The traces

in $z = k$ are $x^2 + 9y^2 = k$, a family of ellipses. The surface is an

elliptic paraboloid.

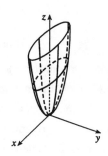

19. $y = z^2 - x^2$. The traces in $x = k$ are the parabolas $y = z^2 - k^2$; the traces in $y = k$ are $k = z^2 - x^2$, which are hyperbolas (note the hyperbolas are oriented differently for $k > 0$ than for $k < 0$); and the traces in $z = k$ are the parabolas $y = k^2 - x^2$. Thus, $\dfrac{y}{1} = \dfrac{z^2}{1^2} - \dfrac{x^2}{1^2}$ is a hyperbolic paraboloid.

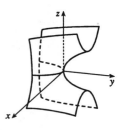

21. (a) In $\mathbb{R}^2$, $x^2 + y^2 = 1$ represents a circle of radius 1 centered at the origin.

(b) In $\mathbb{R}^3$, the equation doesn't involve z, which means that any horizontal plane $z = k$ intersects the surface in a circle $x^2 + y^2 = 1$, $z = k$. Thus the surface is a circular cylinder, made up of infinitely many shifted copies of the circle $x^2 + y^2 = 1$, with axis the z-axis.

(c) In $\mathbb{R}^3$, $x^2 + z^2 = 1$ also represents a circular cylinder of radius 1, this time with axis the y-axis.

23. The traces of $x^2 + y^2 - z^2 = 1$ in $x = k$ are $y^2 - z^2 = 1 - k^2$, a family of hyperbolas (note that the hyperbolas are oriented differently for $-1 < k < 1$ than for $k < -1$ or $k > 1$). The traces in $y = k$ are $x^2 - z^2 = 1 - k^2$, a similar family of hyperbolas. The traces in $z = k$ are $x^2 + y^2 = 1 + k^2$, a family of circles. For $k = 0$, the trace in the xy-plane, the circle is of radius 1. As $|k|$ increases, so does the radius of the circle. This behavior, combined with the hyperbolic vertical traces, gives the graph in Figure 15.

25. $f(x, y) = 3x - x^4 - 4y^2 - 10xy$

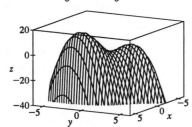

Three-dimensional view

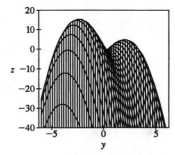

Front view

It does appear that the function has a maximum value, at the higher of the two "hilltops." From the front view graph, the maximum value appears to be approximately 15. Both hilltops could be considered local maximum points, as the values of f there are larger than at the neighboring points. There does not appear to be any local minimum point; although the valley shape between the two peaks looks like a minimum of some kind, some neighboring points have lower function values.

27.

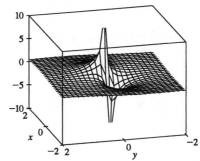

$f(x, y) = \dfrac{x+y}{x^2+y^2}$. As both x and y become large, the function values appear to approach 0, regardless of which direction is considered. As (x, y) approaches the origin, the graph exhibits asymptotic behavior. From some directions, $f(x, y) \to \infty$, while in others $f(x, y) \to -\infty$. (These are the vertical spikes visible in the graph.) If the graph is examined carefully, however, one can see that $f(x, y)$ approaches 0 along the line $y = -x$.

29.

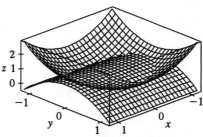

The curve of intersection looks like a bent ellipse. The projection of this curve onto the xy-plane is the set of points $(x, y, 0)$ which satisfy $x^2 + y^2 = 1 - y^2$ $\Leftrightarrow$ $x^2 + 2y^2 = 1$ $\Leftrightarrow$ $x^2 + \dfrac{y^2}{\left(1/\sqrt{2}\right)^2} = 1$.

This is an equation of an ellipse.

31. If (a, b, c) satisfies $z = y^2 - x^2$, then $c = b^2 - a^2$. L_1: $x = a + t$, $y = b + t$, $z = c + 2(b - a)t$, L_2: $x = a + t$, $y = b - t$, $z = c - 2(b + a)t$. Substitute the parametric equations of L_1 into the equation of the hyperbolic paraboloid in order to find the points of intersection: $z = y^2 - x^2$ $\Rightarrow$ $c + 2(b - a)t = (b + t)^2 - (a + t)^2 = b^2 - a^2 + 2(b - a)t$ $\Rightarrow$ $c = b^2 - a^2$. As this is true for all values of t, L_1 lies on $z = y^2 - x^2$. Performing similar operations with L_2 gives: $z = y^2 - x^2$ $\Rightarrow$ $c - 2(b + a)t = (b - t)^2 - (a + t)^2 = b^2 - a^2 - 2(b + a)t$ $\Rightarrow$ $c = b^2 - a^2$. This tells us that all of L_2 also lies on $z = y^2 - x^2$.

Section 9.7 Cylindrical and Spherical Coordinates

1. See Figure 1 and the accompanying discussion on page 692; see the paragraph accompanying Figure 3 on page 693.

3. (a)

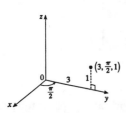

$x = 3 \cos \frac{\pi}{2} = 0$, $y = 3 \sin \frac{\pi}{2} = 3$, and $z = 1$, so the point is $(0, 3, 1)$ in rectangular coordinates.

(b)

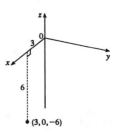

$x = 3 \cos 0 = 3$, $y = 3 \sin 0 = 0$, and $z = -6$, so the point is $(3, 0, -6)$ in rectangular coordinates.

5. (a) $r^2 = (-1)^2 + (0)^2 = 1$ so $r = 1$; $z = 0$; $\tan \theta = 0$ so $\theta = 0$ or π. But $x = -1$ so $\theta = \pi$ and in cylindrical coordinates, the point is $(1, \pi, 0)$.

(b) $r^2 = 4$ so $r = 2$, $\tan \theta = \frac{1}{\sqrt{3}}$ so $\theta = \frac{\pi}{6}$, and $z = 4$. Thus the point is $\left(2, \frac{\pi}{6}, 4\right)$ in cylindrical coordinates.

7. (a)

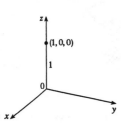

$x = (1) \sin 0 \cos 0 = 0$, $y = (1) \sin 0 \sin 0 = 0$, $z = (1) \cos 0 = 1$ so the point is $(0, 0, 1)$ in rectangular coordinates.

(b)

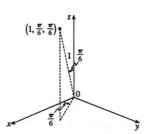

$x = \sin \frac{\pi}{6} \cos \frac{\pi}{6} = \frac{\sqrt{3}}{4}$, $y = \sin \frac{\pi}{6} \sin \frac{\pi}{6} = \frac{1}{4}$, and $z = \cos \frac{\pi}{6} = \frac{\sqrt{3}}{2}$, so the point is $\left(\frac{\sqrt{3}}{4}, \frac{1}{4}, \frac{\sqrt{3}}{2}\right)$ in rectangular coordinates.

9. (a) $\rho = \sqrt{9 + 0 + 0} = 3$, $\cos \phi = \frac{0}{3} = 0$ so $\phi = \frac{\pi}{2}$, and $\cos \theta = \dfrac{-3}{3 \sin \frac{\pi}{2}} = -1$ so $\theta = \pi$, thus the spherical coordinates are $\left(3, \pi, \frac{\pi}{2}\right)$.

(b) $\rho = \sqrt{3 + 1} = 2$, $\cos \phi = \frac{1}{2}$ so $\phi = \frac{\pi}{3}$, and $\cos \theta = \dfrac{\sqrt{3}}{2 \sin \frac{\pi}{3}} = \dfrac{\sqrt{3} \cdot 2}{2 \cdot \sqrt{3}} = 1$ so $\theta = 0$, thus the point is $\left(2, 0, \frac{\pi}{3}\right)$ in spherical coordinates.

Note: It is also apparent that $\theta = 0$ since the point is in the xz-plane and $x > 0$.

11. Since $r = 3$, $x^2 + y^2 = 9$ and the surface is a cylinder with radius 3 and axis the z-axis.

13. Since $\phi = \frac{\pi}{3}$, the surface is one frustum of the right circular cone with vertex at the origin and axis the positive z-axis.

15. $z = r^2 = x^2 + y^2$, so the surface is a circular paraboloid with vertex at the origin and axis the positive z-axis.

17. $r = 2\cos\theta \implies r^2 = x^2 + y^2 = 2r\cos\theta = 2x \iff (x-1)^2 + y^2 = 1$, which is the equation of a circular cylinder with radius 1, whose axis is the vertical line $x = 1$, $y = 0$, $z = z$.

19. Since $r^2 + z^2 = 25$ and $r^2 = x^2 + y^2$, we have $x^2 + y^2 + z^2 = 25$, a sphere with radius 5 and center at the origin.

21. (a) $r^2 = x^2 + y^2$, so $r^2 + z^2 = 16$.

(b) $\rho^2 = x^2 + y^2 + z^2$, so $\rho^2 = 16$ or $\rho = 4$.

23. (a) $r^2 = 2r\sin\theta$ or $r = 2\sin\theta$.

(b) $\rho^2 \sin^2\phi \left(\cos^2\theta + \sin^2\theta\right) = 2\rho\sin\phi\sin\theta$ or $\rho\sin^2\phi = 2\sin\phi\sin\theta$ or $\rho\sin\phi = 2\sin\theta$.

25. $z = r^2 = x^2 + y^2$ is a circular paraboloid with vertex $(0,0,0)$, opening upward. $z = 2 - r^2 \implies z - 2 = -\left(x^2 + y^2\right)$ is a circular paraboloid with vertex $(0,0,2)$ opening downward. Thus $r^2 \le z \le 2 - r^2$ is the solid region enclosed by these two surfaces.

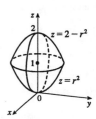

27. $-\frac{\pi}{2} \le \theta \le \frac{\pi}{2}$ restricts the solid to the 4 octants in which x is positive. $\rho = \sec\phi \implies \rho\cos\phi = z = 1$, which is the equation of a horizontal plane. $0 \le \phi \le \frac{\pi}{6}$ describes a cone, opening upward. So the solid lies above the cone $\phi = \frac{\pi}{6}$ and below the plane $z = 1$.

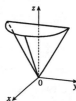

29. $z \ge \sqrt{x^2 + y^2}$ because the solid lies above the cone. Squaring both sides of this inequality gives $z^2 \ge x^2 + y^2 \implies 2z^2 \ge x^2 + y^2 + z^2 = \rho^2 \implies z^2 = \rho^2\cos^2\phi \ge \frac{1}{2}\rho^2 \implies \cos^2\phi \ge \frac{1}{2}$. The cone opens upward so that the inequality is $\cos\phi \ge \frac{1}{\sqrt{2}}$, or equivalently $0 \le \phi \le \frac{\pi}{4}$. In spherical coordinates the sphere $z = x^2 + y^2 + z^2$ is $\rho\cos\phi = \rho^2 \implies \rho = \cos\phi$. $0 \le \rho \le \cos\phi$ because the solid lies below the sphere. The solid can therefore be described as the region in spherical coordinates satisfying $0 \le \rho \le \cos\phi$, $0 \le \phi \le \frac{\pi}{4}$.

31. In cylindrical coordinates, the equation of the cylinder is $r = 3$, $0 \le z \le 10$. The hemisphere is the upper part of the sphere radius 3, center $(0,0,10)$, equation $r^2 + (z - 10)^2 = 3^2$, $z \ge 10$. In Maple, we can use either the `coords=cylindrical` option in a regular `plot` command, or the `plots[cylinderplot]` command. In Mathematica, we can use `ParametricPlot3d`.

Chapter 9 Review

Concept Check

1. A scalar is a real number, while a vector is a quantity that has both a real-valued magnitude and a direction.

2. To add two vectors geometrically, we can use either the Triangle Law or the Parallelogram Law, as illustrated in Figures 4 and 5 in Section 9.2. Algebraically, we add the corresponding components of the vectors.

3. For $c > 0$, $c\mathbf{a}$ is a vector with the same direction as $\mathbf{a}$ and length c times the length of $\mathbf{a}$. If $c < 0$, $c\mathbf{a}$ points in the opposite direction as $\mathbf{a}$ and has length $|c|$ times the length of $\mathbf{a}$. (See Figures 6 and 7 in Section 9.2.) Algebraically, to find $c\mathbf{a}$ we multiply each component of $\mathbf{a}$ by c.

4. See (1) in Section 9.2.

5. See the definition on page 661 and the boxed equation on page 662.

6. The dot product can be used to determine the work done moving an object given the force and displacement vectors. The dot product can also be used to find the angle between two vectors and the scalar projection of one vector onto another. In particular, the dot product can determine if two vectors are orthogonal.

7. See the boxed equations as well as Figures 4 and 5 and the accompanying discussion on page 664.

8. See the definition on page 667; use either (2) or (4) in Section 9.4.

9. The cross product can be used to determine torque if the force and position vectors are known. In addition, the cross product can be used to create a vector orthogonal to two given vectors as well as to determine if two vectors are parallel. The cross product can also be used to determine the area of a parallelogram determined by two vectors.

10. **(a)** The area of the parallelogram determined by $\mathbf{a}$ and $\mathbf{b}$ is the length of the cross product: $|\mathbf{a} \times \mathbf{b}|$.

    (b) The volume of the parallelepiped determined by $\mathbf{a}$, $\mathbf{b}$, and $\mathbf{c}$ is the magnitude of their scalar triple product: $|\mathbf{a} \cdot (\mathbf{b} \times \mathbf{c})|$.

11. If an equation of the plane is known, it can be written as $ax + by + cz = d$. A normal vector, which is perpendicular to the plane, is $\langle a, b, c \rangle$ (or any scalar multiple of $\langle a, b, c \rangle$). If an equation is not known, we can use points on the plane to find two non-parallel vectors which lie in the plane. The cross product of these vectors is a vector perpendicular to the plane.

12. See (1), (2), and (3) in Section 9.5.

13. See (4), (5), and (6) in Section 9.5.

14. (a) Determine the vectors $\overrightarrow{PQ} = \langle a_1, a_2, a_3 \rangle$ and $\overrightarrow{PR} = \langle b_1, b_2, b_3 \rangle$. If there is a scalar t such that $\langle a_1, a_2, a_3 \rangle = t \langle b_1, b_2, b_3 \rangle$, then the vectors are parallel and the points must all lie on the same line. Alternatively, if $\overrightarrow{PQ} \times \overrightarrow{PR} = \mathbf{0}$, then $\overrightarrow{PQ}$ and $\overrightarrow{PR}$ are parallel, so P, Q, and R are collinear. Thirdly, an algebraic method is to determine an equation of the line joining two of the points, and then check whether or not the third point satisfies this equation.

(b) Find the vectors $\overrightarrow{PQ} = \mathbf{a}$, $\overrightarrow{PR} = \mathbf{b}$, $\overrightarrow{PS} = \mathbf{c}$. $\mathbf{a} \times \mathbf{b}$ is normal to the plane formed by P, Q and R, and so S lies on this plane if $\mathbf{a} \times \mathbf{b}$ and $\mathbf{c}$ are orthogonal, that is, if $(\mathbf{a} \times \mathbf{b}) \cdot \mathbf{c} = 0$. (Or use the reasoning in Example 6 in Section 9.4.)

Alternatively, find an equation for the plane determined by three of the points and check whether or not the fourth point satisfies this equation.

15. (a) See Exercise 9.4.25.

(b) See Example 8 in Section 9.5.

(c) See Example 10 in Section 9.5.

16. One method of graphing a function of two variables is to first find traces (see Example 6 in Section 9.6 and the discussion preceding it).

17. (a) See (1) and the discussion accompanying Figure 3 in Section 9.7.

(b) See (3) and Figures 6-8, and the accompanying discussion, in Section 9.7.

True-False Quiz

1. True, by Property 2 of the dot product. (See page 663.)

3. True. If θ is the angle between $\mathbf{u}$ and $\mathbf{v}$, then by the definition of the cross product, $|\mathbf{u} \times \mathbf{v}| = |\mathbf{u}|\,|\mathbf{v}| \sin \theta = |\mathbf{v}|\,|\mathbf{u}| \sin \theta = |\mathbf{v} \times \mathbf{u}|$.
(Or, by Properties 1 and 2 of the cross product, $|\mathbf{u} \times \mathbf{v}| = |-\mathbf{v} \times \mathbf{u}| = |-1|\,|\mathbf{v} \times \mathbf{u}| = |\mathbf{v} \times \mathbf{u}|$.)

5. Property 2 of the cross product tells us that this is true.

7. This is true by (6) in Section 9.4.

9. This is true because $\mathbf{u} \times \mathbf{v}$ is orthogonal to $\mathbf{u}$ (see page 667), and the dot product of two orthogonal vectors is 0.

11. If $|\mathbf{u}| = 1$, $|\mathbf{v}| = 1$ and θ is the angle between these two vectors (so $0 \le \theta \le \pi$), then by the definition of the cross product, $|\mathbf{u} \times \mathbf{v}| = |\mathbf{u}|\,|\mathbf{v}| \sin \theta = \sin \theta$, which is equal to 1 if and only if $\theta = \frac{\pi}{2}$ (that is, if and only if the two vectors are orthogonal). Therefore, the assertion that the cross product of two unit vectors is a unit vector is false.

13. This is false. In $\mathbb{R}^2$, $x^2 + y^2 = 1$ represents a circle, but $\{(x, y, z) \mid x^2 + y^2 = 1\}$ represents a *three-dimensional surface*, namely, a circular cylinder with axis the z-axis.

Exercises

1. (a) By the formula for an equation of a sphere (see page 649), an equation of the sphere with center $(1, -1, 2)$ and radius 3 is $(x - 1)^2 + (y + 1)^2 + (z - 2)^2 = 9$.

(b) Completing the squares gives $(x + 2)^2 + (y + 3)^2 + (z - 5)^2 = -2 + 4 + 9 + 25 = 36$. Thus, the sphere is centered at $(-2, -3, 5)$ and has radius 6.

3. $\mathbf{u} \cdot \mathbf{v} = |\mathbf{u}|\,|\mathbf{v}| \cos 45° = (2)\,(3)\,\frac{\sqrt{2}}{2} = 3\sqrt{2}$. $|\mathbf{u} \times \mathbf{v}| = |\mathbf{u}|\,|\mathbf{v}| \sin 45° = (2)\,(3)\,\frac{\sqrt{2}}{2} = 3\sqrt{2}$. By the right-hand rule, $\mathbf{u} \times \mathbf{v}$ is directed out of the page.

5. The vectors are orthogonal if and only if their dot product is 0. So we require that
$$\langle 2, x, 4 \rangle \cdot \langle 2x, 3, -7 \rangle = 0 \quad \Leftrightarrow \quad 4x + 3x - 28 = 0 \quad \Leftrightarrow \quad x = 4.$$

7. (a) $(\mathbf{u} \times \mathbf{v}) \cdot \mathbf{w} = \mathbf{u} \cdot (\mathbf{v} \times \mathbf{w}) = 2$

(b) $\mathbf{u} \cdot (\mathbf{w} \times \mathbf{v}) = \mathbf{u} \cdot [-(\mathbf{v} \times \mathbf{w})] = -\mathbf{u} \cdot (\mathbf{v} \times \mathbf{w}) = -2$

(c) $\mathbf{v} \cdot (\mathbf{u} \times \mathbf{w}) = (\mathbf{v} \times \mathbf{u}) \cdot \mathbf{w} = -(\mathbf{u} \times \mathbf{v}) \cdot \mathbf{w} = -2$

(d) $(\mathbf{u} \times \mathbf{v}) \cdot \mathbf{v} = \mathbf{u} \cdot (\mathbf{v} \times \mathbf{v}) = \mathbf{u} \cdot \mathbf{0} = 0$

9. For simplicity, consider a unit cube positioned with its back left corner at the origin. Vector representations of the diagonals joining the points $(0, 0, 0)$ to $(1, 1, 1)$ and $(1, 0, 0)$ to $(0, 1, 1)$ are $\langle 1, 1, 1 \rangle$ and $\langle -1, 1, 1 \rangle$. Let θ be the angle between these two vectors.
$\langle 1, 1, 1 \rangle \cdot \langle -1, 1, 1 \rangle = -1 + 1 + 1 = 1 = |\langle 1, 1, 1 \rangle|\,|\langle -1, 1, 1 \rangle| \cos \theta = 3 \cos \theta \quad \Rightarrow \quad \cos \theta = \frac{1}{3} \quad \Rightarrow$
$\theta = \cos^{-1} \frac{1}{3} \approx 71°$.

11. $\overrightarrow{AB} = \langle 1, 0, -1 \rangle$, $\overrightarrow{AC} = \langle 0, 4, 3 \rangle$, so

(a) a vector perpendicular to the plane is $\overrightarrow{AB} \times \overrightarrow{AC} = \langle 0 + 4, -(3 + 0), 4 - 0 \rangle = \langle 4, -3, 4 \rangle$.

(b) $\frac{1}{2} \left| \overrightarrow{AB} \times \overrightarrow{AC} \right| = \frac{1}{2}\sqrt{16 + 9 + 16} = \frac{\sqrt{41}}{2}$.

13. Let F_1 be the magnitude of the force directed $20°$ away from the direction of shore, and let F_2 be the magnitude of the other force. Separating these forces into components parallel to the direction of the resultant force and perpendicular to it gives $F_1 \cos 20° + F_2 \cos 30° = 255$ (1), and
$$F_1 \sin 20° - F_2 \sin 30° = 0 \quad \Rightarrow \quad F_1 = F_2 \frac{\sin 30°}{\sin 20°} \quad (2).\text{ Substituting (2) into (1) gives}$$
$F_2 (\sin 30° \cot 20° + \cos 30°) = 255 \quad \Rightarrow \quad F_2 \approx 114\text{ N.}$ Substituting this into (2) gives $F_1 \approx 166\text{ N.}$

15. $x = 1 + 2t,\ y = 2 - t,\ z = 4 + 3t$

17. $\mathbf{v} = \langle 4, -3, 5 \rangle$, so $x = 1 + 4t,\ y = -3t,\ z = 1 + 5t$.

19. $(x + 4) + 2(y - 1) + 5(z - 2) = 0$ or $x + 2y + 5z = 8$.

21. $\mathbf{n_1} = \langle 1, 0, -1 \rangle$ and $\mathbf{n_2} = \langle 0, 1, 2 \rangle$. Setting $z = 0$, it is easy to see that $(1, 3, 0)$ is a point on the line of intersection of $x - z = 1$ and $y + 2z = 3$. The direction of this line is $\mathbf{v_1} = \mathbf{n_1} \times \mathbf{n_2} = \langle 1, -2, 1 \rangle$. A second vector parallel to the desired plane is $\mathbf{v_2} = \langle 1, 1, -2 \rangle$, since it is perpendicular to $x + y - 2z = 1$. Therefore, the normal of the plane in question is $\mathbf{n} = \mathbf{v_1} \times \mathbf{v_2} = \langle 4 - 1, 1 + 2, 1 + 2 \rangle = 3 \langle 1, 1, 1 \rangle$. Taking $(x_0, y_0, z_0) = (1, 3, 0)$, the equation we are looking for is $(x - 1) + (y - 3) + z = 0 \iff x + y + z = 4$.

23. Since the direction vectors $\langle 2, 3, 4 \rangle$ and $\langle 6, -1, 2 \rangle$ aren't parallel, neither are the lines. For the lines to intersect, the three equations $1 + 2t = -1 + 6s$, $2 + 3t = 3 - s$, $3 + 4t = -5 + 2s$ must be satisfied simultaneously. Solving the first two equations gives $t = \frac{1}{5}$, $s = \frac{2}{5}$ and checking we see these values don't satisfy the third equation. Thus the lines aren't parallel and they don't intersect, so they must be skew.

25. By Exercise 9.5.45, $D = \dfrac{|2 - 24|}{\sqrt{26}} = \dfrac{22}{\sqrt{26}}$.

27. $x \neq 1$ and $x + y + 1 > 0$, so $D = \{(x, y) \mid y > -x - 1, x \neq 1\}$.

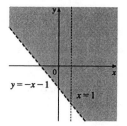

29. The graph is the plane $z = 6 - 2x - 3y \implies 2x + 3y + z = 6$. The intercepts with the coordinate axes are $(3, 0, 0)$, $(0, 2, 0)$, and $(0, 0, 6)$ which enable us to sketch the portion of the plane that lies in the first octant.

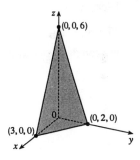

31. The equation is $z = 4 - x^2 - 4y^2$. The traces in $x = k$ are
$z = 4 - k^2 - 4y^2$, a family of parabolas opening downward, as are the
traces in $y = k$, $z = 4 - 4k^2 - x^2$. The traces in $z = k$ are
$x^2 + 4y^2 = 4 - k$, a family of ellipses, so the surface is an elliptic
paraboloid.

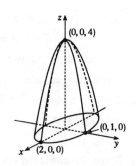

33. Since the equation $y^2 + z^2 = 1$ doesn't involve x, the traces in $x = k$
are each the circle $y^2 + z^2 = 1$, $x = k$, and we have a circular cylinder
with axis the x-axis.

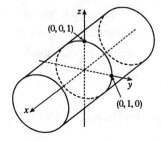

35. $x = 2\cos\frac{\pi}{6} = \sqrt{3}$, $y = 2\sin\frac{\pi}{6} = 1$, $z = 2$, so in rectangular coordinates the point is $\left(\sqrt{3}, 1, 2\right)$.
$\rho = \sqrt{3 + 1 + 4} = 2\sqrt{2}$, $\theta = \frac{\pi}{6}$, and $\cos\phi = z/\rho = \frac{1}{\sqrt{2}}$, so $\phi = \frac{\pi}{4}$ and the spherical coordinates are
$\left(2\sqrt{2}, \frac{\pi}{6}, \frac{\pi}{4}\right)$.

37. $x = 4\sin\frac{\pi}{6}\cos\frac{\pi}{3} = 1$, $y = 4\sin\frac{\pi}{6}\sin\frac{\pi}{3} = \sqrt{3}$, $z = 4\cos\frac{\pi}{6} = 2\sqrt{3}$ so in rectangular coordinates the
point is $\left(1, \sqrt{3}, 2\sqrt{3}\right)$. $r^2 = x^2 + y^2 = 4$, $r = 2$, so the cylindrical coordinates are $\left(2, \frac{\pi}{3}, 2\sqrt{3}\right)$.

39. $x^2 + y^2 + z^2 = 4$. In cylindrical coordinates, this becomes $r^2 + z^2 = 4$. In spherical coordinates, it
becomes $\rho^2 = 4$ or $\rho = 2$.

41. The resulting surface is a circular paraboloid with equation $z = 4x^2 + 4y^2$. Changing to cylindrical
coordinates we have $z = 4\left(x^2 + y^2\right) = 4r^2$.

Focus on Problem Solving

1. Since three-dimensional situations are often difficult to visualize and work with, let us first try to find an analogous problem in two dimensions. The analogue of a cube is a square and the analogue of a sphere is a circle. Thus a similar problem in two dimensions is the following: if five circles with the same radius r are contained in a square of side 1 m so that the circles touch each other and four of the circles touch two sides of the square, find r.

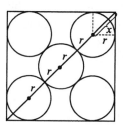

The diagonal of the square is $\sqrt{2}$. The diagonal is also $4r + 2x$. But x is the diagonal of a smaller square of side r. Therefore $x = \sqrt{2}r \;\Rightarrow\; \sqrt{2} = 4r + 2x = 4r + 2\sqrt{2}r = \left(4 + 2\sqrt{2}\right)r \;\Rightarrow\;$ $r = \frac{\sqrt{2}}{4+2\sqrt{2}}$.

Let us use these ideas to solve the original three-dimensional problem. The diagonal of the cube is $\sqrt{1^2 + 1^2 + 1^2} = \sqrt{3}$. The diagonal of the cube is also $4r + 2x$ where x is the diagonal of a smaller cube with edge r. Therefore $x = \sqrt{r^2 + r^2 + r^2} = \sqrt{3}r \;\Rightarrow\;$ $\sqrt{3} = 4r + 2x = 4r + 2\sqrt{3}r = \left(4 + 2\sqrt{3}\right)r$. Thus $r = \frac{\sqrt{3}}{4+2\sqrt{3}} = \frac{2\sqrt{3}-3}{2}$. The radius of each ball is $\left(\sqrt{3} - \frac{3}{2}\right)$ m.

3. (a) We find the line of intersection L as in Example 9.5.7(b). Observe that the point $(-1, c, c)$ lies on both planes. Now since L lies in both planes, it is perpendicular to both of the normal vectors $\mathbf{n_1}$ and $\mathbf{n_2}$, and thus parallel to their cross product $\mathbf{n_1} \times \mathbf{n_2} = \begin{vmatrix} \mathbf{i} & \mathbf{j} & \mathbf{k} \\ c & 1 & 1 \\ 1 & -c & c \end{vmatrix} = \langle 2c, -c^2 + 1, -c^2 - 1 \rangle$. So symmetric equations of L can be written as $\dfrac{x+1}{-2c} = \dfrac{y-c}{c^2 - 1} = \dfrac{z-c}{c^2 + 1}$, provided that $c \neq 0, \pm 1$.

If $c = 0$, then the two planes are given by $y + z = 0$ and $x = -1$, so symmetric equations of L are $x = -1$, $y = -z$. If $c = -1$, then the two planes are given by $-x + y + z = -1$ and $x + y + z = -1$, and they intersect in the line $x = 0$, $y = -z - 1$. If $c = 1$, then the two planes are given by $x + y + z = 1$ and $x - y + z = 1$, and they intersect in the line $y = 0$, $x = 1 - z$.

(b) If we set $z = t$ in the symmetric equations and solve for x and y separately, we get $x + 1 = \dfrac{(t-c)(-2c)}{c^2 + 1}$, $y - c = \dfrac{(t-c)(c^2 - 1)}{c^2 + 1} \;\Rightarrow\; x = \dfrac{-2ct + (c^2 - 1)}{c^2 + 1}$, $y = \dfrac{(c^2 - 1)t + 2c}{c^2 + 1}$. Eliminating c from these equations, we have $x^2 + y^2 = t^2 + 1$. So the curve traced out by L in the plane $z = t$ is a circle with center at $(0, 0, t)$ and radius $\sqrt{t^2 + 1}$.

(c) The area of a horizontal cross-section of the solid is $A(z) = \pi(z^2 + 1)$, so $V = \int_0^1 A(z)\, dz = \pi \left[\frac{1}{3}z^3 + z\right]_0^1 = \frac{4\pi}{3}$.

5. (a) When $\theta = \theta_s$, the block is not moving, so the sum of the forces on the block must be **0**, thus $\mathbf{N} + \mathbf{F} + \mathbf{W} = \mathbf{0}$. This relationship is illustrated geometrically in the figure. Since the vectors form a right triangle, we have $\tan(\theta_s) = \dfrac{|\mathbf{F}|}{|\mathbf{N}|} = \dfrac{\mu_s n}{n} = \mu_s$.

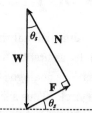

(b) We place the block at the origin and sketch the force vectors acting on the block, including the additional horizontal force **H**, with initial points at the origin. We then rotate this system so that **F** lies along the positive x-axis and the inclined plane is parallel to the x-axis.

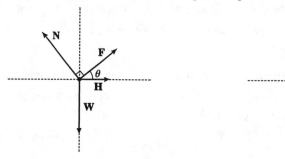

If we assume that $|\mathbf{F}| = \mu_s n$ for $\theta > \theta_s$, then the vectors, in terms of components parallel and perpendicular to the inclined plane, are

$$\mathbf{N} = n\mathbf{j} \qquad \mathbf{F} = (\mu_s n)\,\mathbf{i}$$
$$\mathbf{W} = (-mg\sin\theta)\,\mathbf{i} + (-mg\cos\theta)\,\mathbf{j}$$
$$\mathbf{H} = (h_{\min}\cos\theta)\,\mathbf{i} + (-h_{\min}\sin\theta)\,\mathbf{j}$$

Equating components, we have

$$\mu_s n - mg\sin\theta + h_{\min}\cos\theta = 0 \quad \Rightarrow \quad h_{\min}\cos\theta + \mu_s n = mg\sin\theta \tag{1}$$
$$n - mg\cos\theta - h_{\min}\sin\theta = 0 \quad \Rightarrow \quad h_{\min}\sin\theta + mg\cos\theta = n \tag{2}$$

(c) Since (2) is solved for n, we substitute into (1):

$$h_{\min}\cos\theta + \mu_s\,(h_{\min}\sin\theta + mg\cos\theta) = mg\sin\theta \quad \Rightarrow$$
$$h_{\min}\cos\theta + h_{\min}\mu_s\sin\theta = mg\sin\theta - mg\mu_s\cos\theta \quad \Rightarrow$$
$$h_{\min} = mg\left(\frac{\sin\theta - \mu_s\cos\theta}{\cos\theta + \mu_s\sin\theta}\right) = mg\left(\frac{\tan\theta - \mu_s}{1 + \mu_s\tan\theta}\right)$$

From part (a) we know $\mu_s = \tan\theta_s$, so this becomes $h_{\min} = mg\left(\dfrac{\tan\theta - \tan\theta_s}{1 + \tan\theta_s\tan\theta}\right)$ and using a trigonometric identity, this is $mg\tan(\theta - \theta_s)$ as desired.

Note for $\theta = \theta_s$, $h_{min} = mg \tan 0 = 0$, which makes sense since the block is at rest for θ_s, thus no additional force $\mathbf{H}$ is necessary to prevent it from moving. As θ increases, the factor $\tan(\theta - \theta_s)$, and hence the value of h_{min}, increases slowly for small values of $\theta - \theta_s$ but much more rapidly as $\theta - \theta_s$ becomes significant. This seems reasonable, as the steeper the inclined plane, the less the horizontal components of the various forces affect the movement of the block, so we would need a much larger magnitude of horizontal force to keep the block motionless. If we allow $\theta \to 90°$, corresponding to the inclined plane being placed vertically, the value of h_{min} is quite large; this is to be expected, as it takes a great amount of horizontal force to keep an object from moving vertically. In fact, without friction (so $\theta_s = 0$), we would have $\theta \to 90° \implies h_{min} \to \infty$, and it would be impossible to keep the block from slipping.

(d) Since h_{max} is the largest value of h that keeps the block from slipping, the force of friction is keeping the block from moving *up* the inclined plane; thus, $\mathbf{F}$ is directed *down* the plane. Our system of forces is similar to that in part (b), then, except that we have $\mathbf{F} = -(\mu_s n)\,\mathbf{i}$. Following our procedure in parts (b) and (c), we equate components:

$$-\mu_s n - mg \sin\theta + h_{max} \cos\theta = 0 \quad \Rightarrow \quad h_{max} \cos\theta - \mu_s n = mg \sin\theta$$

$$n - mg \cos\theta - h_{max} \sin\theta = 0 \quad \Rightarrow \quad h_{max} \sin\theta + mg \cos\theta = n$$

Then substituting,

$$h_{max} \cos\theta - \mu_s (h_{max} \sin\theta + mg \cos\theta) = mg \sin\theta \quad \Rightarrow$$

$$h_{max} \cos\theta - h_{max}\mu_s \sin\theta = mg \sin\theta + mg\mu_s \cos\theta \quad \Rightarrow$$

$$h_{max} = mg \left(\frac{\sin\theta + \mu_s \cos\theta}{\cos\theta - \mu_s \sin\theta} \right) = mg \left(\frac{\tan\theta + \mu_s}{1 - \mu_s \tan\theta} \right)$$

$$= mg \left(\frac{\tan\theta + \tan\theta_s}{1 - \tan\theta_s \tan\theta} \right) = mg \tan(\theta + \theta_s)$$

We would expect h_{max} to increase as θ increases, with similar behavior as we established for h_{min}, but with h_{max} values always larger than h_{min}. We can see that this is the case if we graph h_{max} as a function of θ, as the curve is the graph of h_{min} translated $2\theta_s$ to the left, so the equation does seem reasonable. Notice that the equation predicts $h_{max} \to \infty$ as $\theta \to 90° - \theta_s$. In fact, as θ increases, the normal force increases as well. When $90° - \theta_s \le \theta \le 90°$, the horizontal force is completely counteracted by the sum of the normal and frictional forces, so no part of the horizontal force contributes to moving the block up the plane no matter how large its magnitude.

Chapter 10   Vector Functions

Section 10.1   Vector Functions and Space Curves

1. The component functions t^2, $\sqrt{t-1}$, and $\sqrt{5-t}$ are all defined when $t-1 \geq 0$ $\Rightarrow$ $t \geq 1$ and $5 - t \geq 0$ $\Rightarrow$ $t \leq 5$, so the domain of $\mathbf{r}\,(t)$ is $[1, 5]$.

3. $\displaystyle\lim_{t\to 1} \sqrt{t+3} = 2,\ \lim_{t\to 1}\frac{t-1}{t^2-1} = \lim_{t\to 1}\frac{1}{t+1} = \frac{1}{2},\ \lim_{t\to 1}\left(\frac{\tan t}{t}\right) = \tan 1$

Thus the given limit equals $\langle 2, \frac{1}{2}, \tan 1 \rangle$.

5. $x = \cos 4t$, $y = t$, $z = \sin 4t$. At any point (x, y, z) on the curve, $x^2 + z^2 = \cos^2 4t + \sin^2 4t = 1$. So the curve lies on a circular cylinder with axis the y-axis. Since $y = t$, this is a helix. So the graph is V.

7. $x = t$, $y = 1/\left(1+t^2\right)$, $z = t^2$. Note that y and z are positive for all t. The curve passes through $(0, 1, 0)$ when $t = 0$. As $t \to \infty$, $(x, y, z) \to (\infty, 0, \infty)$, and as $t \to -\infty$, $(x, y, z) \to (-\infty, 0, \infty)$. So the graph is I.

9. $x = \cos t$, $y = \sin t$, $z = \sin 5t$. $x^2 + y^2 = \cos^2 t + \sin^2 t = 1$, so the curve lies on a circular cylinder with axis the z-axis. Each of x, y and z is periodic, and at $t = 0$ and $t = 2\pi$ the curve passes through the same point, so the curve repeats itself and the graph is IV.

11. The parametric equations for this curve are $x = t^4 + 1$, $y = t$. We can make a table of values, or we can eliminate the parameter: $t = y$ $\Rightarrow$ $x = y^4 + 1$, with $y \in \mathbb{R}$. By comparing different values of t, we find the direction in which t increases as indicated in the graph.

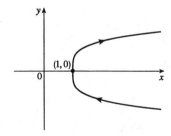

13. The corresponding parametric equations are $x = t$, $y = -t$, $z = 2t$, which are the parametric equations of a line through the origin and with direction vector $\langle 1, -1, 2 \rangle$.

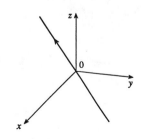

15. The parametric equations give

$x^2 + z^2 = \sin^2 t + \cos^2 t = 1$, $y = 3$, which is a circle
of radius 1, center $(0, 3, 0)$ in the plane $y = 3$.

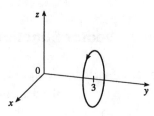

17. The parametric equations are $x = t^2$, $y = t^4$, $z = t^6$.
These are positive for $t \neq 0$ and 0 when $t = 0$. So the
curve lies entirely in the first quadrant. The projection
of the graph onto the xy-plane is $y = x^2$, $y > 0$, a half
parabola. On the xz-plane $z = x^3$, $z > 0$, a half cubic,
and the yz-plane, $y^3 = z^2$.

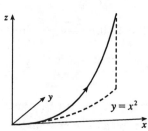

19. If $x = t \cos t$, $y = t \sin t$, and $z = t$, then

$x^2 + y^2 = t^2 \cos^2 t + t^2 \sin^2 t = t^2 = z^2$, so the curve
lies on the cone $z^2 = x^2 + y^2$. Thus the curve is a
spiral on this cone.

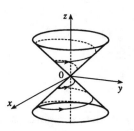

21. $\mathbf{r}(t) = \langle \sin t, \cos t, t^2 \rangle$

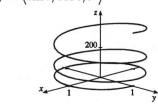

23. $\mathbf{r}(t) = \langle \sqrt{t}, t, t^2 - 2 \rangle$

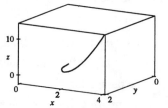

25.

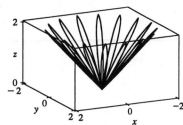

$x = (1 + \cos 16t) \cos t$, $y = (1 + \cos 16t) \sin t$,
$z = 1 + \cos 16t$. At any point on the graph,

$x^2 + y^2 = (1 + \cos 16t)^2 \cos^2 t + (1 + \cos 16t)^2 \sin^2 t$
$\qquad = (1 + \cos 16t)^2 = z^2$, so the graph lies on the cone

$x^2 + y^2 = z^2$. From the graph at left, we see that this curve
looks like the projection of a leaved two-dimensional curve
onto a cone.

27. If $t = -1$, then $x = 1, y = 4, z = 0$, so the curve passes through the point $(1, 4, 0)$. If $t = 3$, then $x = 9, y = -8, z = 28$, so the curve passes through the point $(9, -8, 28)$. For the point $(4, 7, -6)$ to be on the curve, we require $y = 1 - 3t = 7 \Rightarrow t = -2$. But then $z = 1 + (-2)^3 = -7 \neq -6$, so $(4, 7, -6)$ is not on the curve.

29. Both equations are solved for z, so we can substitute to eliminate z: $\sqrt{x^2 + y^2} = 1 + y \Rightarrow$ $x^2 + y^2 = 1 + 2y + y^2 \Rightarrow x^2 = 1 + 2y \Rightarrow y = \frac{1}{2}(x^2 - 1)$. We can form parametric equations for the curve C of intersection by choosing a parameter $x = t$, then $y = \frac{1}{2}(t^2 - 1)$ and $z = 1 + y = 1 + \frac{1}{2}(t^2 - 1) = \frac{1}{2}(t^2 + 1)$. Thus a vector function representing C is $\mathbf{r}(t) = t\,\mathbf{i} + \frac{1}{2}(t^2 - 1)\,\mathbf{j} + \frac{1}{2}(t^2 + 1)\,\mathbf{k}$.

31.

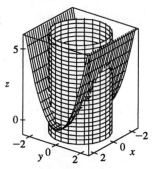 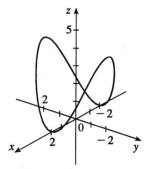

$x = 2\cos t \Rightarrow y^2 = 4 - 4\cos^2 t = 4\sin^2 t \Rightarrow y = \pm 2\sin t$ and $z = x^2 = 4\cos^2 t$. We choose the $+$ sign, so parametric equations for the curve of intersection are $x = 2\cos t$, $y = 2\sin t$, $z = 4\cos^2 t$.

33. Let $\mathbf{u}(t) = \langle u_1(t), u_2(t), u_3(t) \rangle$ and $\mathbf{v}(t) = \langle v_1(t), v_2(t), v_3(t) \rangle$. In each part of this problem the basic procedure is to use Equation 1 and then analyze the individual component functions using the limit properties we have already developed for real-valued functions.

(a) $\displaystyle \lim_{t \to a} \mathbf{u}(t) + \lim_{t \to a} \mathbf{v}(t) = \left\langle \lim_{t \to a} u_1(t), \lim_{t \to a} u_2(t), \lim_{t \to a} u_3(t) \right\rangle + \left\langle \lim_{t \to a} v_1(t), \lim_{t \to a} v_2(t), \lim_{t \to a} v_3(t) \right\rangle$

and the limits of these component functions must each exist since the vector functions both possess limits as $t \to a$. Then adding the two vectors and using the addition property of limits for real-valued functions, we have that

$$
\begin{aligned}
\lim_{t \to a} \mathbf{u}(t) + \lim_{t \to a} \mathbf{v}(t) &= \left\langle \lim_{t \to a} u_1(t) + \lim_{t \to a} v_1(t), \lim_{t \to a} u_2(t) + \lim_{t \to a} v_2(t), \lim_{t \to a} u_3(t) + \lim_{t \to a} v_3(t) \right\rangle \\
&= \left\langle \lim_{t \to a} [u_1(t) + v_1(t)], \lim_{t \to a} [u_2(t) + v_2(t)], \lim_{t \to a} [u_3(t) + v_3(t)] \right\rangle \\
&= \lim_{t \to a} \langle u_1(t) + v_1(t), u_2(t) + v_2(t), u_3(t) + v_3(t) \rangle \quad \text{[using (1) backward]} \\
&= \lim_{t \to a} [\mathbf{u}(t) + \mathbf{v}(t)]
\end{aligned}
$$

(b) $\lim\limits_{t \to a} c\mathbf{u}\,(t) = \lim\limits_{t \to a} \langle cu_1\,(t)\,, cu_2\,(t)\,, cu_3\,(t) \rangle = \left\langle \lim\limits_{t \to a} cu_1\,(t)\,, \lim\limits_{t \to a} cu_2\,(t)\,, \lim\limits_{t \to a} cu_3\,(t) \right\rangle$

$$= \left\langle c \lim_{t \to a} u_1\,(t)\,, c \lim_{t \to a} u_2\,(t)\,, c \lim_{t \to a} u_3\,(t) \right\rangle = c \left\langle \lim_{t \to a} u_1\,(t)\,, \lim_{t \to a} u_2\,(t)\,, \lim_{t \to a} u_3\,(t) \right\rangle$$

$$= c \lim_{t \to a} \langle u_1\,(t)\,, u_2\,(t)\,, u_3\,(t) \rangle = c \lim_{t \to a} \mathbf{u}\,(t)$$

(c) $\lim\limits_{t \to a} \mathbf{u}\,(t) \cdot \lim\limits_{t \to a} \mathbf{v}\,(t) = \left\langle \lim\limits_{t \to a} u_1\,(t)\,, \lim\limits_{t \to a} u_2\,(t)\,, \lim\limits_{t \to a} u_3\,(t) \right\rangle \cdot \left\langle \lim\limits_{t \to a} v_1\,(t)\,, \lim\limits_{t \to a} v_2\,(t)\,, \lim\limits_{t \to a} v_3\,(t) \right\rangle$

$$= \left[\lim_{t \to a} u_1\,(t) \right] \left[\lim_{t \to a} v_1\,(t) \right] + \left[\lim_{t \to a} u_2\,(t) \right] \left[\lim_{t \to a} v_2\,(t) \right]$$

$$+ \left[\lim_{t \to a} u_3\,(t) \right] \left[\lim_{t \to a} v_3\,(t) \right]$$

$$= \lim_{t \to a} u_1\,(t)\, v_1\,(t) + \lim_{t \to a} u_2\,(t)\, v_2\,(t) + \lim_{t \to a} u_3\,(t)\, v_3\,(t)$$

$$= \lim_{t \to a} [u_1\,(t)\, v_1\,(t) + u_2\,(t)\, v_2\,(t) + u_3\,(t)\, v_3\,(t)] = \lim_{t \to a} [\mathbf{u}\,(t) \cdot \mathbf{v}\,(t)]$$

(d) $\lim\limits_{t \to a} \mathbf{u}\,(t) \times \lim\limits_{t \to a} \mathbf{v}\,(t) = \left\langle \lim\limits_{t \to a} u_1\,(t)\,, \lim\limits_{t \to a} u_2\,(t)\,, \lim\limits_{t \to a} u_3\,(t) \right\rangle \times \left\langle \lim\limits_{t \to a} v_1\,(t)\,, \lim\limits_{t \to a} v_2\,(t)\,, \lim\limits_{t \to a} v_3\,(t) \right\rangle$

$$= \left\langle \left[\lim_{t \to a} u_2\,(t) \right] \left[\lim_{t \to a} v_3\,(t) \right] - \left[\lim_{t \to a} u_3\,(t) \right] \left[\lim_{t \to a} v_2\,(t) \right] \,, \right.$$

$$\left[\lim_{t \to a} u_3\,(t) \right] \left[\lim_{t \to a} v_1\,(t) \right] - \left[\lim_{t \to a} u_1\,(t) \right] \left[\lim_{t \to a} v_3\,(t) \right] \,,$$

$$\left. \left[\lim_{t \to a} u_1\,(t) \right] \left[\lim_{t \to a} v_2\,(t) \right] - \left[\lim_{t \to a} u_2\,(t) \right] \left[\lim_{t \to a} v_1\,(t) \right] \right\rangle$$

$$= \left\langle \lim_{t \to a} [u_2\,(t)\, v_3\,(t) - u_3\,(t)\, v_2\,(t)]\,, \lim_{t \to a} [u_3\,(t)\, v_1\,(t) - u_1\,(t)\, v_3\,(t)]\,, \right.$$

$$\left. \lim_{t \to a} [u_1\,(t)\, v_2\,(t) - u_2\,(t)\, v_1\,(t)] \right\rangle$$

$$= \lim_{t \to a} \langle u_2\,(t)\, v_3\,(t) - u_3\,(t)\, v_2\,(t)\,, u_3\,(t)\, v_1\,(t) - u_1\,(t)\, v_3\,(t)\,,$$

$$u_1\,(t)\, v_2\,(t) - u_2\,(t)\, v_1\,(t) \rangle$$

$$= \lim_{t \to a} [\mathbf{u}\,(t) \times \mathbf{v}\,(t)]$$

Section 10.2 Derivatives and Integrals of Vector Functions

1. (a)

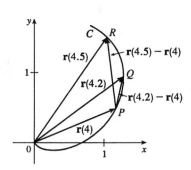

(b) $\dfrac{\mathbf{r}\,(4.5) - \mathbf{r}\,(4)}{0.5} = 2\,[\mathbf{r}\,(4.5) - \mathbf{r}\,(4)]$, so we draw a vector in

the same direction but with twice the length of the vector

$\mathbf{r}\,(4.5) - \mathbf{r}\,(4)$. $\dfrac{\mathbf{r}\,(4.2) - \mathbf{r}\,(4)}{0.2} = 5\,[\mathbf{r}\,(4.2) - \mathbf{r}\,(4)]$, so we

draw a vector in the same direction but with 5 times the

length of the vector $\mathbf{r}\,(4.2) - \mathbf{r}\,(4)$.

(c) By Definition 1, $\mathbf{r}'\,(4) = \lim\limits_{h \to 0} \dfrac{\mathbf{r}\,(4 + h) - \mathbf{r}\,(4)}{h}$.

$\mathbf{T}\,(4) = \dfrac{\mathbf{r}'\,(4)}{|\mathbf{r}'\,(4)|}$.

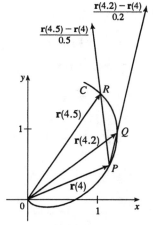

(d) $\mathbf{T}\,(4)$ is a unit vector in the same direction as $\mathbf{r}'\,(4)$, that is, parallel to the tangent line to the curve at
$\mathbf{r}\,(4)$ with length 1.

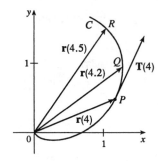

3. (a),(c)

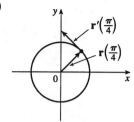

(b) $\mathbf{r}'(t) = \langle -\sin t, \cos t \rangle$

5. Since $(x-1)^2 = t^2 = y$, the curve is a parabola.

(a),(c)

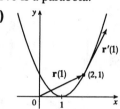

(b) $\mathbf{r}'(t) = \mathbf{i} + 2t\mathbf{j}$

7. $x^{-2} = e^{-2t} = y$, so $y = 1/x^2$, $x > 0$.

(a),(c)

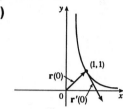

(b) $\mathbf{r}'(t) = e^t\mathbf{i} - 2e^{-2t}\mathbf{j}$

9. $\mathbf{r}'(t) = \left(\sec^2 t\right)\mathbf{j} + \left(\sec t \tan t\right)\mathbf{k}$.

11. $\mathbf{r}'(t) = -\dfrac{2t}{4-t^2}\mathbf{i} + \dfrac{1}{2\sqrt{1+t}}\mathbf{j} - 12e^{3t}\mathbf{k}$.

13. $\mathbf{r}'(t) = \mathbf{0} + \mathbf{b} + 2\,\mathbf{c}\,t = \mathbf{b} + 2t\,\mathbf{c}$ by Formulas 1 and 3 of Theorem 3.

15. $\mathbf{r}'(t) = \mathbf{i} + 2\cos t\,\mathbf{j} - 3\sin t\,\mathbf{k}$, $\mathbf{r}'\left(\frac{\pi}{6}\right) = \mathbf{i} + \sqrt{3}\,\mathbf{j} - \frac{3}{2}\mathbf{k}$. Thus

$\mathbf{T}\left(\frac{\pi}{6}\right) = \dfrac{1}{\sqrt{25/4}}\left(\mathbf{i} + \sqrt{3}\,\mathbf{j} - \frac{3}{2}\mathbf{k}\right) = \frac{2}{5}\mathbf{i} + \frac{2\sqrt{3}}{5}\mathbf{j} - \frac{3}{5}\mathbf{k}$.

17. $\mathbf{r}(t) = \langle t, t^2, t^3 \rangle \Rightarrow \mathbf{r}'(t) = \langle 1, 2t, 3t^2 \rangle$. Then $\mathbf{r}'(1) = \langle 1, 2, 3 \rangle$ and $|\mathbf{r}'(1)| = \sqrt{1^2 + 2^2 + 3^2} = \sqrt{14}$,

so $\mathbf{T}(1) = \dfrac{\mathbf{r}'(1)}{|\mathbf{r}'(1)|} = \frac{1}{\sqrt{14}}\langle 1, 2, 3 \rangle = \left\langle \frac{1}{\sqrt{14}}, \frac{2}{\sqrt{14}}, \frac{3}{\sqrt{14}} \right\rangle$. $\mathbf{r}''(t) = \langle 0, 2, 6t \rangle$, so

$\mathbf{r}'(t) \times \mathbf{r}''(t) = \begin{vmatrix} \mathbf{i} & \mathbf{j} & \mathbf{k} \\ 1 & 2t & 3t^2 \\ 0 & 2 & 6t \end{vmatrix} = \begin{vmatrix} 2t & 3t^2 \\ 2 & 6t \end{vmatrix}\mathbf{i} - \begin{vmatrix} 1 & 3t^2 \\ 0 & 6t \end{vmatrix}\mathbf{j} + \begin{vmatrix} 1 & 2t \\ 0 & 2 \end{vmatrix}\mathbf{k}$

$= \left(12t^2 - 6t^2\right)\mathbf{i} - (6t - 0)\mathbf{j} + (2 - 0)\mathbf{k} = \langle 6t^2, -6t, 2 \rangle$.

19. The vector equation of the curve is $\mathbf{r}(t) = t\mathbf{i} + t^2\mathbf{j} + t^3\mathbf{k}$, so $\mathbf{r}'(t) = \mathbf{i} + 2t\mathbf{j} + 3t^2\mathbf{k}$. At the point $(1, 1, 1)$, $t = 1$, so the tangent vector here is $\mathbf{i} + 2\mathbf{j} + 3\mathbf{k}$. The tangent line goes through the point $(1, 1, 1)$ and has direction vector $\mathbf{i} + 2\mathbf{j} + 3\mathbf{k}$. Thus, parametric equations are $x = 1 + t$, $y = 1 + 2t$, $z = 1 + 3t$.

21. $\mathbf{r}(t) = \langle t\cos 2\pi t, t\sin 2\pi t, 4t \rangle$, $\mathbf{r}'(t) = \langle \cos 2\pi t - 2\pi t\sin 2\pi t, \sin 2\pi t + 2\pi t\cos 2\pi t, 4 \rangle$. At $\left(0, \frac{1}{4}, 1\right)$, $t = \frac{1}{4}$ and $\mathbf{r}'\left(\frac{1}{4}\right) = \left\langle 0 - \frac{\pi}{2}, 1 + 0, 4 \right\rangle = \left\langle -\frac{\pi}{2}, 1, 4 \right\rangle$. Thus, parametric equations of the tangent line are $x = -\frac{\pi}{2}t$, $y = \frac{1}{4} + t$, $z = 1 + 4t$.

23. $\mathbf{r}(t) = \langle t, \sqrt{2}\cos t, \sqrt{2}\sin t \rangle \Rightarrow$

$\mathbf{r}'(t) = \langle 1, -\sqrt{2}\sin t, \sqrt{2}\cos t \rangle$. At $\left(\frac{\pi}{4}, 1, 1\right)$, $t = \frac{\pi}{4}$ and $\mathbf{r}'\left(\frac{\pi}{4}\right) = \langle 1, -1, 1 \rangle$. Thus, parametric equations of the tangent line are $x = \frac{\pi}{4} + t$, $y = 1 - t$, $z = 1 + t$.

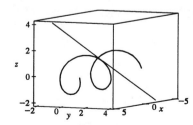

25. (a) $\mathbf{r}\,(t) = \langle t^3, t^4, t^5 \rangle \;\Rightarrow\; \mathbf{r}'\,(t) = \langle 3t^2, 4t^3, 5t^4 \rangle$, and since $\mathbf{r}'\,(0) = \langle 0,0,0 \rangle = \mathbf{0}$, the curve is not smooth.

(b) $\mathbf{r}\,(t) = \langle t^3 + t, t^4, t^5 \rangle \;\Rightarrow\; \mathbf{r}'\,(t) = \langle 3t^2 + 1, 4t^3, 5t^4 \rangle$. $\mathbf{r}'\,(t)$ is continuous since its component functions are continuous. Also, $\mathbf{r}'\,(t) \neq \mathbf{0}$, as the y- and z-components are 0 only for $t = 0$, but $\mathbf{r}'\,(0) = \langle 1,0,0 \rangle \neq \mathbf{0}$. Thus, the curve is smooth.

(c) $\mathbf{r}\,(t) = \langle \cos^3 t, \sin^3 t \rangle \;\Rightarrow\; \mathbf{r}'\,(t) = \langle -3\cos^2 t \sin t, 3\sin^2 t \cos t \rangle$. Since $\mathbf{r}'\,(0) = \langle -3\cos^2 0 \sin 0, 3\sin^2 0 \cos 0 \rangle = \langle 0,0 \rangle = \mathbf{0}$, the curve is not smooth.

27. The angle of intersection of the two curves is the angle between the two tangent vectors to the curves at the point of intersection. Since $\mathbf{r}_1'\,(t) = \langle 1, 2t, 3t^2 \rangle$ and $t = 0$ at $(0,0,0)$, $\mathbf{r}_1'\,(0) = \langle 1,0,0 \rangle$ is a tangent vector to $\mathbf{r}_1$ at $(0,0,0)$. Also, $\mathbf{r}_2' = \langle \cos t, 2\cos t, 1 \rangle$ so $\mathbf{r}_2'\,(0) = \langle 1,2,1 \rangle$ is a tangent vector to $\mathbf{r}_2$ at $(0,0,0)$. If θ is the angle between these two tangent vectors, then $\cos \theta = \frac{1}{\sqrt{1}\sqrt{6}} \langle 1,0,0 \rangle \cdot \langle 1,2,1 \rangle = \frac{1}{\sqrt{6}}$ and $\theta = \cos^{-1} \frac{1}{\sqrt{6}} \approx 66°$.

29. $\int_0^1 \left(t\,\mathbf{i} + t^2\,\mathbf{j} + t^3\,\mathbf{k} \right) dt = \left(\int_0^1 t\,dt \right) \mathbf{i} + \left(\int_0^1 t^2\,dt \right) \mathbf{j} + \left(\int_0^1 t^3\,dt \right) \mathbf{k}$

$$= \left[\frac{t^2}{2} \right]_0^1 \mathbf{i} + \left[\frac{t^3}{3} \right]_0^1 \mathbf{j} + \left[\frac{t^4}{4} \right]_0^1 \mathbf{k}$$

$$= \tfrac{1}{2}\mathbf{i} + \tfrac{1}{3}\mathbf{j} + \tfrac{1}{4}\mathbf{k}$$

31. $\int_0^{\pi/4} \left(\cos 2t\,\mathbf{i} + \sin 2t\,\mathbf{j} + t \sin t\,\mathbf{k} \right) dt$

$$= \left[\tfrac{1}{2}\sin 2t\,\mathbf{i} - \tfrac{1}{2}\cos 2t\,\mathbf{j} \right]_0^{\pi/4} + \left[[-t\cos t]_0^{\pi/4} + \int_0^{\pi/4} \cos t\,dt \right]\mathbf{k}$$

$$= \tfrac{1}{2}\mathbf{i} + \tfrac{1}{2}\mathbf{j} + \left[-\tfrac{\pi}{4}\cos \tfrac{\pi}{4} + \sin \tfrac{\pi}{4} \right]\mathbf{k}$$

$$= \tfrac{1}{2}\mathbf{i} + \tfrac{1}{2}\mathbf{j} + \tfrac{1}{\sqrt{2}}\left(1 - \tfrac{\pi}{4} \right)\mathbf{k}$$

$$= \tfrac{1}{2}\mathbf{i} + \tfrac{1}{2}\mathbf{j} + \tfrac{4-\pi}{4\sqrt{2}}\mathbf{k}$$

33. $\int \left(e^t\,\mathbf{i} + 2t\,\mathbf{j} + \ln t\,\mathbf{k} \right) dt = \left(\int e^t\,dt \right)\mathbf{i} + \left(\int 2t\,dt \right)\mathbf{j} + \left(\int \ln t\,dt \right)\mathbf{k} = e^t\,\mathbf{i} + t^2\,\mathbf{j} + (t \ln t - t)\,\mathbf{k} + \mathbf{C}$, where $\mathbf{C}$ is a vector constant of integration.

35. $\mathbf{r}'\,(t) = t^2\,\mathbf{i} + 4t^3\,\mathbf{j} - t^2\,\mathbf{k} \;\Rightarrow\; \mathbf{r}\,(t) = \tfrac{1}{3}t^3\,\mathbf{i} + t^4\,\mathbf{j} - \tfrac{1}{3}t^3\,\mathbf{k} + \mathbf{C}$, where $\mathbf{C}$ is a constant vector. But $\mathbf{j} = \mathbf{r}\,(0) = (0)\,\mathbf{i} + (0)\,\mathbf{j} - (0)\,\mathbf{k} + \mathbf{C}$. Thus $\mathbf{C} = \mathbf{j}$ and $\mathbf{r}\,(t) = \tfrac{1}{3}t^3\,\mathbf{i} + (t^4 + 1)\,\mathbf{j} - \tfrac{1}{3}t^3\,\mathbf{k}$.

37. $\dfrac{d}{dt}\left[\mathbf{u}\,(t) + \mathbf{v}\,(t) \right] = \dfrac{d}{dt} \langle u_1\,(t) + v_1\,(t), u_2\,(t) + v_2\,(t), u_3\,(t) + v_3\,(t) \rangle$

$$= \left\langle \dfrac{d}{dt}\left[u_1\,(t) + v_1\,(t) \right], \dfrac{d}{dt}\left[u_2\,(t) + v_2\,(t) \right], \dfrac{d}{dt}\left[u_3\,(t) + v_3\,(t) \right] \right\rangle$$

$$= \langle u_1'\,(t) + v_1'\,(t), u_2'\,(t) + v_2'\,(t), u_3'\,(t) + v_3'\,(t) \rangle$$

$$= \langle u_1'\,(t), u_2'\,(t), u_3'\,(t) \rangle + \langle v_1'\,(t), v_2'\,(t), v_3'\,(t) \rangle = \mathbf{u}'\,(t) + \mathbf{v}'\,(t).$$

39. $\dfrac{d}{dt}\left[\mathbf{u}\left(t\right)\times\mathbf{v}\left(t\right)\right]$

$$= \frac{d}{dt}\left\langle u_2\left(t\right)v_3\left(t\right)-u_3\left(t\right)v_2\left(t\right),\,u_3\left(t\right)v_1\left(t\right)-u_1\left(t\right)v_3\left(t\right),\,u_1\left(t\right)v_2\left(t\right)-u_2\left(t\right)v_1\left(t\right)\right\rangle$$

$$= \langle u_2'v_3\left(t\right)+u_2\left(t\right)v_3'\left(t\right)-u_3'\left(t\right)v_2\left(t\right)-u_3\left(t\right)v_2'\left(t\right),$$
$$u_3'\left(t\right)v_1\left(t\right)+u_3\left(t\right)v_1'\left(t\right)-u_1'\left(t\right)v_3\left(t\right)-u_1\left(t\right)v_3'\left(t\right),$$
$$u_1'\left(t\right)v_2\left(t\right)+u_1\left(t\right)v_2'\left(t\right)-u_2'\left(t\right)v_1\left(t\right)-u_2\left(t\right)v_1'\left(t\right)\rangle$$

$$= \langle u_2'\left(t\right)v_3\left(t\right)-u_3'\left(t\right)v_2\left(t\right),\,u_3'\left(t\right)v_1\left(t\right)-u_1'\left(t\right)v_3\left(t\right),\,u_1'\left(t\right)v_2\left(t\right)-u_2'\left(t\right)v_1\left(t\right)\rangle$$

$$\quad+ \langle u_2\left(t\right)v_3'\left(t\right)-u_3\left(t\right)v_2'\left(t\right),\,u_3\left(t\right)v_1'\left(t\right)-u_1\left(t\right)v_3'\left(t\right),\,u_1\left(t\right)v_2'\left(t\right)-u_2\left(t\right)v_1'\left(t\right)\rangle$$

$$= \mathbf{u}'\left(t\right)\times\mathbf{v}\left(t\right)+\mathbf{u}\left(t\right)\times\mathbf{v}'\left(t\right)$$

Alternate Solution: Let $\mathbf{r}\left(t\right)=\mathbf{u}\left(t\right)\times\mathbf{v}\left(t\right)$. Then

$$\mathbf{r}\left(t+h\right)-\mathbf{r}\left(t\right)=\left[\mathbf{u}\left(t+h\right)\times\mathbf{v}\left(t+h\right)\right]-\left[\mathbf{u}\left(t\right)\times\mathbf{v}\left(t\right)\right]$$
$$=\left[\mathbf{u}\left(t+h\right)\times\mathbf{v}\left(t+h\right)\right]-\left[\mathbf{u}\left(t\right)\times\mathbf{v}\left(t\right)\right]+\left[\mathbf{u}\left(t+h\right)\times\mathbf{v}\left(t\right)\right]-\left[\mathbf{u}\left(t+h\right)\times\mathbf{v}\left(t\right)\right]$$
$$=\mathbf{u}\left(t+h\right)\times\left[\mathbf{v}\left(t+h\right)-\mathbf{v}\left(t\right)\right]+\left[\mathbf{u}\left(t+h\right)-\mathbf{u}\left(t\right)\right]\times\mathbf{v}\left(t\right)$$

(Be careful of the order of the cross product.)

Dividing through by h and taking the limit as $h\to 0$ we have

$$\mathbf{r}'\left(t\right)=\lim_{h\to 0}\frac{\mathbf{u}\left(t+h\right)\times\left[\mathbf{v}\left(t+h\right)-\mathbf{v}\left(t\right)\right]}{h}+\lim_{h\to 0}\frac{\left[\mathbf{u}\left(t+h\right)-\mathbf{u}\left(t\right)\right]\times\mathbf{v}\left(t\right)}{h}$$

$$=\mathbf{u}\left(t\right)\times\mathbf{v}'\left(t\right)+\mathbf{u}'\left(t\right)\times\mathbf{v}\left(t\right)$$

by Exercise 10.1.33(a) and Definition 1.

41. $D_t\left[\mathbf{u}\left(t\right)\cdot\mathbf{v}\left(t\right)\right]=\mathbf{u}'\left(t\right)\cdot\mathbf{v}\left(t\right)+\mathbf{u}\left(t\right)\cdot\mathbf{v}'\left(t\right)$ by Formula 4 of Theorem 3

$$=\left(-4t\,\mathbf{j}+9t^2\,\mathbf{k}\right)\cdot\left(t\,\mathbf{i}+\cos t\,\mathbf{j}+\sin t\,\mathbf{k}\right)$$

$$\quad+\left(\mathbf{i}-2t^2\,\mathbf{j}+3t^3\,\mathbf{k}\right)\cdot\left(\mathbf{i}-\sin t\,\mathbf{j}+\cos t\,\mathbf{k}\right)$$

$$=-4t\cos t+9t^2\sin t+1+2t^2\sin t+3t^3\cos t$$

$$=1-4t\cos t+11t^2\sin t+3t^3\cos t$$

43. $\dfrac{d}{dt}\left[\mathbf{r}\left(t\right)\times\mathbf{r}'\left(t\right)\right]=\mathbf{r}'\left(t\right)\times\mathbf{r}'\left(t\right)+\mathbf{r}\left(t\right)\times\mathbf{r}''\left(t\right)$ by Formula 5 of Theorem 3. But $\mathbf{r}'\left(t\right)\times\mathbf{r}'\left(t\right)=\mathbf{0}$

(by the margin note on page 667). Thus, $\dfrac{d}{dt}\left[\mathbf{r}\left(t\right)\times\mathbf{r}'\left(t\right)\right]=\mathbf{r}\left(t\right)\times\mathbf{r}''\left(t\right)$.

45. $\dfrac{d}{dt}\left|\mathbf{r}\left(t\right)\right|=\dfrac{d}{dt}\left[\mathbf{r}\left(t\right)\cdot\mathbf{r}\left(t\right)\right]^{1/2}=\tfrac{1}{2}\left[\mathbf{r}\left(t\right)\cdot\mathbf{r}\left(t\right)\right]^{-1/2}\left[2\mathbf{r}\left(t\right)\cdot\mathbf{r}'\left(t\right)\right]=\dfrac{\mathbf{r}\left(t\right)\cdot\mathbf{r}'\left(t\right)}{\left|\mathbf{r}\left(t\right)\right|}$

47. Since $\mathbf{u}\left(t\right)=\mathbf{r}\left(t\right)\cdot\left[\mathbf{r}'\left(t\right)\times\mathbf{r}''\left(t\right)\right]$,

$$\mathbf{u}'\left(t\right)=\mathbf{r}'\left(t\right)\cdot\left[\mathbf{r}'\left(t\right)\times\mathbf{r}''\left(t\right)\right]+\mathbf{r}\left(t\right)\cdot\frac{d}{dt}\left[\mathbf{r}'\left(t\right)\times\mathbf{r}''\left(t\right)\right]$$

$$=0+\mathbf{r}\left(t\right)\cdot\left[\mathbf{r}''\left(t\right)\times\mathbf{r}''\left(t\right)+\mathbf{r}'\left(t\right)\times\mathbf{r}'''\left(t\right)\right]\qquad\qquad\text{[since }\mathbf{r}'(t)\perp\mathbf{r}'\left(t\right)\times\mathbf{r}''\left(t\right)\text{]}$$

$$=\mathbf{r}\left(t\right)\cdot\left[\mathbf{r}'\left(t\right)\times\mathbf{r}'''\left(t\right)\right]\qquad\qquad\qquad\qquad\qquad\text{[since }\mathbf{r}''\left(t\right)\times\mathbf{r}''\left(t\right)=\mathbf{0}\text{]}$$

Section 10.3   Arc Length and Curvature

1. $\mathbf{r}'(t) = \langle 2, 3\cos t, -3\sin t \rangle$, $|\mathbf{r}'(t)| = \sqrt{4 + 9\cos^2 t + 9\sin^2 t} = \sqrt{13}$

Then by (3), $L = \int_a^b \sqrt{13}\, dt = \sqrt{13}\,(b - a)$

3. $\mathbf{r}'(t) = \langle 6, 6\sqrt{2}t, 6t^2 \rangle$, $|\mathbf{r}'(t)| = 6\sqrt{1 + 2t^2 + t^4} = 6\,(1 + t^2)$

$L = \int_0^1 6\,(1 + t^2)\, dt = \left[6\,(t + \tfrac{1}{3}t^3)\right]_0^1 = \tfrac{24}{3} = 8$

5. The point $(2, 4, 8)$ corresponds to $t = 2$, so by Equation 2, $L = \int_0^2 \sqrt{(1)^2 + (2t)^2 + (3t^2)^2}\, dt$.

If $f(t) = \sqrt{1 + 4t^2 + 9t^4}$, then Simpson's Rule gives

$L \approx \dfrac{2 - 0}{10 \cdot 3}\,[f(0) + 4f(0.2) + 2f(0.4) + \cdots + 4f(1.8) + f(2)] \approx 9.5706.$

7. $\mathbf{r}'(t) = e^t\,(\cos t + \sin t)\,\mathbf{i} + e^t\,(\cos t - \sin t)\,\mathbf{j}$,

$ds/dt = |\mathbf{r}'(t)| = e^t \sqrt{(\cos t + \sin t)^2 + (\cos t - \sin t)^2} = e^t \sqrt{2\cos^2 t + 2\sin^2 t} = \sqrt{2}\,e^t$

$s(t) = \int_0^t |\mathbf{r}'(u)|\, du = \int_0^t \sqrt{2}\,e^u\, du = \sqrt{2}\,(e^t - 1) \quad \Rightarrow \quad \tfrac{1}{\sqrt{2}}s + 1 = e^t \quad \Rightarrow \quad t(s) = \ln\left(\tfrac{1}{\sqrt{2}}s + 1\right).$

Therefore, $\mathbf{r}(t(s)) = \left(\tfrac{1}{\sqrt{2}}s + 1\right)\left[\sin\left(\ln\left(\tfrac{1}{\sqrt{2}}s + 1\right)\right)\mathbf{i} + \cos\left(\ln\left(\tfrac{1}{\sqrt{2}}s + 1\right)\right)\mathbf{j}\right].$

9. $|\mathbf{r}'(t)| = \sqrt{(3\cos t)^2 + 16 + (-3\sin t)^2} = \sqrt{9 + 16} = 5$ and $s(t) = \int_0^t |\mathbf{r}'(u)|\, du = \int_0^t 5\, du = 5t$

$\Rightarrow \quad t(s) = \tfrac{1}{5}s$. Therefore, $\mathbf{r}(t(s)) = 3\sin\left(\tfrac{1}{5}s\right)\mathbf{i} + \tfrac{4}{5}s\,\mathbf{j} + 3\cos\left(\tfrac{1}{5}s\right)\mathbf{k}.$

11. (a) $\mathbf{T}(t) = \dfrac{\mathbf{r}'(t)}{|\mathbf{r}'(t)|} = \dfrac{\langle 4\cos 4t, 3, -4\sin 4t \rangle}{|\langle 4\cos 4t, 3, -4\sin 4t \rangle|} = \dfrac{1}{\sqrt{16 + 9}}\,\langle 4\cos 4t, 3, -4\sin 4t \rangle$

$= \tfrac{1}{5}\,\langle 4\cos 4t, 3, -4\sin 4t \rangle$

$\mathbf{N}(t) = \dfrac{\mathbf{T}'(t)}{|\mathbf{T}'(t)|} = \dfrac{\tfrac{1}{5}\langle -16\sin 4t, 0, -16\cos 4t \rangle}{|\tfrac{1}{5}\langle -16\sin 4t, 0, -16\cos 4t \rangle|} = \dfrac{1/5}{16/5}\,\langle -16\sin 4t, 0, -16\cos 4t \rangle$

$= \langle -\sin 4t, 0, -\cos 4t \rangle$

(b) $\kappa(t) = \dfrac{|\mathbf{T}'(t)|}{|\mathbf{r}'(t)|} = \dfrac{16/5}{5} = \dfrac{16}{25}$

13. (a) $\mathbf{T}(t) = \dfrac{\mathbf{r}'(t)}{|\mathbf{r}'(t)|} = \dfrac{1}{\sqrt{2\sin^2 t + 2\cos^2 t}}\,\langle -\sqrt{2}\sin t, \cos t, \cos t \rangle = \tfrac{1}{\sqrt{2}}\,\langle -\sqrt{2}\sin t, \cos t, \cos t \rangle$

$\mathbf{N}(t) = \dfrac{\mathbf{T}'(t)}{|\mathbf{T}'(t)|} = \dfrac{1}{\sqrt{2\cos^2 t + 2\sin^2 t}}\,\langle -\sqrt{2}\cos t, -\sin t, -\sin t \rangle$

$= \tfrac{1}{\sqrt{2}}\,\langle -\sqrt{2}\cos t, -\sin t, -\sin t \rangle$

(b) $\kappa(t) = \dfrac{|\mathbf{T}'(t)|}{|\mathbf{r}'(t)|} = \dfrac{1}{\sqrt{2}}$

15. $\mathbf{r}'(t) = \mathbf{j} - 2t\,\mathbf{k}$, $\mathbf{r}''(t) = -2\,\mathbf{k}$, $|\mathbf{r}'(t)|^3 = \left(4t^2 + 1\right)^{3/2}$, $|\mathbf{r}'(t) \times \mathbf{r}''(t)| = |-2\,\mathbf{i}| = 2$,

$$\kappa(t) = \frac{|\mathbf{r}'(t) \times \mathbf{r}''(t)|}{|\mathbf{r}'(t)|^3} = \frac{2}{\left(4t^2 + 1\right)^{3/2}}$$

17. $\mathbf{r}'(t) = \langle \cos t, -\sin t, \cos t \rangle$, $\mathbf{r}''(t) = \langle -\sin t, -\cos t, -\sin t \rangle$, $|\mathbf{r}'(t)|^3 = \left(\sqrt{\cos^2 t + 1}\right)^3$,

$|\mathbf{r}'(t) \times \mathbf{r}''(t)| = |\langle 1, 0, -1 \rangle| = \sqrt{2}$, $\kappa(t) = \dfrac{|\mathbf{r}'(t) \times \mathbf{r}''(t)|}{|\mathbf{r}'(t)|^3} = \dfrac{\sqrt{2}}{\left(1 + \cos^2 t\right)^{3/2}}$

19. $f(x) = x^3$, $f'(x) = 3x^2$, $f''(x) = 6x$, $\kappa(x) = \dfrac{|f''(x)|}{\left[1 + (f'(x))^2\right]^{3/2}} = \dfrac{6\,|x|}{\left(1 + 9x^4\right)^{3/2}}$

21. $y' = \cos x$, $y'' = -\sin x$, $\kappa(x) = \dfrac{|y''(x)|}{\left[1 + (y'(x))^2\right]^{3/2}} = \dfrac{|\sin x|}{\left(1 + \cos^2 x\right)^{3/2}}$

23. Since $y' = y'' = e^x$, the curvature is $\kappa(x) = \dfrac{|y''(x)|}{\left[1 + (y'(x))^2\right]^{3/2}} = \dfrac{e^x}{\left(1 + e^{2x}\right)^{3/2}} = e^x \left(1 + e^{2x}\right)^{-3/2}$.

To find the maximum curvature, we first find the critical numbers of $\kappa(x)$:

$\kappa'(x) = e^x \left(1 + e^{2x}\right)^{-3/2} + e^x \left(-\frac{3}{2}\right) \left(1 + e^{2x}\right)^{-5/2} \left(2e^{2x}\right) = e^x \dfrac{1 + e^{2x} - 3e^{2x}}{\left(1 + e^{2x}\right)^{5/2}} = e^x \dfrac{1 - 2e^{2x}}{\left(1 + e^{2x}\right)^{5/2}}$.

$\kappa'(x) = 0$ when $1 - 2e^{2x} = 0$, so $e^{2x} = \frac{1}{2}$ or $x = -\frac{1}{2}\ln 2$. And since $1 - 2e^{2x} > 0$ for $x < -\frac{1}{2}\ln 2$

and $1 - 2e^{2x} < 0$ for $x > -\frac{1}{2}\ln 2$, the maximum curvature is attained at the point

$\left(-\frac{1}{2}\ln 2,\, e^{(-\ln 2)/2}\right) = \left(-\frac{1}{2}\ln 2,\, \frac{1}{\sqrt{2}}\right)$. Since $\lim\limits_{x \to \infty} e^x \left(1 + e^{2x}\right)^{-3/2} = 0$, $\kappa(x)$ approaches 0 as

$x \to \infty$.

25. (a) C appears to be changing direction more quickly at P than Q, so we would expect the curvature to be greater at P.

(b) First we sketch approximate osculating circles at P and
Q. Using the axes scale as a guide, we measure the
radius of the osculating circle at P to be approximately

0.8 units, thus $\rho = \dfrac{1}{\kappa} \;\Rightarrow\; \kappa = \dfrac{1}{\rho} \approx \dfrac{1}{0.8} \approx 1.3$.

Similarly, we estimate the radius of the osculating

circle at Q to be 1.4 units, so $\kappa = \dfrac{1}{\rho} \approx \dfrac{1}{1.4} \approx 0.7$.

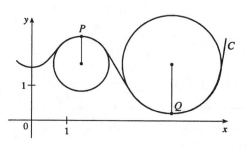

27. $y = x^4 \Rightarrow y' = 4x^3$, $y'' = 12x^2$, and

$$\kappa(x) = \frac{|y''|}{\left[1 + (y')^2\right]^{3/2}} = \frac{12x^2}{(1 + 16x^6)^{3/2}}.$$ The appearance of

the two humps in this graph is perhaps a little surprising, but it
is explained by the fact that $y = x^4$ is very flat around the
origin, and so here the curvature is zero.

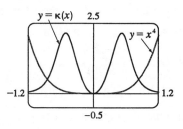

29. $\kappa(t) = \dfrac{|\dot{x}\ddot{y} - \ddot{x}\dot{y}|}{(\dot{x}^2 + \dot{y}^2)^{3/2}} = \dfrac{\left|(3t^2)(2) - (6t)(2t)\right|}{(9t^4 + 4t^2)^{3/2}} = \dfrac{6t^2}{(t^2)^{3/2}(9t^2 + 4)^{3/2}} = \dfrac{6t^2}{|t|^3(9t^2 + 4)^{3/2}}$

$$= \dfrac{6}{|t|(9t^2 + 4)^{3/2}}$$

31. $\left(1, \frac{2}{3}, 1\right)$ corresponds to $t = 1$. $\mathbf{T}(t) = \dfrac{\mathbf{r}'(t)}{|\mathbf{r}'(t)|} = \dfrac{\langle 2t, 2t^2, 1\rangle}{\sqrt{4t^2 + 4t^4 + 1}} = \dfrac{\langle 2t, 2t^2, 1\rangle}{2t^2 + 1}$, so

$\mathbf{T}(1) = \left\langle \frac{2}{3}, \frac{2}{3}, \frac{1}{3}\right\rangle$.

$\mathbf{T}'(t) = -4t(2t^2 + 1)^{-2}\langle 2t, 2t^2, 1\rangle + (2t^2 + 1)^{-1}\langle 2, 4t, 0\rangle$ (By Theorem 10.2.3)

$\qquad = (2t^2 + 1)^{-2}\langle -8t^2 + 4t^2 + 2, -8t^3 + 8t^3 + 4t, -4t\rangle = 2(2t^2 + 1)^{-2}\langle 1 - 2t^2, 2t, -2t\rangle$

$\mathbf{N}(t) = \dfrac{\mathbf{T}'(t)}{|\mathbf{T}'(t)|} = \dfrac{2(2t^2 + 1)^{-2}\langle 1 - 2t^2, 2t, -2t\rangle}{2(2t^2 + 1)^{-2}\sqrt{(1 - 2t^2)^2 + (2t)^2 + (-2t)^2}} = \dfrac{\langle 1 - 2t^2, 2t, -2t\rangle}{\sqrt{1 - 4t^2 + 4t^4 + 8t^2}}$

$\qquad = \dfrac{\langle 1 - 2t^2, 2t, -2t\rangle}{1 + 2t^2}$

$\mathbf{N}(1) = \left\langle -\frac{1}{3}, \frac{2}{3}, -\frac{2}{3}\right\rangle$ and $\mathbf{B}(1) = \mathbf{T}(1) \times \mathbf{N}(1) = \left\langle -\frac{4}{9} - \frac{2}{9}, -\left(-\frac{4}{9} + \frac{1}{9}\right), \frac{4}{9} + \frac{2}{9}\right\rangle = \left\langle -\frac{2}{3}, \frac{1}{3}, \frac{2}{3}\right\rangle$.

33. $t = \pi$ corresponds to $(0, \pi, -2)$.

$\mathbf{T}(t) = \dfrac{\mathbf{r}'(t)}{|\mathbf{r}'(t)|} = \dfrac{\langle 6\cos 3t, 1, -6\sin 3t\rangle}{\sqrt{36\cos^2 3t + 1 + 36\sin^2 3t}} = \dfrac{1}{\sqrt{37}}\langle 6\cos 3t, 1, -6\sin 3t\rangle$. $\mathbf{T}(\pi) = \dfrac{1}{\sqrt{37}}\langle -6, 1, 0\rangle$

is a normal vector for the normal plane, and so $\langle -6, 1, 0\rangle$ is also normal. Thus an equation for the plane
is $-6(x - 0) + 1(y - \pi) + 0(z + 2) = 0$ or $y - 6x = \pi$. $\mathbf{T}'(t) = \dfrac{1}{\sqrt{37}}\langle -18\sin 3t, 0, -18\cos 3t\rangle \Rightarrow$

$|\mathbf{T}'(t)| = \dfrac{\sqrt{18^2\sin^2 3t + 18^2\cos^2 3t}}{\sqrt{37}} = \dfrac{18}{\sqrt{37}} \Rightarrow \mathbf{N}(t) = \langle -\sin 3t, 0, -\cos 3t\rangle$. So

$\mathbf{B}(\pi) = \dfrac{1}{\sqrt{37}}\langle -6, 1, 0\rangle \times \langle 0, 0, 1\rangle = \dfrac{1}{\sqrt{37}}\langle 1, 6, 0\rangle$. Since $\mathbf{B}(\pi)$ is a normal to the osculating plane, so is
$\langle 1, 6, 0\rangle$ and an equation for the plane is $1(x - 0) + 6(y - \pi) + 0(z + 2) = 0$ or $x + 6y = 6\pi$.

35. The ellipse is given by the parametric equations

$x = 2\cos t$, $y = 3\sin t$, so using the result from Exercise 28,

$$\kappa(t) = \frac{|\dot{x}\ddot{y} - \ddot{x}\dot{y}|}{(\dot{x}^2 + \dot{y}^2)^{3/2}} = \frac{|(-2\sin t)(-3\sin t) - (3\cos t)(-2\cos t)|}{(4\sin^2 t + 9\cos^2 t)^{3/2}} = \frac{6}{(4\sin^2 t + 9\cos^2 t)^{3/2}}.$$

At $(2,0)$, $t = 0$. Now $\kappa(0) = \frac{6}{27} = \frac{2}{9}$, so the radius of the

osculating circle is $1/\kappa(0) = \frac{9}{2}$ and its center is $\left(-\frac{5}{2}, 0\right)$. Its

equation is therefore $\left(x + \frac{5}{2}\right)^2 + y^2 = \frac{81}{4}$. At $(0,3)$, $t = \frac{\pi}{2}$,

and $\kappa\left(\frac{\pi}{2}\right) = \frac{6}{8} = \frac{3}{4}$. So the radius of the osculating circle is $\frac{4}{3}$

and its center is $\left(0, \frac{5}{3}\right)$. Hence its equation is

$$x^2 + \left(y - \frac{5}{3}\right)^2 = \frac{16}{9}.$$

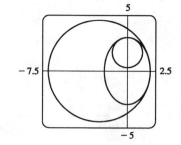

37. The tangent vector is normal to the normal plane, and the vector $\langle 6, 6, -8 \rangle$ is normal to the given plane.
But $\mathbf{T}(t) \parallel \mathbf{r}'(t)$ and $\langle 6, 6, -8 \rangle \parallel \langle 3, 3, -4 \rangle$, so we need to find t such that $\mathbf{r}'(t) \parallel \langle 3, 3, -4 \rangle$.
$\mathbf{r}(t) = \langle t^3, 3t, t^4 \rangle \;\Rightarrow\; \mathbf{r}'(t) = \langle 3t^2, 3, 4t^3 \rangle \parallel \langle 3, 3, -4 \rangle$ when $t = -1$. So the planes are parallel at
the point $\mathbf{r}(-1) = (-1, -3, 1)$.

39. $\kappa = \left|\dfrac{d\mathbf{T}}{ds}\right| = \left|\dfrac{d\mathbf{T}/dt}{ds/dt}\right| = \dfrac{|d\mathbf{T}/dt|}{ds/dt}$ and $\mathbf{N} = \dfrac{d\mathbf{T}/dt}{|d\mathbf{T}/dt|}$, so $\kappa\mathbf{N} = \dfrac{\left|\dfrac{d\mathbf{T}}{dt}\right|\dfrac{d\mathbf{T}}{dt}}{\left|\dfrac{d\mathbf{T}}{dt}\right|\dfrac{ds}{dt}} = \dfrac{d\mathbf{T}/dt}{ds/dt} = \dfrac{d\mathbf{T}}{ds}$ by the

Chain Rule.

41. (a) $|\mathbf{B}| = 1 \;\Rightarrow\; \mathbf{B} \cdot \mathbf{B} = 1 \;\Rightarrow\; \dfrac{d}{ds}(\mathbf{B} \cdot \mathbf{B}) = 0 \;\Rightarrow\; 2\dfrac{d\mathbf{B}}{ds} \cdot \mathbf{B} = 0 \;\Rightarrow\; \dfrac{d\mathbf{B}}{ds} \perp \mathbf{B}$

(b) $\mathbf{B} = \mathbf{T} \times \mathbf{N} \;\Rightarrow\;$

$$\frac{d\mathbf{B}}{ds} = \frac{d}{ds}(\mathbf{T} \times \mathbf{N}) = \frac{d}{dt}(\mathbf{T} \times \mathbf{N})\frac{1}{ds/dt} = \frac{d}{dt}(\mathbf{T} \times \mathbf{N})\frac{1}{|\mathbf{r}'(t)|}$$

$$= [(\mathbf{T}' \times \mathbf{N}) + (\mathbf{T} \times \mathbf{N}')]\frac{1}{|\mathbf{r}'(t)|} = \left[\left(\mathbf{T}' \times \frac{\mathbf{T}'}{|\mathbf{T}'|}\right) + (\mathbf{T} \times \mathbf{N}')\right]\frac{1}{|\mathbf{r}'(t)|} = \frac{\mathbf{T} \times \mathbf{N}'}{|\mathbf{r}'(t)|}$$

$$\Rightarrow\; \frac{d\mathbf{B}}{ds} \perp \mathbf{T}$$

(c) $\mathbf{B} = \mathbf{T} \times \mathbf{N} \;\Rightarrow\; \mathbf{T} \perp \mathbf{N}$, $\mathbf{B} \perp \mathbf{T}$ and $\mathbf{B} \perp \mathbf{N}$. So $\mathbf{B}$, $\mathbf{T}$ and $\mathbf{N}$ form an orthogonal set of vectors
in the three-dimensional space $\mathbb{R}^3$. From parts (a) and (b), $d\mathbf{B}/ds$ is perpendicular to both $\mathbf{B}$ and $\mathbf{T}$,
so $d\mathbf{B}/ds$ is parallel to $\mathbf{N}$. Therefore, $d\mathbf{B}/ds = -\tau(s)\mathbf{N}$, where $\tau(s)$ is a scalar.

(d) Since $\mathbf{B} = \mathbf{T} \times \mathbf{N}$, $\mathbf{T} \perp \mathbf{N}$ and both $\mathbf{T}$ and $\mathbf{N}$ are unit vectors, $\mathbf{B}$ is a unit vector mutually
perpendicular to both $\mathbf{T}$ and $\mathbf{N}$. For a plane curve, $\mathbf{T}$ and $\mathbf{N}$ always lie in the plane of the curve, so
that $\mathbf{B}$ is a constant unit vector always perpendicular to the plane. Thus $d\mathbf{B}/ds = \mathbf{0}$, but
$d\mathbf{B}/ds = -\tau(s)\mathbf{N}$ and $\mathbf{N} \neq \mathbf{0}$, so $\tau(s) = 0$.

43. (a) $\mathbf{r}' = s'\mathbf{T} \Rightarrow \mathbf{r}'' = s''\mathbf{T} + s'\mathbf{T}' = s''\mathbf{T} + s'\dfrac{d\mathbf{T}}{ds}s' = s''\mathbf{T} + \kappa\,(s')^2\,\mathbf{N}$ by the first

Serret-Frenet formula.

(b) Using part (a), we have

$$\mathbf{r}' \times \mathbf{r}'' = (s'\mathbf{T}) \times \left[s''\mathbf{T} + \kappa\,(s')^2\,\mathbf{N} \right]$$

$$= \left[(s'\mathbf{T}) \times (s''\mathbf{T}) \right] + \left[(s'\mathbf{T}) \times \left(\kappa\,(s')^2\,\mathbf{N} \right) \right] \quad \text{(By Property 3 of the cross product)}$$

$$= (s's'')\,(\mathbf{T} \times \mathbf{T}) + \kappa\,(s')^3\,(\mathbf{T} \times \mathbf{N}) = 0 + \kappa\,(s')^3\,\mathbf{B} = \kappa\,(s')^3\,\mathbf{B}$$

(c) Using part (a), we have

$$\mathbf{r}''' = \left[s''\mathbf{T} + \kappa\,(s')^2\,\mathbf{N} \right]' = s'''\mathbf{T} + s''\mathbf{T}' + \kappa'\,(s')^2\,\mathbf{N} + 2\kappa s's''\mathbf{N} + \kappa\,(s')^2\,\mathbf{N}'$$

$$= s'''\mathbf{T} + s''\frac{d\mathbf{T}}{ds}s' + \kappa'\,(s')^2\,\mathbf{N} + 2\kappa s's''\mathbf{N} + \kappa\,(s')^2\frac{d\mathbf{N}}{ds}s'$$

$$= s'''\mathbf{T} + s''s'\kappa\mathbf{N} + \kappa'\,(s')^2\,\mathbf{N} + 2\kappa s's''\mathbf{N} + \kappa\,(s')^3\,(-\kappa\mathbf{T} + \tau\mathbf{B}) \quad \text{(by the second formula)}$$

$$= \left[s''' - \kappa^2\,(s')^3 \right]\mathbf{T} + \left[3\kappa s's'' + \kappa'\,(s')^2 \right]\mathbf{N} + \kappa\tau\,(s')^3\,\mathbf{B}$$

(d) Using parts (b) and (c) and the facts that $\mathbf{B} \cdot \mathbf{T} = 0$, $\mathbf{B} \cdot \mathbf{N} = 0$, and $\mathbf{B} \cdot \mathbf{B} = 1$, we get

$$\frac{(\mathbf{r}' \times \mathbf{r}'') \cdot \mathbf{r}'''}{|\mathbf{r}' \times \mathbf{r}''|^2} = \frac{\kappa\,(s')^3\,\mathbf{B} \cdot \left\{ \left[s''' - \kappa^2\,(s')^3 \right]\mathbf{T} + \left[3\kappa s's'' + \kappa'\,(s')^2 \right]\mathbf{N} + \kappa\tau\,(s')^3\,\mathbf{B} \right\}}{\left| \kappa\,(s')^3\,\mathbf{B} \right|^2}$$

$$= \frac{\kappa\,(s')^3\,\kappa\tau\,(s')^3}{\left[\kappa\,(s')^3 \right]^2} = \tau$$

45. For one helix, the vector equation is $\mathbf{r}\,(t) = \langle 10\cos t,\ 10\sin t,\ 34t/\,(2\pi) \rangle$ (measuring in angstroms), because the radius of each helix is 10 angstroms, and z increases by 34 angstroms for each increase of 2π in t. Using the arc length formula, letting t go from 0 to $2.9 \times 10^8 \times 2\pi$, we find the approximate length of each helix to be

$$L = \int_0^{2.9 \times 10^8 \times 2\pi} |\mathbf{r}'\,(t)|\ dt$$

$$= \int_0^{2.9 \times 10^8 \times 2\pi} \sqrt{(-10\sin t)^2 + (10\cos t)^2 + \left(\tfrac{34}{2\pi} \right)^2}\ dt$$

$$= \sqrt{100 + \left(\tfrac{34}{2\pi} \right)^2\,t}\ \Bigg]_0^{2.9 \times 10^8 \times 2\pi}$$

$$= 2.9 \times 10^8 \times 2\pi\sqrt{100 + \left(\tfrac{34}{2\pi} \right)^2}$$

$$\approx 2.07 \times 10^{10}\ \text{Å} \,-\, \text{more than two meters!}$$

Section 10.4   Motion in Space

1. (a) If $\mathbf{r}(t) = x(t)\mathbf{i} + y(t)\mathbf{j} + z(t)\mathbf{k}$ is the position vector of the particle at time t, then the average velocity over the time interval $[0, 1]$ is

$$\mathbf{v}_{ave} = \frac{\mathbf{r}(1) - \mathbf{r}(0)}{1 - 0} = \frac{(4.5\mathbf{i} + 6.0\mathbf{j} + 3.0\mathbf{k}) - (2.7\mathbf{i} + 9.8\mathbf{j} + 3.7\mathbf{k})}{1} = 1.8\mathbf{i} - 3.8\mathbf{j} - 0.7\mathbf{k}.$$

Similarly, over the other intervals we have

$$[0.5, 1]: \quad \mathbf{v}_{ave} = \frac{\mathbf{r}(1) - \mathbf{r}(0.5)}{1 - 0.5} = \frac{(4.5\mathbf{i} + 6.0\mathbf{j} + 3.0\mathbf{k}) - (3.5\mathbf{i} + 7.2\mathbf{j} + 3.3\mathbf{k})}{0.5}$$

$$= 2.0\mathbf{i} - 2.4\mathbf{j} - 0.6\mathbf{k}$$

$$[1, 2]: \quad \mathbf{v}_{ave} = \frac{\mathbf{r}(2) - \mathbf{r}(1)}{2 - 1} = \frac{(7.3\mathbf{i} + 7.8\mathbf{j} + 2.7\mathbf{k}) - (4.5\mathbf{i} + 6.0\mathbf{j} + 3.0\mathbf{k})}{1}$$

$$= 2.8\mathbf{i} + 1.8\mathbf{j} - 0.3\mathbf{k}$$

$$[1, 1.5]: \quad \mathbf{v}_{ave} = \frac{\mathbf{r}(1.5) - \mathbf{r}(1)}{1.5 - 1} = \frac{(5.9\mathbf{i} + 6.4\mathbf{j} + 2.8\mathbf{k}) - (4.5\mathbf{i} + 6.0\mathbf{j} + 3.0\mathbf{k})}{0.5}$$

$$= 2.8\mathbf{i} + 0.8\mathbf{j} - 0.4\mathbf{k}$$

(b) We can estimate the velocity at $t = 1$ by averaging the average velocities over the time intervals $[0.5, 1]$ and $[1, 1.5]$: $\mathbf{v}(1) \approx \frac{1}{2}[(2\mathbf{i} - 2.4\mathbf{j} - 0.6\mathbf{k}) + (2.8\mathbf{i} + 0.8\mathbf{j} - 0.4\mathbf{k})] = 2.4\mathbf{i} - 0.8\mathbf{j} - 0.5\mathbf{k}$.

Then the speed is $|\mathbf{v}(1)| \approx \sqrt{(2.4)^2 + (-0.8)^2 + (-0.5)^2} \approx 2.58$.

3. $\mathbf{r}(t) = \langle t^2 - 1, t \rangle \quad \Rightarrow$
$\mathbf{v}(t) = \mathbf{r}'(t) = \langle 2t, 1 \rangle$,
$\mathbf{a}(t) = \mathbf{r}''(t) = \langle 2, 0 \rangle$,
$|\mathbf{v}(t)| = \sqrt{4t^2 + 1}$

At $t = 1$:
$\mathbf{v}(1) = \langle 2, 1 \rangle$
$\mathbf{a}(1) = \langle 2, 0 \rangle$

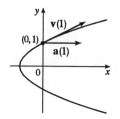

5. $\mathbf{r}(t) = e^t\mathbf{i} + e^{-t}\mathbf{j} \quad \Rightarrow$
$\mathbf{v}(t) = e^t\mathbf{i} - e^{-t}\mathbf{j}$,
$\mathbf{a}(t) = e^t\mathbf{i} + e^{-t}\mathbf{j}$

At $t = 0$:
$\mathbf{v}(0) = \mathbf{i} - \mathbf{j}$,
$\mathbf{a}(0) = \mathbf{i} + \mathbf{j}$

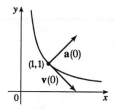

$|\mathbf{v}(t)| = \sqrt{e^{2t} + e^{-2t}} = e^{-t}\sqrt{e^{4t} + 1}$

Since $x = e^t$, $t = \ln x$ and $y = e^{-t} = e^{-\ln x} = 1/x$, and $x > 0, y > 0$.

7. $\mathbf{r}(t) = \langle \sin t, t, \cos t \rangle \Rightarrow$

$\mathbf{v}(t) = \langle \cos t, 1, -\sin t \rangle,\ \mathbf{v}(0) = \langle 1, 1, 0 \rangle$

$\mathbf{a}(t) = \langle -\sin t, 0, -\cos t \rangle,\ \mathbf{a}(0) = \langle 0, 0, -1 \rangle$

$|\mathbf{v}(t)| = \sqrt{\cos^2 t + 1 + \sin^2 t} = \sqrt{2}$

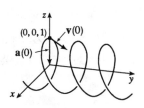

Since $x^2 + z^2 = 1$, $y = t$, the path of the particle is a helix about the y-axis.

9. $\mathbf{r}(t) = e^t \langle \cos t, \sin t, t \rangle \Rightarrow$

$\mathbf{v}(t) = e^t \langle \cos t, \sin t, t \rangle + e^t \langle -\sin t, \cos t, 1 \rangle = e^t \langle \cos t - \sin t, \sin t + \cos t, t + 1 \rangle$

$\mathbf{a}(t) = e^t \langle \cos t - \sin t - \sin t - \cos t, \sin t + \cos t + \cos t - \sin t, t + 1 + 1 \rangle$

$\quad = e^t \langle -2 \sin t, 2 \cos t, t + 2 \rangle$

$|\mathbf{v}(t)| = e^t \sqrt{\cos^2 t + \sin^2 t - 2 \cos t \sin t + \sin^2 t + \cos^2 t + 2 \sin t \cos t + t^2 + 2t + 1}$

$\quad = e^t \sqrt{t^2 + 2t + 3}$

11. $\mathbf{a}(t) = \mathbf{k} \Rightarrow \mathbf{v}(t) = \int \mathbf{k}\,dt = t\mathbf{k} + \mathbf{c}_1$ and $\mathbf{i} - \mathbf{j} = \mathbf{v}(0) = 0\mathbf{k} + \mathbf{c}_1$, so $\mathbf{c}_1 = \mathbf{i} - \mathbf{j}$ and

$\mathbf{v}(t) = \mathbf{i} - \mathbf{j} + t\mathbf{k}.\ \mathbf{r}(t) = \int (\mathbf{i} - \mathbf{j} + t\mathbf{k})\,dt = t\mathbf{i} - t\mathbf{j} + \frac{1}{2}t^2\mathbf{k} + \mathbf{c}_2.$ But $\mathbf{0} = \mathbf{r}(0) = \mathbf{0} + \mathbf{c}_2$, so

$\mathbf{c}_2 = \mathbf{0}$ and $\mathbf{r}(t) = t\mathbf{i} - t\mathbf{j} + \frac{1}{2}t^2\mathbf{k}.$

13. (a) $\mathbf{a}(t) = \mathbf{i} + 2\mathbf{j} + 2t\mathbf{k} \Rightarrow$

(b)

$\quad \mathbf{v}(t) = \int (\mathbf{i} + 2\mathbf{j} + 2t\mathbf{k})\,dt = t\mathbf{i} + 2t\mathbf{j} + t^2\mathbf{k} + \mathbf{c}_1$, and

$\quad \mathbf{0} = \mathbf{v}(0) = \mathbf{0} + \mathbf{c}_1$, so $\mathbf{c}_1 = \mathbf{0}$ and $\mathbf{v}(t) = \mathbf{i} + 2t\mathbf{j} + t^2\mathbf{k}.$

$\quad \mathbf{r}(t) = \int (t\mathbf{i} + 2t\mathbf{j} + t^2\mathbf{k})\,dt = \frac{1}{2}t^2\mathbf{i} + t^2\mathbf{j} + \frac{1}{3}t^3\mathbf{k} + \mathbf{c}_2.$

$\quad$ But $\mathbf{i} + \mathbf{k} = \mathbf{r}(0) = \mathbf{0} + \mathbf{c}_2$, so $\mathbf{c}_2 = \mathbf{i} + \mathbf{k}$ and

$\quad \mathbf{r}(t) = \left(1 + \frac{1}{2}t^2\right)\mathbf{i} + t^2\mathbf{j} + \left(1 + \frac{1}{3}t^3\right)\mathbf{k}.$

15. $\mathbf{r}(t) = \langle t^2, 5t, t^2 - 16t \rangle \Rightarrow \mathbf{v}(t) = \langle 2t, 5, 2t - 16 \rangle,$

$|\mathbf{v}(t)| = \sqrt{4t^2 + 25 + 4t^2 - 64t + 256} = \sqrt{8t^2 - 64t + 281}$ and

$\frac{d}{dt}|\mathbf{v}(t)| = \frac{1}{2}(8t^2 - 64t + 281)^{-1/2}(16t - 64).$ This is zero if and only if the numerator is zero, that

is, $16t - 64 = 0$ or $t = 4$. Since $\frac{d}{dt}|\mathbf{v}(t)| < 0$ for $t < 4$ and $\frac{d}{dt}|\mathbf{v}(t)| > 0$ for $t > 4$, the minimum

speed of $\sqrt{153}$ is attained at $t = 4$ units of time.

17. $|\mathbf{F}(t)| = 20\,\text{N}$ in the direction of the positive z-axis, so $\mathbf{F}(t) = 20\mathbf{k}$. Also $m = 4\,\text{kg}$, $\mathbf{r}(0) = \mathbf{0}$ and

$\mathbf{v}(0) = \mathbf{i} - \mathbf{j}.$ Since $20\mathbf{k} = \mathbf{F}(t) = 4\mathbf{a}(t)$, $\mathbf{a}(t) = 5\mathbf{k}.$ Then $\mathbf{v}(t) = 5t\mathbf{k} + \mathbf{c}_1$ where $\mathbf{c}_1 = \mathbf{i} - \mathbf{j}$ so

$\mathbf{v}(t) = \mathbf{i} - \mathbf{j} + 5t\mathbf{k}$ and the speed is $|\mathbf{v}(t)| = \sqrt{1 + 1 + 25t^2} = \sqrt{25t^2 + 2}.$ Also

$\mathbf{r}(t) = t\mathbf{i} - t\mathbf{j} + \frac{5}{2}t^2\mathbf{k} + \mathbf{c}_2$ and $\mathbf{0} = \mathbf{r}(0)$, so $\mathbf{c}_2 = \mathbf{0}$ and $\mathbf{r}(t) = t\mathbf{i} - t\mathbf{j} + \frac{5}{2}t^2\mathbf{k}.$

19. $|v(0)| = 500$ m/s and since the angle of elevation is $30°$, the direction of the velocity is $\frac{1}{2}\left(\sqrt{3}\,i + j\right)$.

Thus $v(0) = 250\left(\sqrt{3}\,i + j\right)$ and if we set up the axes so the projectile starts at the origin, then $r(0) = 0$. Ignoring air resistance, the only force is that due to gravity, so $F(t) = -mg\,j$ where $g \approx 9.8$ m/s^2. Thus $a(t) = -g\,j$ and $v(t) = -gt\,j + c_1$. But $250\left(\sqrt{3}\,i + j\right) = v(0) = c_1$, so $v(t) = 250\sqrt{3}\,i + (250 - gt)\,j$ and $r(t) = 250\sqrt{3}t\,i + \left(250t - \frac{1}{2}gt^2\right)j + c_2$ where $0 = r(0) = c_2$. Thus $r(t) = 250\sqrt{3}t\,i + \left(250t - \frac{1}{2}gt^2\right)j$.

(a) Setting $250t - \frac{1}{2}gt^2 = 0$ gives $t = 0$ or $t = \frac{500}{g} \approx 51.0$ s. So the range is $250\sqrt{3} \cdot \frac{500}{g} \approx 22$ km.

(b) $0 = \frac{d}{dt}\left(250t - \frac{1}{2}gt^2\right) = 250 - gt$ implies that the maximum height is attained when $t = 250/g \approx 25.5$ s. Thus, the maximum height is

$(250)(250/g) - g(250/g)^2\,\frac{1}{2} = (250)^2/(2g) \approx 3.2$ km.

(c) From part (a), impact occurs at $t = 500/g \approx 51.0$. Thus, the velocity at impact is $v(500/g) = 250\sqrt{3}\,i + [250 - g(500/g)]\,j = 250\sqrt{3}\,i - 250\,j$ and the speed is $|v(500/g)| = 250\sqrt{3+1} = 500$ m/s.

21. As in Example 5, $r(t) = (v_0 \cos 45°)\,t\,i + \left[(v_0 \sin 45°)\,t - \frac{1}{2}gt^2\right]j = \frac{1}{2}\left[v_0\sqrt{2}t\,i + \left(v_0\sqrt{2}t - gt^2\right)j\right]$.

Then the ball lands at $t = \dfrac{v_0\sqrt{2}}{g}$ s. Now since it lands 90 m away, $90 = \frac{1}{2}v_0\sqrt{2}\dfrac{v_0\sqrt{2}}{g}$ or $v_0^2 = 90g$ and the initial velocity is $v_0 = \sqrt{90g} \approx 30$ m/s.

23. From (4), $x = (v_0 \cos \alpha)\,t$ or $t = \dfrac{x}{v_0 \cos \alpha}$. Thus

$$y = (v_0 \sin \alpha)\frac{x}{v_0 \cos \alpha} - \frac{g}{2}\left(\frac{x}{v_0 \cos \alpha}\right)^2 = (\tan \alpha)x - \frac{g}{2v_0^2 \cos^2 \alpha}x^2.$$ Thus the trajectory is a parabola. Continuing by completing the square, we see that

$$y - \frac{(\tan^2 \alpha)v_0^2 \cos^2 \alpha}{2g} = -\frac{g}{2v_0^2 \cos^2 \alpha}\left[x - \frac{(\tan \alpha)v_0^2(\cos^2 \alpha)}{g}\right]^2 \text{ or}$$

$$y - \frac{v_0^2 \sin^2 \alpha}{2g} = -\frac{g}{2v_0^2 \cos^2 \alpha}\left(x - \frac{v_0^2 \sin \alpha \cos \alpha}{g}\right)^2. \text{ Thus the vertex of the parabola lies at}$$

$\left(\dfrac{v_0^2 \sin \alpha \cos \alpha}{g}, \dfrac{v_0^2 \sin^2 \alpha}{2g}\right)$, so the maximum height is $y = \dfrac{v_0^2 \sin^2 \alpha}{2g}$.

25. $r'(t) = (1 - \cos t)\,i + (\sin t)\,j$, $|r'(t)| = \sqrt{1 - 2\cos t + 1} = \sqrt{2(1 - \cos t)}$,

$r''(t) = (\sin t)\,i + (\cos t)\,j$. Thus $a_T = \dfrac{\sin t}{\sqrt{2(1 - \cos t)}}$ and

$$a_N = \frac{|(\cos t - \cos^2 t - \sin^2 t)\,k|}{\sqrt{2(1 - \cos t)}} = \frac{\sqrt{[(\cos t) - 1]^2}}{\sqrt{2}\sqrt{1 - \cos t}} = \frac{1}{\sqrt{2}}\sqrt{\frac{(1 - \cos t)^2}{1 - \cos t}} = \frac{\sqrt{1 - \cos t}}{\sqrt{2}}.$$

27. $\mathbf{r}'(t) = e^t\mathbf{i} + \sqrt{2}\mathbf{j} - e^{-t}\mathbf{k}$, $|\mathbf{r}(t)| = \sqrt{e^{2t} + 2 + e^{-2t}} = e^t + e^{-t}$,

$\mathbf{r}''(t) = e^t\mathbf{i} + e^{-t}\mathbf{k}$. Then $a_T = \dfrac{e^{2t} - e^{-2t}}{e^t + e^{-t}} = e^t - e^{-t} = 2\sinh t$ and

$a_N = \dfrac{\left|\sqrt{2}e^{-t}\mathbf{i} - 2\mathbf{j} - \sqrt{2}e^t\mathbf{k}\right|}{e^t + e^{-t}} = \dfrac{\sqrt{2\left(e^{-2t} + 2 + e^{2t}\right)}}{e^t + e^{-t}} = \sqrt{2}\dfrac{e^t + e^{-t}}{e^t + e^{-t}} = \sqrt{2}.$

29. If the engines are turned off at time t, then the spacecraft will continue to travel
in the direction of $\mathbf{v}(t)$, so we need a t such that for some scalar $s > 0$,

$\mathbf{r}(t) + s\,\mathbf{v}(t) = \langle 6, 4, 9\rangle$. $\mathbf{v}(t) = \mathbf{r}'(t) = \mathbf{i} + \dfrac{1}{t}\mathbf{j} + \dfrac{8t}{(t^2+1)^2}\mathbf{k} \quad \Rightarrow$

$\mathbf{r}(t) + s\,\mathbf{v}(t) = \left\langle 3 + t + s, 2 + \ln t + \dfrac{s}{t}, 7 - \dfrac{4}{t^2+1} + \dfrac{8st}{(t^2+1)^2}\right\rangle \quad \Rightarrow \quad 3 + t + s = 6 \quad \Rightarrow$

$s = 3 - t$, so $7 - \dfrac{4}{t^2+1} + \dfrac{8(3-t)t}{(t^2+1)^2} = 9 \quad \Leftrightarrow \quad \dfrac{24t - 12t^2 - 4}{(t^2+1)^2} = 2 \quad \Leftrightarrow \quad t^4 + 8t^2 - 12t + 3 = 0$. It

is easily seen that $t = 1$ is a root of this polynomial. Also $2 + \ln 1 + \dfrac{3-1}{1} = 4$, so $t = 1$ is the desired

solution.

Section 10.5 Parametric Surfaces

1. $\mathbf{r}(u, v) = u \cos v \, \mathbf{i} + u \sin v \, \mathbf{j} + u^2 \, \mathbf{k}$. The parametric equations for the surface are
$x = u \cos v, y = u \sin v, z = u^2$. For any point (x, y, z) on the surface, we have
$x^2 + y^2 = u^2 \cos^2 v + u^2 \sin^2 v = u^2 = z$. Since no restrictions are placed on the parameters, the
surface is $z = x^2 + y^2$, which we recognize as a circular paraboloid opening upward whose axis is the
z-axis.

3. $\mathbf{r}(x, \theta) = \langle x, \cos \theta, \sin \theta \rangle$. The parametric equations for the surface are $x = x, y = \cos \theta, z = \sin \theta$. For
any point (x, y, z) on the surface, we have $y^2 + z^2 = \cos^2 \theta + \sin^2 \theta = 1$, so any vertical trace in $x = k$
is the circle $y^2 + z^2 = 1, x = k$. Since $x = x$ with no restriction, the surface is a circular cylinder with
radius 1 whose axis is the x-axis.

5. $\mathbf{r}(u, v) = \langle u^2 + 1, v^3 + 1, u + v \rangle, -1 \le u \le 1, -1 \le v \le 1$

The surface has parametric equations $x = u^2 + 1, y = v^3 + 1$,
$z = u + v, -1 \le u \le 1, -1 \le v \le 1$. If we keep u constant at
$u_0, x = u_0^2 + 1$, a constant, so the corresponding grid curves
must be the curves parallel to the yz-plane. If v is constant, we
have $y = v_0^3 + 1$, a constant, so these grid curves are the curves
parallel to the xz-plane.

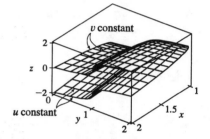

7. $x = \cos u \sin 2v, y = \sin u \sin 2v, z = \sin v$

The complete graph of the surface is given by the parametric
domain $0 \le u \le \pi, 0 \le v \le 2\pi$. Note that if $v = v_0$ is constant,
the parametric equations become $x = \cos u \sin 2v_0$,
$y = \sin u \sin 2v_0, z = \sin v_0$ which represent a circle of radius
$\sin 2v_0$ in the plane $z = \sin v_0$. So the circular grid curves we see
lying horizontally are the grid curves which have v constant. The
vertical grid curves, then, correspond to $u = u_0$ being held
constant, giving $x = \cos u_0 \sin 2v$ and $y = \sin u_0 \sin 2v$ with
$z = \sin v$ which has a "figure-eight" shape.

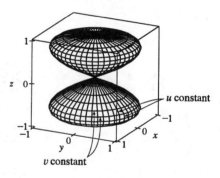

9. $\mathbf{r}(u, v) = \cos v \, \mathbf{i} + \sin v \, \mathbf{j} + u \, \mathbf{k}$. The parametric equations for the surface are $x = \cos v, y = \sin v$,
$z = u$. Then $x^2 + y^2 = \cos^2 v + \sin^2 v = 1$ and $z = u$ with no restriction on u, so we have a circular
cylinder, graph IV. The grid curves with u constant are the horizontal circles we see in the plane $z = u$.
If v is constant, both x and y are constant with z free to vary, so the corresponding grid curves are the
lines on the cylinder parallel to the z-axis.

11. $\mathbf{r}(u, v) = u \cos v \, \mathbf{i} + u \sin v \, \mathbf{j} + v \, \mathbf{k}$. The parametric equations for the surface are
$x = u \cos v$, $y = u \sin v$, $z = v$. We look at the grid curves first; if we fix v, then x and y parametrize a
straight line in the plane $z = v$ which intersects the z-axis. If u is held constant, the projection onto the
xy-plane is circular; with $z = v$, each grid curve is a helix. The surface is a spiraling ramp, graph I.

13. $x = (u - \sin u) \cos v$, $y = (1 - \cos u) \sin v$, $z = u$. If u is held constant, x and y give an equation of an
ellipse in the plane $z = u$, thus the grid curves are horizontally oriented ellipses. Note that when $u = 0$,
the "ellipse" is the single point $(0, 0, 0)$, and when $u = \pi$, we have $y = 0$ while x ranges from $-\pi$ to π, a
line segment parallel to the x-axis in the plane $z = \pi$. This is the upper "seam" we see in graph II. When
v is held constant, $z = u$ is free to vary, so the corresponding grid curves are the curves we see running
up and down along the surface.

15. Letting x and y be the parameters, parametric equations are $x = x$, $y = y$, $z = \sqrt{1 - 3x^2 - 2y^2}$ where
$-\frac{1}{\sqrt{3}} \le x \le \frac{1}{\sqrt{3}}$ and $-\frac{1}{\sqrt{2}} \le y \le \frac{1}{\sqrt{2}}$. Then a vector equation of the surface is
$\mathbf{r}(x, y) = x \, \mathbf{i} + y \, \mathbf{j} + \sqrt{1 - 3x^2 - 2y^2} \, \mathbf{k}$.
Alternate Solution: Letting ϕ and θ be the parameters, parametric equations are $x = \frac{1}{\sqrt{3}} \sin \phi \cos \theta$,
$y = \frac{1}{\sqrt{2}} \sin \phi \sin \theta$, $z = \cos \phi$ where $0 \le \phi \le \frac{\pi}{2}$ and $0 \le \theta \le 2\pi$.
Note: There are many parametric representations of a given surface.

17. $x = x$, $y = 6 - 3x^2 - 2z^2$, $z = z$ where $3x^2 + 2z^2 \le 6$ since $y \ge 0$. Then the associated vector
equation is $\mathbf{r}(x, y) = x \, \mathbf{i} + (6 - 3x^2 - 2z^2) \, \mathbf{j} + z \, \mathbf{k}$.

19. Since the cone intersects the sphere in the circle $x^2 + y^2 = 2$, $z = 2$ and we want the portion of the
sphere above this, we can parametrize the surface as $x = x$, $y = y$, $z = \sqrt{4 - x^2 - y^2}$ where
$2 \le x^2 + y^2 \le 4$.
Alternate Solution: Using spherical coordinates, $x = 2 \sin \phi \cos \theta$, $y = 2 \sin \phi \sin \theta$, $z = 2 \cos \phi$ where
$0 \le \phi \le \frac{\pi}{4}$ and $0 \le \theta \le 2\pi$.

21. The surface is a disc with radius 4 and center $(0, 0, 5)$. Thus, $x = r \cos \theta$, $y = r \sin \theta$, $z = 5$ where
$0 \le r \le 4$, $0 \le \theta \le 2\pi$ is a parametric representation of the surface.
Alternate Solution: In rectangular coordinates we could represent the surface as $x = x$, $y = y$, $z = 5$
where $0 \le x^2 + y^2 \le 16$.

23. Using Equations 3, we have the
parametrization $x = x$, $y = e^{-x} \cos \theta$,
$z = e^{-x} \sin \theta$, $0 \le x \le 3$, $0 \le \theta \le 2\pi$.

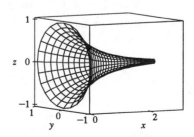

25. (a) $x = a \sin u \cos v$, $y = b \sin u \sin v$, $z = c \cos u$ $\Rightarrow$ **(b)**

$$\frac{x^2}{a^2} + \frac{y^2}{b^2} + \frac{z^2}{c^2} = (\sin u \cos v)^2 + (\sin u \sin v)^2 + (\cos u)^2$$

$$= \sin^2 u + \cos^2 u = 1$$

and since the ranges of u and v are sufficient to generate the entire graph, the parametric equations represent an ellipsoid.

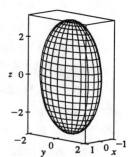

27. (a) Replacing $\cos u$ by $\sin u$ and $\sin u$ by $\cos u$ gives parametric equations
$x = (2 + \sin v) \sin u$, $y = (2 + \sin v) \cos u$, $z = u + \cos v$. From the
graph, it appears that the direction of the spiral is reversed. We can verify
this observation by noting that the projection of the spiral grid curves onto
the xy-plane, given by $x = (2 + \sin v) \sin u$, $y = (2 + \sin v) \cos u$, $z = 0$,
draws a circle in the clockwise direction for each value of v. The original
equations, on the other hand, give circular projections drawn in the
counterclockwise direction. The equation for z is identical in both
surfaces, so as z increases, these grid curves spiral up in opposite
directions for the two surfaces.

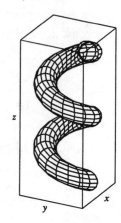

(b) Replacing $\cos u$ by $\cos 2u$ and $\sin u$ by $\sin 2u$ gives parametric equations
$x = (2 + \sin v) \cos 2u$, $y = (2 + \sin v) \sin 2u$, $z = u + \cos v$. From the
graph, it appears that the number of coils in the surface doubles within the
same parametric domain. We can verify this observation by noting that the
projection of the spiral grid curves onto the xy-plane, given by
$x = (2 + \sin v) \cos 2u$, $y = (2 + \sin v) \sin 2u$, $z = 0$ (where v is
constant), complete circular revolutions for $0 \le u \le \pi$ while the original
surface requires $0 \le u \le 2\pi$ for a complete revolution. Thus, the new
surface winds around twice as fast as the original surface, and since the
equation for z is identical in both surfaces, we observe twice as many
circular coils in the same z-interval.

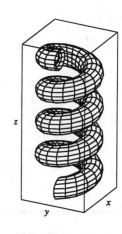

Chapter 10 Review

Concept Check

1. A vector function is a function whose domain is a set of real numbers and whose range is a set of vectors. To find the derivative or integral, we can differentiate or integrate each component of the vector function.

2. The tip of the moving vector $\mathbf{r}(t)$ of a continuous vector function traces out a space curve.

3. (a) A curve represented by the vector function $\mathbf{r}(t)$ is smooth if $\mathbf{r}'(t)$ is continuous and $\mathbf{r}'(t) \neq \mathbf{0}$ on its parametric domain (except possibly at the endpoints).

(b) The tangent vector to a smooth curve at a point P with position vector $\mathbf{r}(t)$ is the vector $\mathbf{r}'(t)$. The tangent line at P is the line through P parallel to the tangent vector $\mathbf{r}'(t)$. The unit tangent vector is

$$\mathbf{T}(t) = \frac{\mathbf{r}'(t)}{|\mathbf{r}'(t)|}.$$

4. (a)–(f) See Theorem 3 in Section 10.2.

5. Use (2), or equivalently (3), in Section 10.3.

6. (a) The curvature of a curve is $\kappa = \left|\dfrac{d\mathbf{T}}{ds}\right|$ where $\mathbf{T}$ is the unit tangent vector.

(b) $\kappa(t) = \left|\dfrac{\mathbf{T}'(t)}{\mathbf{r}'(t)}\right|$

(c) $\kappa(t) = \dfrac{|\mathbf{r}'(t) \times \mathbf{r}''(t)|}{|\mathbf{r}'(t)|^3}$

(d) $\kappa(x) = \dfrac{|f''(x)|}{\left[1 + (f'(x))^2\right]^{3/2}}$

7. (a) The unit normal vector: $\mathbf{N}(t) = \dfrac{\mathbf{T}'(t)}{|\mathbf{T}'(t)|}$

The binormal vector: $\mathbf{B}(t) = \mathbf{T}(t) \times \mathbf{N}(t)$

(b) See the discussion preceding Example 7 on page 721.

8. (a) If $\mathbf{r}(t)$ is the position vector of the particle on the space curve, the velocity $\mathbf{v}(t) = \mathbf{r}'(t)$, the speed is given by $|\mathbf{v}(t)|$, and the acceleration $\mathbf{a}(t) = \mathbf{v}'(t) = \mathbf{r}''(t)$.

(b) $\mathbf{a} = a_T\mathbf{T} + a_N\mathbf{N}$ where $a_T = v'$ and $a_N = \kappa v^2$.

9. See the statement of Kepler's Laws on page 729.

10. See the discussion on pages 734 and 735.

True-False Quiz

1. True. If we reparametrize the curve by replacing $u = t^3$, we have $\mathbf{r}(u) = u\,\mathbf{i} + 2u\,\mathbf{j} + 3u\,\mathbf{k}$, which is a line through the origin with direction vector $\mathbf{i} + 2\mathbf{j} + 3\mathbf{k}$.

3. False. $\mathbf{r}'(t) = \langle -\sin t, 2t, 4t^3 \rangle$, and since $\mathbf{r}'(0) = \langle 0, 0, 0 \rangle = \mathbf{0}$, the curve is not smooth.

5. False. By Formula 5 of Theorem 10.2.3, $\frac{d}{dt}[\mathbf{u}(t) \times \mathbf{v}(t)] = \mathbf{u}'(t) \times \mathbf{v}(t) + \mathbf{u}(t) \times \mathbf{v}'(t)$.

7. False. κ is the magnitude of the rate of change of the unit tangent vector $\mathbf{T}$ with respect to arc length s, not with respect to t.

9. True. See the discussion preceding Example 7 on page 721.

Exercises

1. (a) Since $x = 2$ and $y^2 + z^2 = 1$, the curve is a circle in the plane $x = 2$ with center $(2, 0, 0)$ and radius 1.

(b) $\mathbf{r}'(t) = \cos t\,\mathbf{j} - \sin t\,\mathbf{k} \quad \Rightarrow \quad \mathbf{r}''(t) = -\sin t\,\mathbf{j} - \cos t\,\mathbf{k}$

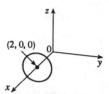

3. The projection of the curve C of intersection onto the xy-plane is the circle $x^2 + y^2 = 16$, $z = 0$. So we can write $x = 4\cos t$, $y = 4\sin t$, $0 \le t \le 2\pi$. From the equation of the plane, we have $z = 5 - x = 5 - 4\cos t$, so parametric equations for C are $x = 4\cos t$, $y = 4\sin t$, $z = 5 - 4\cos t$, $0 \le t \le 2\pi$, and the corresponding vector function is $\mathbf{r}(t) = 4\cos t\,\mathbf{i} + 4\sin t\,\mathbf{j} + (5 - 4\cos t)\,\mathbf{k}$, $0 \le t \le 2\pi$.

5. $\int_0^1 \left[(t + t^2)\,\mathbf{i} + (2 + t^3)\,\mathbf{j} + t^4\,\mathbf{k} \right] dt = \left[\left(\frac{1}{2}t^2 + \frac{1}{3}t^3 \right)\mathbf{i} + \left(2t + \frac{1}{4}t^4 \right)\mathbf{j} + \left(\frac{1}{5}t^5 \right)\mathbf{k} \right]_0^1$

$$= \tfrac{5}{6}\,\mathbf{i} + \tfrac{9}{4}\,\mathbf{j} + \tfrac{1}{5}\,\mathbf{k}$$

7. $t = 1$ at $(1, 4, 2)$ and $t = 4$ at $(2, 1, 17)$, so

$$L = \int_1^4 \sqrt{\frac{1}{4t} + \frac{16}{t^4} + 4t^2}\, dt$$

$$\approx \frac{4 - 1}{3 \cdot 4} \left[\sqrt{\tfrac{1}{4} + 16 + 4} + 4 \cdot \sqrt{\frac{1}{4 \cdot \frac{7}{4}} + \frac{16}{\left(\frac{7}{4}\right)^4} + 4\left(\tfrac{7}{4}\right)^2} + 2 \cdot \sqrt{\frac{1}{4 \cdot \frac{10}{4}} + \frac{16}{\left(\frac{10}{4}\right)^4} + 4\left(\tfrac{10}{4}\right)^2} \right.$$

$$\left. + 4 \cdot \sqrt{\frac{1}{4 \cdot \frac{13}{4}} + \frac{16}{\left(\frac{13}{4}\right)^4} + 4\left(\tfrac{13}{4}\right)^2} + \sqrt{\frac{1}{4 \cdot 4} + \frac{16}{4^4} + 4 \cdot 4^2} \right]$$

$$\approx 15.9241$$

9. The angle of intersection of the two curves, θ, is the angle between their respective tangents at the point of intersection. For both curves the point $(1, 0, 0)$ occurs when $t = 0$. $\mathbf{r}_1'(t) = -\sin t\, \mathbf{i} + \cos t\, \mathbf{j} + \mathbf{k} \Rightarrow$ $\mathbf{r}_1'(0) = \mathbf{j} + \mathbf{k}$ and $\mathbf{r}_2'(t) = \mathbf{i} + 2t\, \mathbf{j} + 3t^2\, \mathbf{k} \quad \Rightarrow \quad \mathbf{r}_2'(0) = \mathbf{i}$. $\mathbf{r}_1'(0) \cdot \mathbf{r}_2'(0) = (\mathbf{j} + \mathbf{k}) \cdot \mathbf{i} = 0$. Therefore, the curves intersect in a right angle, that is, $\theta = \frac{\pi}{2}$.

11. (a) $\mathbf{T}(t) = \dfrac{\langle t^2, t, 1\rangle}{\sqrt{t^4 + t^2 + 1}}$

(b) $\mathbf{T}'(t) = -\frac{1}{2}\left(t^4 + t^2 + 1\right)^{-3/2}\left(4t^3 + 2t\right)\langle t^2, t, 1\rangle + \left(t^4 + t^2 + 1\right)^{-1/2}\langle 2t, 1, 0\rangle$

$\quad = \dfrac{-2t^3 - t}{\left(t^4 + t^2 + 1\right)^{3/2}}\langle t^2, t, 1\rangle + \dfrac{1}{\left(t^4 + t^2 + 1\right)^{1/2}}\langle 2t, 1, 0\rangle$

$\quad = \dfrac{\langle -2t^5 - t^3, -2t^4 - t^2, -2t^3 - t\rangle + \langle 2t^5 + 2t^3 + 2t, t^4 + t^2 + 1, 0\rangle}{\left(t^4 + t^2 + 1\right)^{3/2}}$

$\quad = \dfrac{\langle 2t, -t^4 + 1, -2t^3 - t\rangle}{\left(t^4 + t^2 + 1\right)^{3/2}}$

$|\mathbf{T}'(t)| = \dfrac{\sqrt{4t^2 + t^8 - 2t^4 + 1 + 4t^6 + 4t^4 + t^2}}{\left(t^4 + t^2 + 1\right)^{3/2}} = \dfrac{\sqrt{t^8 + 4t^6 + 2t^4 + 5t^2}}{\left(t^4 + t^2 + 1\right)^{3/2}}$, and

$\mathbf{N}(t) = \dfrac{\langle 2t, 1 - t^4, -2t^3 - t\rangle}{\sqrt{t^8 + 4t^6 + 2t^4 + 5t^2}}$.

(c) $\kappa(t) = \dfrac{|\mathbf{T}'(t)|}{|\mathbf{r}'(t)|} = \dfrac{\sqrt{t^8 + 4t^6 + 2t^4 + 5t^2}}{\left(t^4 + t^2 + 1\right)^2}$

13. $y' = 4x^3$, $y'' = 12x^2$ and $\kappa(x) = \dfrac{|12x^2|}{\left(1 + 16x^6\right)^{3/2}}$, so $\kappa(1) = \dfrac{12}{(17)^{3/2}}$.

15. $\mathbf{r}(t) = \langle \sin 2t, t, \cos 2t\rangle \Rightarrow \mathbf{r}'(t) = \langle 2\cos 2t, 1, -2\sin 2t\rangle \Rightarrow \mathbf{T}(t) = \frac{1}{\sqrt{5}}\langle 2\cos 2t, 1, -2\sin 2t\rangle$
$\Rightarrow \mathbf{T}'(t) = \frac{1}{\sqrt{5}}\langle -4\sin 2t, 0, -4\cos 2t\rangle \Rightarrow \mathbf{N}(t) = \langle -\sin 2t, 0, -\cos 2t\rangle$. So
$\mathbf{N} = \mathbf{N}(\pi) = \langle 0, 0, -1\rangle$ and $\mathbf{B} = \mathbf{T} \times \mathbf{N} = \frac{1}{\sqrt{5}}\langle -1, 2, 0\rangle$. So a normal to the osculating plane is
$\langle -1, 2, 0\rangle$ and an equation is $-1(x - 0) + 2(y - \pi) + 0(z - 1) = 0$ or $x - 2y + 2\pi = 0$.

17. $\mathbf{v}(t) = 2\sqrt{2}\,\mathbf{i} + 2e^{2t}\,\mathbf{j} - 2e^{-2t}\,\mathbf{k}$, $|\mathbf{v}(t)| = \sqrt{8 + 4e^{4t} + 4e^{-4t}} = 2\left(e^{2t} + e^{-2t}\right)$,
$\mathbf{a}(t) = 4e^{2t}\,\mathbf{j} + 4e^{-2t}\,\mathbf{k}$

19. We set up the axes so that the shot leaves the athlete's hand 7 ft above the origin. Then we are given $\mathbf{r}(0) = 7\mathbf{j}$, $|\mathbf{v}(0)| = 43$ ft/s, and $\mathbf{v}(0)$ has direction given by a $45°$ angle of elevation. Then a unit vector in the direction of $\mathbf{v}(0)$ is $\frac{1}{\sqrt{2}}(\mathbf{i}+\mathbf{j})$ $\Rightarrow$ $\mathbf{v}(0) = \frac{43}{\sqrt{2}}(\mathbf{i}+\mathbf{j})$. Assuming air resistance is negligible, the only external force is due to gravity, so as in Example 5 in Section 10.4 we have $\mathbf{a} = -g\mathbf{j}$ where here $g \approx 32$ ft/s². Since $\mathbf{v}'(t) = \mathbf{a}(t)$, we integrate, giving $\mathbf{v}(t) = -gt\mathbf{j} + \mathbf{C}$ where $\mathbf{C} = \mathbf{v}(0) = \frac{43}{\sqrt{2}}(\mathbf{i}+\mathbf{j})$ $\Rightarrow$ $\mathbf{v}(t) = \frac{43}{\sqrt{2}}\mathbf{i} + \left(\frac{43}{\sqrt{2}} - gt\right)\mathbf{j}$. Since $\mathbf{r}'(t) = \mathbf{v}(t)$ we integrate again, so $\mathbf{r}(t) = \frac{43}{\sqrt{2}}t\,\mathbf{i} + \left(\frac{43}{\sqrt{2}}t - \frac{1}{2}gt^2\right)\mathbf{j} + \mathbf{D}$. But $\mathbf{D} = \mathbf{r}(0) = 7\mathbf{j}$ $\Rightarrow$ $\mathbf{r}(t) = \frac{43}{\sqrt{2}}t\,\mathbf{i} + \left(\frac{43}{\sqrt{2}}t - \frac{1}{2}gt^2 + 7\right)\mathbf{j}$.

(a) At 2 seconds, the shot is at $\mathbf{r}(2) = \frac{43}{\sqrt{2}}(2)\,\mathbf{i} + \left(\frac{43}{\sqrt{2}}(2) - \frac{1}{2}g(2)^2 + 7\right)\mathbf{j} \approx 60.8\,\mathbf{i} + 3.8\,\mathbf{j}$, so the shot is about 3.8 ft above the ground, at a horizontal distance of 60.8 ft from the athlete.

(b) The shot reaches its maximum height when the vertical component of velocity is 0: $\frac{43}{\sqrt{2}} - gt = 0$ $\Rightarrow$ $t = \dfrac{43}{\sqrt{2g}} \approx 0.95$ s. Then $\mathbf{r}(0.95) \approx 28.9\,\mathbf{i} + 21.4\,\mathbf{j}$, so the maximum height is approximately 21.4 ft.

(c) The shot hits the ground when the vertical component of $\mathbf{r}(t)$ is 0, so $\frac{43}{\sqrt{2}}t - \frac{1}{2}gt^2 + 7 = 0$ $\Rightarrow$ $-16t^2 + \frac{43}{\sqrt{2}}t + 7 = 0$ $\Rightarrow$ $t \approx 2.11$ s. $\mathbf{r}(2.11) \approx 64.2\,\mathbf{i} - 0.08\,\mathbf{j}$, thus the shot lands approximately 64.2 ft from the athlete.

21. From Example 3 in Section 10.5, a parametric representation of the sphere $x^2 + y^2 + z^2 = 4$ is $x = 2\sin\phi\cos\theta$, $y = 2\sin\phi\sin\theta$, $z = 2\cos\phi$ with $0 \le \theta \le 2\pi$ and $0 \le \phi \le \pi$. We can restrict the surface to that portion between the planes $z = 1$ and $z = -1$ by restricting $-1 \le z \le 1$ $\Rightarrow$ $-1 \le 2\cos\phi \le 1$ $\Rightarrow$ $\frac{\pi}{3} \le \phi \le \frac{2\pi}{3}$.

23. (a) Instead of proceeding directly, we use Formula 3 of Theorem 10.2.3: $\mathbf{r}(t) = t\,\mathbf{R}(t)$ $\Rightarrow$ $\mathbf{v} = \mathbf{r}'(t) = \mathbf{R}(t) + t\,\mathbf{R}'(t) = \cos\omega t\,\mathbf{i} + \sin\omega t\,\mathbf{j} + t\,\mathbf{v}_d$.

(b) Using the same method as in part (a) and starting with $\mathbf{v} = \mathbf{R}(t) + t\,\mathbf{R}'(t)$, we have $\mathbf{a} = \mathbf{v}' = \mathbf{R}'(t) + \mathbf{R}'(t) + t\,\mathbf{R}''(t) = 2\,\mathbf{R}'(t) + t\,\mathbf{R}''(t) = 2\,\mathbf{v}_d + t\,\mathbf{a}_d$.

(c) Here we have $\mathbf{r}(t) = e^{-t}\cos\omega t\,\mathbf{i} + e^{-t}\sin\omega t\,\mathbf{j} = e^{-t}\,\mathbf{R}(t)$. So, as in parts (a) and (b),
$\mathbf{v} = \mathbf{r}'(t) = e^{-t}\,\mathbf{R}'(t) - e^{-t}\,\mathbf{R}(t) = e^{-t}\,[\mathbf{R}'(t) - \mathbf{R}(t)]$ $\Rightarrow$
$\mathbf{a} = \mathbf{v}' = e^{-t}\,[\mathbf{R}''(t) - \mathbf{R}'(t)] - e^{-t}\,[\mathbf{R}'(t) - \mathbf{R}(t)] = e^{-t}\,[\mathbf{R}''(t) - 2\,\mathbf{R}'(t) + \mathbf{R}(t)]$
$= e^{-t}\,\mathbf{a}_d - 2e^{-t}\,\mathbf{v}_d + e^{-t}\,\mathbf{R}$

Thus, the Coriolis acceleration (the sum of the "extra" terms not involving $\mathbf{a}_d$) is $-2e^{-t}\,\mathbf{v}_d + e^{-t}\,\mathbf{R}$.

25. (a) $F(x) = \begin{cases} 1 & \text{if } x \le 0 \\ \sqrt{1-x^2} & \text{if } 0 < x < \frac{1}{\sqrt{2}} \\ -x + \sqrt{2} & \text{if } x \ge \frac{1}{\sqrt{2}} \end{cases} \Rightarrow F'(x) = \begin{cases} 0 & \text{if } x < 0 \\ -x/\sqrt{1-x^2} & \text{if } 0 < x < \frac{1}{\sqrt{2}} \\ -1 & \text{if } x > \frac{1}{\sqrt{2}} \end{cases} \Rightarrow$

$F''(x) = \begin{cases} 0 & \text{if } x < 0 \\ -1/(1-x^2)^{3/2} & \text{if } 0 < x < \frac{1}{\sqrt{2}} \\ 0 & \text{if } x > \frac{1}{\sqrt{2}} \end{cases}$ since

$\frac{d}{dx}\left[-x(1-x^2)^{-1/2}\right] = -(1-x^2)^{-1/2} - x^2(1-x^2)^{-3/2} = -(1-x^2)^{-3/2}.$

Now $\lim_{x \to 0^+} \sqrt{1-x^2} = 1 = F(0)$ and $\lim_{x \to (1/\sqrt{2})^-} \sqrt{1-x^2} = \frac{1}{\sqrt{2}} = F\left(\frac{1}{\sqrt{2}}\right)$, so F is continuous.

Also, since $\lim_{x \to 0^+} F'(x) = 0 = \lim_{x \to 0^-} F'(x)$ and $\lim_{x \to (1/\sqrt{2})^-} F'(x) = -1 = \lim_{x \to (1/\sqrt{2})^+} F'(x)$, F' is

continuous. But $\lim_{x \to 0^+} F''(x) = -1 \ne 0 = \lim_{x \to 0^-} F''(x)$, so F'' is not continuous at $x = 0$. (The

same is true at $x = \frac{1}{\sqrt{2}}$.) So F does not have continuous curvature.

(b) Set $P(x) = ax^5 + bx^4 + cx^3 + dx^2 + ex + f$. The continuity conditions on P are $P(0) = 0$,
$P(1) = 1$, $P'(0) = 0$ and $P'(1) = 1$. Also the curvature must be continuous. For $x \le 0$ and $x \ge 1$,

$\kappa(x) = 0$; elsewhere $\kappa(x) = \dfrac{|P''(x)|}{\left(1 + [P'(x)]^2\right)^{3/2}}$, so we need $P''(0) = 0$ and $P''(1) = 0$.

The conditions $P(0) = P'(0) = P''(0) = 0$ imply that $d = e = f = 0$. The other conditions imply
that $a + b + c = 1$, $5a + 4b + 3c = 1$, and $10a + 6b + 3c = 0$. From these, we find that $a = 3$,
$b = -8$, and $c = 6$. Therefore $P(x) = 3x^5 - 8x^4 + 6x^3$. Since there was no solution with $a = 0$,
this could not have been done with a polynomial of degree 4.

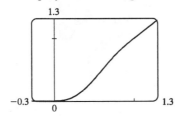

Focus on Problem Solving

1. (a) $r(t) = R\cos\omega t\,\mathbf{i} + R\sin\omega t\,\mathbf{j} \;\Rightarrow\; \mathbf{v} = \mathbf{r}'(t) = -\omega R\sin\omega t\,\mathbf{i} + \omega R\cos\omega t\,\mathbf{j}$,

so $\mathbf{r} = R(\cos\omega t\,\mathbf{i} + \sin\omega t\,\mathbf{j})$ and $\mathbf{v} = \omega R(-\sin\omega t\,\mathbf{i} + \cos\omega t\,\mathbf{j})$.

$\mathbf{v}\cdot\mathbf{r} = \omega R^2(-\cos\omega t\sin\omega t + \sin\omega t\cos\omega t) = 0$, so $\mathbf{v}\perp\mathbf{r}$. Since $\mathbf{r}$ points along a radius of the circle, and $\mathbf{v}\perp\mathbf{r}$, $\mathbf{v}$ is tangent to the circle. Because it is a velocity vector, $\mathbf{v}$ points in the direction of motion.

(b) In (a), we wrote $\mathbf{v}$ in the form $\omega R\mathbf{u}$, where $\mathbf{u}$ is the unit vector $-\sin\omega t\,\mathbf{i} + \cos\omega t\,\mathbf{j}$. Clearly $|\mathbf{v}| = \omega R|\mathbf{u}| = \omega R$. At speed ωR, the particle completes one revolution, a distance $2\pi R$, in time

$$T = \frac{2\pi R}{\omega R} = \frac{2\pi}{\omega}.$$

(c) $\mathbf{a} = \dfrac{d\mathbf{v}}{dt} = -\omega^2 R\cos\omega t\,\mathbf{i} - \omega^2 R\sin\omega t\,\mathbf{j} = -\omega^2 R(\cos\omega t\,\mathbf{i} + \sin\omega t\,\mathbf{j})$, so $\mathbf{a} = -\omega^2\mathbf{r}$. This shows that $\mathbf{a}$ is proportional to $\mathbf{r}$ and points in the opposite direction (toward the origin). Also, $|\mathbf{a}| = \omega^2|\mathbf{r}| = \omega^2 R$.

(d) By Newton's Second Law (see Section 10.4), $\mathbf{F} = m\mathbf{a}$, so

$$|\mathbf{F}| = m|\mathbf{a}| = mR\omega^2 = \frac{m(\omega R)^2}{R} = \frac{m|\mathbf{v}|^2}{R}.$$

3. (a) The projectile reaches maximum height when $0 = \dfrac{dy}{dt} = \dfrac{d}{dt}\left[(v_0\sin\alpha)t - \tfrac{1}{2}gt^2\right] = v_0\sin\alpha - gt$;

that is, when $t = \dfrac{v_0\sin\alpha}{g}$ and $y = (v_0\sin\alpha)\left(\dfrac{v_0\sin\alpha}{g}\right) - \dfrac{1}{2}g\left(\dfrac{v_0\sin\alpha}{g}\right)^2 = \dfrac{v_0^2\sin^2\alpha}{2g}$. This is the maximum height attained when the projectile is fired with an angle of elevation α. This maximum height is largest when $\alpha = \tfrac{\pi}{2}$. In that case, $\sin\alpha = 1$ and the maximum height is $\dfrac{v_0^2}{2g}$.

(b) Let $R = v_0^2/g$. We are asked to consider the parabola $x^2 + 2Ry - R^2 = 0$ which can be rewritten as

$y = -\dfrac{1}{2R}x^2 + \dfrac{R}{2}$. The points on or inside this parabola are those for which $-R \le x \le R$ and

$0 \le y \le \dfrac{-1}{2R}x^2 + \dfrac{R}{2}$. When the projectile is fired at angle of elevation α, the points (x, y) along its path satisfy the relations $x = (v_0\cos\alpha)t$ and $y = (v_0\sin\alpha)t - \tfrac{1}{2}gt^2$, where $0 \le t \le (2v_0\sin\alpha)/g$ (as in Example 5 in Section 10.4). Thus $|x| \le \left|v_0\cos\alpha\left(\dfrac{2v_0\sin\alpha}{g}\right)\right| = \left|\dfrac{v_0^2}{g}\sin2\alpha\right| \le \left|\dfrac{v_0^2}{g}\right| = |R|$.

This shows that $-R \le x \le R$.

For t in the specified range, we also have $y = t\left(v_0 \sin \alpha - \frac{1}{2}gt\right) = \frac{1}{2}gt\left(\dfrac{2v_0 \sin \alpha}{g} - t\right) \geq 0$ and

$$y = (v_0 \sin \alpha)\frac{x}{v_0 \cos \alpha} - \frac{g}{2}\left(\frac{x}{v_0 \cos \alpha}\right)^2 = (\tan \alpha)\,x - \frac{g}{2v_0^2 \cos^2 \alpha}x^2 = -\frac{1}{2R\cos^2 \alpha}x^2 + (\tan \alpha)\,x.$$

Thus

$$\begin{aligned}
y - \left(\frac{-1}{2R}x^2 + \frac{R}{2}\right) &= \frac{-1}{2R\cos^2 \alpha}x^2 + \frac{1}{2R}x^2 + (\tan \alpha)\,x - \frac{R}{2}\\[2mm]
&= \frac{x^2}{2R}\left(1 - \frac{1}{\cos^2 \alpha}\right) + (\tan \alpha)\,x - \frac{R}{2} = \frac{x^2\left(1 - \sec^2 \alpha\right) + 2R\,(\tan \alpha)\,x - R^2}{2R}\\[2mm]
&= \frac{-\left(\tan^2 \alpha\right)x^2 + 2R\,(\tan \alpha)\,x - R^2}{2R} = \frac{-\left[(\tan \alpha)\,x - R\right]^2}{2R} \leq 0
\end{aligned}$$

We have shown that every target that can be hit by the projectile lies on or inside the parabola $y = -\dfrac{1}{2R}x^2 + \dfrac{R}{2}$. Now let (a, b) be any point on or inside the parabola $y = -\dfrac{1}{2R}x^2 + \dfrac{R}{2}$. Then $-R \leq a \leq R$ and $0 \leq b \leq -\dfrac{1}{2R}a^2 + \dfrac{R}{2}$. We seek an angle α such that (a, b) lies in the path of the projectile; that is, we wish to find an angle α such that $b = -\dfrac{1}{2R\cos^2 \alpha}a^2 + (\tan \alpha)\,a$ or equivalently $b = \dfrac{-1}{2R}\left(\tan^2 \alpha + 1\right)a^2 + (\tan \alpha)\,a$. Rearranging this equation we get

$$\frac{a^2}{2R}\tan^2 \alpha - a\tan \alpha + \left(\frac{a^2}{2R} + b\right) = 0 \text{ or } a^2\,(\tan \alpha)^2 - 2aR\,(\tan \alpha) + \left(a^2 + 2bR\right) = 0 \;\;(\bigstar).$$

This quadratic equation for $\tan \alpha$ has real solutions exactly when the discriminant is nonnegative.
Now $B^2 - 4AC \geq 0 \;\Leftrightarrow\; (-2aR)^2 - 4a^2\left(a^2 + 2bR\right) \geq 0 \;\Leftrightarrow\; 4a^2\left(R^2 - a^2 - 2bR\right) \geq 0$

$\Leftrightarrow\; -a^2 - 2bR + R^2 \geq 0 \;\Leftrightarrow\; b \leq \dfrac{1}{2R}\left(R^2 - a^2\right) \;\Leftrightarrow\; b \leq \dfrac{-1}{2R}a^2 + \dfrac{R}{2}$. This condition is

satisfied since (a, b) is on or inside the parabola $y = -\dfrac{1}{2R}x^2 + \dfrac{R}{2}$. It follows that (a, b) lies
in the path of the projectile when $\tan \alpha$ satisfies $(\bigstar)$, that is, when

$$\tan \alpha = \frac{2aR \pm \sqrt{4a^2\left(R^2 - a^2 - 2bR\right)}}{2a^2} = \frac{R \pm \sqrt{R^2 - 2bR - a^2}}{a}.$$

(c)

If the gun is pointed at a target with height h at a distance D downrange, then $\tan \alpha = h/D$. When the projectile reaches a distance D downrange (remember we are assuming that it doesn't hit the ground first), we have $D = x = (v_0 \cos \alpha)\,t$, so $t = \dfrac{D}{v_0 \cos \alpha}$ and

$$y = (v_0 \sin \alpha)\,t - \tfrac{1}{2}gt^2 = D\tan \alpha - \frac{gD^2}{2v_0^2 \cos^2 \alpha}.$$ Meanwhile, the target, whose x-coordinate is also

D, has fallen from height h to height $h - \tfrac{1}{2}gt^2 = D\tan \alpha - \dfrac{gD^2}{2v_0^2 \cos^2 \alpha}$. Thus the projectile hits the
target.

5. (a) $m\dfrac{d^2\mathbf{R}}{dt^2} = -mg\mathbf{j} - k\dfrac{d\mathbf{R}}{dt} \ \Rightarrow \ \dfrac{d}{dt}\left(m\dfrac{d\mathbf{R}}{dt} + k\mathbf{R} + mgt\,\mathbf{j}\right) = 0 \ \Rightarrow \ m\dfrac{d\mathbf{R}}{dt} + k\mathbf{R} + mgt\,\mathbf{j} = \mathbf{c}$

(c is a constant vector in the xy-plane). At $t = 0$, this says that $m\mathbf{v}\,(0) + k\mathbf{R}\,(0) = \mathbf{c}$. Since

$\mathbf{v}\,(0) = \mathbf{v_0}$ and $\mathbf{R}\,(0) = 0$, we have $\mathbf{c} = m\mathbf{v_0}$. Therefore $\dfrac{d\mathbf{R}}{dt} + \dfrac{k}{m}\mathbf{R} + gt\,\mathbf{j} = \mathbf{v_0}$, or

$\dfrac{d\mathbf{R}}{dt} + \dfrac{k}{m}\mathbf{R} = \mathbf{v_0} - gt\,\mathbf{j}.$

(b) Multiplying by $e^{(k/m)t}$ gives $e^{(k/m)t}\dfrac{d\mathbf{R}}{dt} + \dfrac{k}{m}e^{(k/m)t}\mathbf{R} = e^{(k/m)t}\mathbf{v_0} - gte^{(k/m)t}\,\mathbf{j}$ or

$\dfrac{d}{dt}\left(e^{(k/m)t}\mathbf{R}\right) = e^{(k/m)t}\mathbf{v_0} - gte^{(k/m)t}\,\mathbf{j}.$ Integrating gives

$e^{(k/m)t}\mathbf{R} = \dfrac{m}{k}e^{(k/m)t}\mathbf{v_0} - \left[\dfrac{mg}{k}te^{(k/m)t} - \dfrac{m^2 g}{k^2}e^{(k/m)t}\right]\mathbf{j} + \mathbf{b}$ for some constant vector $\mathbf{b}$.

Setting $t = 0$ yields the relation $\mathbf{R}\,(0) = \dfrac{m}{k}\mathbf{v_0} + \dfrac{m^2 g}{k^2}\mathbf{j} + \mathbf{b}$, so $\mathbf{b} = -\dfrac{m}{k}\mathbf{v_0} - \dfrac{m^2 g}{k^2}\mathbf{j}$. Thus

$e^{(k/m)t}\mathbf{R} = \dfrac{m}{k}\left[e^{(k/m)t} - 1\right]\mathbf{v_0} - \left[\dfrac{mg}{k}te^{(k/m)t} - \dfrac{m^2 g}{k^2}\left(e^{(k/m)t} - 1\right)\right]\mathbf{j}$ and

$\mathbf{R}\,(t) = \dfrac{m}{k}\left[1 - e^{-kt/m}\right]\mathbf{v_0} + \dfrac{mg}{k}\left[\dfrac{m}{k}\left(1 - e^{-kt/m}\right) - t\right]\mathbf{j}.$

7. (a) $\mathbf{a} = -g\mathbf{j} \ \Rightarrow \ \mathbf{v} = \mathbf{v_0} - gt\,\mathbf{j} = 2\mathbf{i} - gt\,\mathbf{j} \ \Rightarrow \ \mathbf{s} = \mathbf{s_0} + 2t\,\mathbf{i} - \tfrac{1}{2}gt^2\,\mathbf{j} = 3.5\mathbf{j} + 2t\,\mathbf{i} - \tfrac{1}{2}gt^2\,\mathbf{j} \ \Rightarrow$

$\mathbf{s} = 2t\,\mathbf{i} + \left(3.5 - \tfrac{1}{2}gt^2\right)\mathbf{j}$. Therefore $y = 0$ when $t = \sqrt{7/g}$ seconds. At that instant, the ball is

$2\sqrt{7/g} \approx 0.94$ ft to the right of the table top. Its coordinates (relative to an origin on the floor

directly under the table's edge) are $(0.94, 0)$. At impact, the velocity is $\mathbf{v} = 2\mathbf{i} - \sqrt{7g}\,\mathbf{j}$, so the

speed is $|\mathbf{v}| = \sqrt{4 + 7g} \approx 15$ ft/s.

(b) The slope of the curve when $t = \sqrt{\dfrac{7}{g}}$ is $\dfrac{dy}{dx} = \dfrac{dy/dt}{dx/dt} = \dfrac{-gt}{2} = \dfrac{-g\sqrt{7/g}}{2} = \dfrac{-\sqrt{7g}}{2}$. Thus

$\cot\theta = \sqrt{7g}/2$ and $\theta \approx 7.6°.$

(c) From (a), $|\mathbf{v}| = \sqrt{4 + 7g}$. So the ball rebounds with speed $0.8\sqrt{4 + 7g} \approx 12.08$ ft/s at angle of

inclination $90° - \theta \approx 82.3886°$. By Example 10.4.5, the horizontal distance traveled between

bounces is $d = \dfrac{v_0^2 \sin 2\alpha}{g}$, where $v_0 \approx 12.08$ ft/s and $\alpha \approx 82.3886°$. Therefore, $d \approx 1.197$ ft. So

the ball strikes the floor at about $2\sqrt{7/g} + 1.197 \approx 2.13$ ft to the right of the table's edge.

Chapter 11   Partial Derivatives

Section 11.1   Functions of Several Variables

1. (a) From Table 1, $f(8, 60) = -7$, which means that if the temperature is $8\,^\circ$C and the wind speed is 60 km/h, then the air would feel equivalent to approximately $-7\,^\circ$C without wind.

(b) The question is asking: when the temperature is $-12\,^\circ$C, what wind speed gives a wind-chill index of $-26\,^\circ$C? From Table 1, the speed is 20 km/h.

(c) The question is asking: when the wind speed is 80 km/h, what temperature gives a wind-chill index of $-14\,^\circ$C? From Table 1, the temperature is $4\,^\circ$C.

(d) The function $I = f(-4, v)$ means that we fix T at -4 and allow v to vary, resulting in a function of one variable. In other words, the function gives wind-chill index values for different wind speeds when the temperature is $-4\,^\circ$C. From Table 1 (look at the row corresponding to $T = -4$), the function decreases and appears to approach a constant value as v increases.

(e) The function $I = f(T, 50)$ means that we fix v at 50 and allow T to vary, again giving a function of one variable. In other words, the function gives wind-chill index values for different temperatures when the wind speed is 50 km/h. From Table 1 (look at the column corresponding to $v = 50$), the function increases almost linearly as T increases.

3. If the amounts of labor and capital are both doubled, we replace L, K in the function with $2L, 2K$, giving

$$P(2L, 2K) = 1.01\,(2L)^{0.75}\,(2K)^{0.25} = 1.01\,\left(2^{0.75}\right)\left(2^{0.25}\right) L^{0.75}K^{0.25} = \left(2^{1}\right)1.01 L^{0.75}K^{0.25}$$
$$= 2P(L, K)$$

Thus, the production is doubled. It is also true for the general case $P(L, K) = bL^\alpha K^{1-\alpha}$:
$$P(2L, 2K) = b\,(2L)^\alpha\,(2K)^{1-\alpha} = b\left(2^\alpha\right)\left(2^{1-\alpha}\right)L^\alpha K^{1-\alpha} = \left(2^{\alpha+1-\alpha}\right)bL^\alpha K^{1-\alpha} = 2P(L, K).$$

5. $f(x, y) = \sqrt{x - y}$. The domain is $\{(x, y) \mid x - y \geq 0\} = \{(x, y) \mid x \geq y\}$ and the range is $\{z \mid z \geq 0\}$.

7. (a) $f(x, y, z) = x^2 \ln(x - y + z)$, so $f(3, 6, 4) = 3^2 \ln(3 - 6 + 4) = 9 \ln 1 = 0$.

(b) For the logarithmic function to be defined, we need $x - y + z > 0$. Thus the domain of f is $\{(x, y, z) \mid x + z > y\}$.

(c) The range of f is $\mathbb{R}$.

9. The point $(-3, 3)$ lies between the level curves with z-values 50 and 60. Since the point is a little closer to the level curve with $z = 60$, we estimate that $f(-3, 3) \approx 56$. The point $(3, -2)$ appears to be just about halfway between the level curves with z-values 30 and 40, so we estimate $f(3, -2) \approx 35$.

11. Near A, the level curves are very close together, indicating that the terrain is quite steep. At B, the level curves are much farther apart, so we would expect the terrain to be much less steep than near A, perhaps almost flat.

13. The level curves are $xy = k$. For $k = 0$ the curves are the coordinate axis; if $k > 0$, they are hyperbolas in the first and third quadrants; if $k < 0$, they are hyperbolas in the second and fourth quadrants.

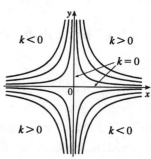

15. $k = x/y$ or $x = ky$ is a family of lines without the point $(0,0)$.

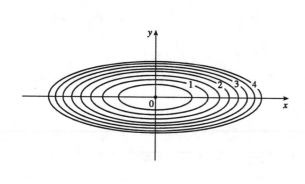

17. $k = \sqrt{x+y}$ or for $x + y \geq 0$, $k^2 = x + y$, or $y = -x + k^2$.
Note: $k \geq 0$ since $k = \sqrt{x+y}$.

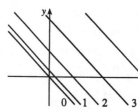

19. $k = x - y^2$, or $x - k = y^2$, a family of parabolas with vertex $(k, 0)$.

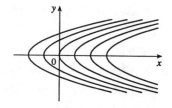

21. The contour map consists of the level curves $k = x^2 + 9y^2$, a family of ellipses with major axis the x-axis. (Or, if $k = 0$, the origin.)

The graph of $f(x, y)$ is the surface $z = x^2 + 9y^2$, an elliptic paraboloid.

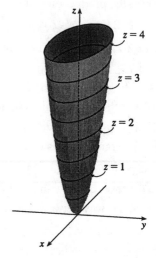

If we visualize lifting each ellipse $k = x^2 + 9y^2$ of the contour map to the plane $z = k$, we have horizontal traces that indicate the shape of the graph of f.

23. The isothermals are given by $k = 100/\left(1 + x^2 + 2y^2\right)$ or
$x^2 + 2y^2 = (100 - k)/k \; (0 < k \leq 100)$, a family of ellipses.

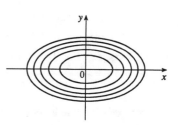

25. $f(x, y) = x^3 + y^3$

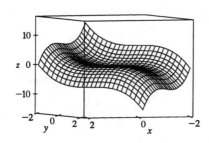
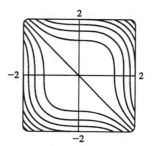

Note that the function is 0 along the line $y = -x$.

27. $f(x, y) = xy^2 - x^3$

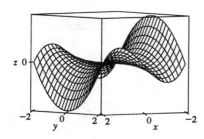
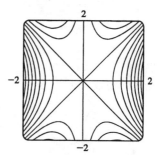

The cross-sections parallel to the yz-plane (such as the left-front trace in the graph above) are parabolas; those parallel to the xz-plane (such as the right-front trace) are cubic curves. The surface is called a monkey saddle because a monkey sitting on the surface near the origin has places for both legs and tail to rest.

29. (a) B   *Reasons:* This function is constant on any circle centered at the origin, a description which
     (b) III   matches only B and III.

31. (a) F   *Reasons:* $f(x, y) \to \infty$ as $(x, y) \to (0, 0)$, a condition satisfied only by F and V.
     (b) V

33. (a) D   *Reasons:* This function is periodic in both x and y, with period 2π in each variable.
     (b) IV

35. $k = x + 3y + 5z$ is a family of parallel planes with normal vector $\langle 1, 3, 5 \rangle$.

37. $k = x^2 - y^2 + z^2$ are the equations of the level surfaces. For $k = 0$, the surface is a right circular cone with vertex the origin and axis the y-axis. For $k > 0$, we have a family of hyperboloids of one sheet with axis the y-axis. For $k < 0$, we have a family of hyperboloids of two sheets with axis the y-axis.

39. $f(x, y) = e^{cx^2 + y^2}$. First, if $c = 0$, the graph is the cylindrical

surface $z = e^{y^2}$ (whose level curves are parallel lines).

When $c > 0$, the vertical trace above the y-axis remains fixed

while the sides of the surface in the x-direction "curl" upward,

giving the graph a shape resembling an elliptic paraboloid. The

level curves of the surface are ellipses centered at the origin.

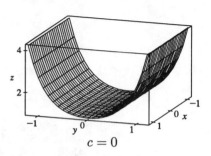

$c = 0$

For $0 < c < 1$, the ellipses have major axis the x-axis and the eccentricity increases as $c \to 0$.

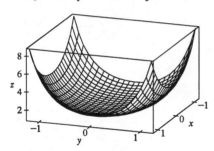
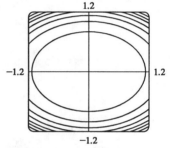

$c = 0.5$ (level curves in increments of 1)

For $c = 1$ the level curves are circles centered at the origin.

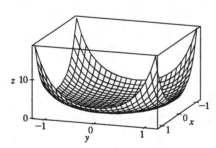
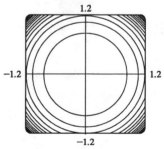

$c = 1$ (level curves in increments of 1)

When $c > 1$, the level curves are ellipses with major axis the y-axis, and the eccentricity increases as c increases.

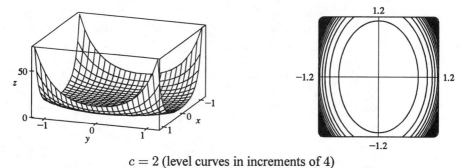

$c = 2$ (level curves in increments of 4)

For values of $c < 0$, the sides of the surface in the x-direction curl downward and approach the xy-plane (while the vertical trace $x = 0$ remains fixed), giving a saddle-shaped appearance to the graph near the point $(0, 0, 1)$. The level curves consist of a family of hyperbolas. As c decreases, the surface becomes flatter in the x-direction and the surface's approach to the curve in the trace $x = 0$ becomes steeper, as the graphs demonstrate.

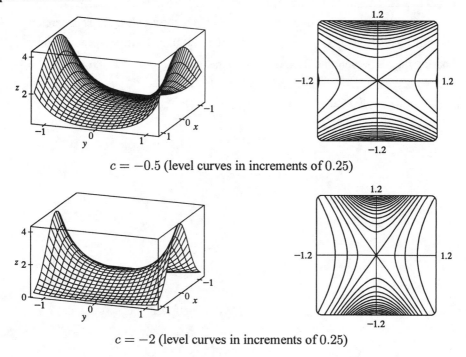

$c = -0.5$ (level curves in increments of 0.25)

$c = -2$ (level curves in increments of 0.25)

41. (a) $P = bL^\alpha K^{1-\alpha} \Rightarrow \dfrac{P}{K} = bL^\alpha K^{-\alpha} \Rightarrow \dfrac{P}{K} = b\left(\dfrac{L}{K}\right)^\alpha \Rightarrow \ln\dfrac{P}{K} = \ln\left(b\left(\dfrac{L}{K}\right)^\alpha\right) \Rightarrow$

$\ln\dfrac{P}{K} = \ln b + \alpha \ln\left(\dfrac{L}{K}\right)$

(b) We list the values for $\ln(L/K)$ and $\ln(P/K)$ for the years 1899-1922. (Historically, these values were rounded to 2 decimal places.)

Year	$x = \ln(L/K)$	$y = \ln(P/K)$	Year	$x = \ln(L/K)$	$y = \ln(P/K)$
1899	0	0	1911	−0.38	−0.34
1900	−0.02	−0.06	1912	−0.38	−0.24
1901	−0.04	−0.02	1913	−0.41	−0.25
1902	−0.04	0	1914	−0.47	−0.37
1903	−0.07	−0.05	1915	−0.53	−0.34
1904	−0.13	−0.12	1916	−0.49	−0.28
1905	−0.18	−0.04	1917	−0.53	−0.39
1906	−0.20	−0.07	1918	−0.60	−0.50
1907	−0.23	−0.15	1919	−0.68	−0.57
1908	−0.41	−0.38	1920	−0.74	−0.57
1909	−0.33	−0.24	1921	−1.05	−0.85
1910	−0.35	−0.27	1922	−0.98	−0.59

After entering the (x, y) pairs into a calculator or CAS, the resulting least squares regression line through the points is approximately $y = 0.75136x + 0.01053$, which we round to $y = 0.75x + 0.01$.

(c) Comparing the regression line from part (b) to the equation $y = \ln b + \alpha x$ with $x = \ln(L/K)$ and $y = \ln(P/K)$, we have $\alpha = 0.75$ and $\ln b = 0.01 \Rightarrow b = e^{0.01} \approx 1.01$. Thus, the Cobb-Douglas production function is $P = bL^\alpha K^{1-\alpha} = 1.01L^{0.75}K^{0.25}$.

Section 11.2   Limits and Continuity

1. In general, we can't say anything about $f(3,1)$! Definition 1 says that the values of $f(x,y)$ approach 6 as (x,y) approaches, but is not equal to, $(3,1)$. If f is continuous, we know that

$\lim\limits_{(x,y)\to(a,b)} f(x,y) = f(a,b)$, so $\lim\limits_{(x,y)\to(3,1)} f(x,y) = f(3,1) = 6$.

3. We make a table of values of $f(x,y) = \dfrac{x^2y^3 + x^3y^2 - 5}{2 - xy}$ for a set of (x,y) points near the origin.

x \ y	−0.2	−0.1	−0.05	0	0.05	0.1	0.2
−0.2	−2.551	−2.525	−2.513	−2.500	−2.488	−2.475	−2.451
−0.1	−2.525	−2.513	−2.506	−2.500	−2.494	−2.488	−2.475
−0.05	−2.513	−2.506	−2.503	−2.500	−2.497	−2.494	−2.488
0	−2.500	−2.500	−2.500		−2.500	−2.500	−2.500
0.05	−2.488	−2.494	−2.497	−2.500	−2.503	−2.506	−2.513
0.1	−2.475	−2.488	−2.494	−2.500	−2.506	−2.513	−2.525
0.2	−2.451	−2.475	−2.488	−2.500	−2.513	−2.525	−2.551

As the table shows, the values of $f(x,y)$ seem to approach -2.5 as (x,y) approaches the origin from a variety of different directions. This suggests that $\lim\limits_{(x,y)\to(0,0)} f(x,y) = -2.5$.

Since f is a rational function, it is continuous on its domain. f is defined at $(0,0)$, so we can use direct substitution to establish that $\lim\limits_{(x,y)\to(0,0)} f(x,y) = \dfrac{0^2 0^3 + 0^3 0^2 - 5}{2 - 0 \cdot 0} = -\dfrac{5}{2}$, verifying our guess.

5. $f(x,y) = x^2y^2 - 2xy^5 + 3y$ is a polynomial, and hence continuous, so

$\lim\limits_{(x,y)\to(2,3)} f(x,y) = f(2,3) = (2^2)(3^2) - 2(2)(3^5) + 3(3) = -927$.

7. $f(x,y) = x^2/(x^2 + y^2)$. First approach $(0,0)$ along the x-axis. Then $f(x,0) = x^2/x^2 = 1$ for $x \neq 0$, so $f(x,y) \to 1$. Now approach $(0,0)$ along the y-axis. Then for $y \neq 0$, $f(0,y) = 0$, so $f(x,y) \to 0$. Since f has two different limits along two different lines, the limit does not exist.

9. $f(x,y) = 8x^2y^2/(x^4 + y^4)$. Approaching $(0,0)$ along the x-axis gives $f(x,y) \to 0$. Approaching $(0,0)$ along the line $y = x$, $f(x,x) = 8x^4/2x^4 = 4$ for $x \neq 0$, so along this line $f(x,y) \to 4$ as $(x,y) \to (0,0)$. Thus the limit doesn't exist.

11. $f(x,y) = \dfrac{xy}{\sqrt{x^2+y^2}}$. We can see that the limit along any line through $(0,0)$ is 0, as well as along other

paths through $(0,0)$ such as $x = y^2$ and $y = x^2$. So we suspect that the limit exists and equals 0; we use

the Squeeze Theorem to prove our assertion. $0 \le \left| \dfrac{xy}{\sqrt{x^2+y^2}} \right| \le |x|$ since $|y| \le \sqrt{x^2+y^2}$, and

$|x| \to 0$ as $(x,y) \to (0,0)$. So $\displaystyle\lim_{(x,y)\to(0,0)} f(x,y) = 0$.

13. Let $f(x,y) = \dfrac{2x^2 y}{x^4 + y^2}$. Then $f(x,0) = 0$ for $x \ne 0$, so $f(x,y) \to 0$ as $(x,y) \to (0,0)$ along the

x-axis. But $f(x,x^2) = \dfrac{2x^4}{2x^4} = 1$ for $x \ne 0$, so $f(x,y) \to 1$ as $(x,y) \to (0,0)$ along the parabola

$y = x^2$. Thus the limit doesn't exist.

15. $\displaystyle\lim_{(x,y)\to(0,0)} \dfrac{x^2+y^2}{\sqrt{x^2+y^2+1}-1} = \lim_{(x,y)\to(0,0)} \dfrac{\left(x^2+y^2\right)\left(\sqrt{x^2+y^2+1}+1\right)}{x^2+y^2}$

$= \displaystyle\lim_{(x,y)\to(0,0)} \left(\sqrt{x^2+y^2+1}+1\right) = 2$

17. $f(x,y,z) = \dfrac{xy + yz^2 + xz^2}{x^2 + y^2 + z^4}$. Then $f(x,0,0) = 0/x^2 = 0$ for $x \ne 0$, so as $(x,y,z) \to (0,0,0)$ along

the x-axis, $f(x,y,z) \to 0$. But $f(x,x,0) = x^2/\left(2x^2\right) = \frac{1}{2}$ for $x \ne 0$, so as $(x,y,z) \to (0,0,0)$ along

the line $y = x$, $z = 0$, $f(x,y,z) \to \frac{1}{2}$. Thus the limit does not exist.

19.

From the ridges on the graph, we see that as $(x,y) \to (0,0)$ along the lines under the two ridges, $f(x,y)$ approaches different values. So the limit does not exist.

21. $h(x,y) = g(f(x,y)) = (2x + 3y - 6)^2 + \sqrt{2x + 3y - 6}$. Since f is a polynomial, it is continuous on
$\mathbb{R}^2$ and g is continuous on its domain $\{t \mid t \ge 0\}$. Thus h is continuous on its domain
$D = \{(x,y) \mid 2x + 3y - 6 \ge 0\} = \{(x,y) \mid y \ge -\frac{2}{3}x + 2\}$, which consists of all points on and above
the line $y = -\frac{2}{3}x + 2$.

23.

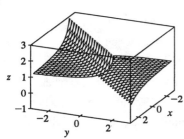

From the graph, it appears that f is discontinuous along the line $y = x$. If we consider $f(x, y) = e^{1/(x-y)}$ as a composition of functions, $g(x, y) = 1/(x - y)$ is a rational function and therefore continuous except where $x - y = 0$ $\Rightarrow$ $y = x$. Since the function $h(t) = e^t$ is continuous everywhere, the composition

$h(g(x, y)) = e^{1/(x-y)} = f(x, y)$ is continuous except along the line $y = x$, as we suspected.

25. $F(x, y) = \dfrac{1}{x^2 - y}$ is a rational function and thus is continuous on its domain

$\{(x, y) \mid x^2 - y \neq 0\} = \{(x, y) \mid y \neq x^2\}$, so F is continuous on $\mathbb{R}^2$ except the parabola $y = x^2$.

27. $G(x, y) = e^{xy} \sin(x + y) = g(x, y) f(x, y)$ where $g(x, y) = e^{xy}$ and $f(x, y) = \sin(x + y)$, both of which are continuous on $\mathbb{R}^2$. Thus, G is continuous on $\mathbb{R}^2$.

29. $f(x, y, z) = \dfrac{xyz}{x^2 + y^2 - z}$ is a rational function and thus is continuous on its domain

$\{(x, y, z) \mid x^2 + y^2 - z \neq 0\} = \{(x, y, z) \mid z \neq x^2 + y^2\}$, so f is continuous on $\mathbb{R}^3$ except on the circular paraboloid $z = x^2 + y^2$.

31. $f(x, y) = \begin{cases} \dfrac{x^2 y^3}{2x^2 + y^2} & \text{if } (x, y) \neq (0, 0) \\ 1 & \text{if } (x, y) = (0, 0) \end{cases}$ The first piece of f is a rational function defined

everywhere except at the origin, so f is continuous on $\mathbb{R}^2$ except possibly at the origin. Since $x^2 \leq 2x^2 + y^2$, we have $\left| x^2 y^3 / (2x^2 + y^2) \right| \leq \left| y^3 \right|$. We know that $\left| y^3 \right| \to 0$ as $(x, y) \to (0, 0)$. So, by the Squeeze Theorem, $\displaystyle\lim_{(x,y)\to(0,0)} f(x, y) = \lim_{(x,y)\to(0,0)} \dfrac{x^2 y^3}{2x^2 + y^2} = 0$. But $f(0, 0) = 1$, so f is discontinuous at $(0, 0)$. Therefore, f is continuous on the set $\{(x, y) \mid (x, y) \neq (0, 0)\}$.

33. $\displaystyle\lim_{(x,y)\to(0,0)} \dfrac{x^3 + y^3}{x^2 + y^2} = \lim_{r\to 0^+} \dfrac{(r \cos\theta)^3 + (r \sin\theta)^3}{r^2} = \lim_{r\to 0^+} (r \cos^3\theta + r \sin^3\theta) = 0$

35. $\displaystyle\lim_{(x,y,z)\to(0,0,0)} \dfrac{xyz}{x^2 + y^2 + z^2} = \lim_{\rho\to 0^+} \dfrac{(\rho \sin\phi \cos\theta)(\rho \sin\phi \sin\theta)(\rho \cos\phi)}{\rho^2}$

$= \displaystyle\lim_{\rho\to 0^+} (\rho \sin^2\phi \cos\phi \sin\theta \cos\theta) = 0$

Section 11.3 Partial Derivatives

1. (a) $\partial T/\partial x$ represents the rate of change of T when we fix y and t and consider T as a function of the single variable x, which describes how quickly the temperature changes when longitude changes but latitude and time are constant. $\partial T/\partial y$ represents the rate of change of T when we fix x and t and consider T as a function of y, which describes how quickly the temperature changes when latitude changes but longitude and time are constant. $\partial T/\partial t$ represents the rate of change of T when we fix x and y and consider T as a function of t, which describes how quickly the temperature changes over time for a constant longitude and latitude.

(b) $f_x(158, 21, 9)$ represents the rate of change of temperature at longitude $158°$ W, latitude $21°$ N at 9:00 A.M. when only longitude varies. Since the air is warmer to the west than to the east, increasing longitude results in an increased air temperature, so we would expect $f_x(158, 21, 9)$ to be positive. $f_y(158, 21, 9)$ represents the rate of change of temperature at the same time and location when only latitude varies. Since the air is warmer to the south and cooler to the north, increasing latitude results in a decreased air temperature, so we would expect $f_y(158, 21, 9)$ to be negative. $f_t(158, 21, 9)$ represents the rate of change of temperature at the same time and location when only time varies. Since typically air temperature increases from the morning to the afternoon as the sun warms it, we would expect $f_t(158, 21, 9)$ to be positive.

3. (a) By Definition 4, $f_T(12, 20) = \lim\limits_{h \to 0} \dfrac{f(12 + h, 20) - f(12, 20)}{h}$, which we can approximate by considering $h = 4$ and $h = -4$ and using the values given in the table: $f_T(12, 20) \approx \dfrac{f(16, 20) - f(12, 20)}{4} = \dfrac{11 - 5}{4} = 1.5$,

$f_T(12, 20) \approx \dfrac{f(8, 20) - f(12, 20)}{-4} = \dfrac{0 - 5}{-4} = 1.25$. Averaging these values, we estimate $f_T(12, 20)$ to be approximately 1.375. Thus, when the actual temperature is $12\,°$C and the wind speed is 20 km/h, the apparent temperature rises by about $1.375\,°$C for every degree that the actual temperature rises.

Similarly, $f_v(12, 20) = \lim\limits_{h \to 0} \dfrac{f(12, 20 + h) - f(12, 20)}{h}$ which we can approximate by considering

$h = 10$ and $h = -10$: $f_v(12, 20) \approx \dfrac{f(12, 30) - f(12, 20)}{10} = \dfrac{3 - 5}{10} = -0.2$,

$f_v(12, 20) \approx \dfrac{f(12, 10) - f(12, 20)}{-10} = \dfrac{9 - 5}{-10} = -0.4$. Averaging these values, we estimate $f_v(12, 20)$ to be approximately -0.3. Thus, when the actual temperature is $12\,°$C and the wind speed is 20 km/h, the apparent temperature decreases by about $0.3\,°$C for every km/h that the wind speed increases.

(b) For a fixed wind speed v, the values of the wind-chill index I increase as temperature T increases (look at a column of the table), so $\dfrac{\partial I}{\partial T}$ is positive. For a fixed temperature T, the values of I decrease (or remain constant) as v increases (look at a row of the table), so $\dfrac{\partial I}{\partial v}$ is negative (or perhaps 0).

(c) For fixed values of T, the function values $f(T, v)$ appear to become constant (or nearly constant) as v increases, so the corresponding rate of change is 0 or near 0 as v increases. This suggests that

$$\lim_{v \to \infty} \left(\frac{\partial I}{\partial v} \right) = 0.$$

5. First of all, if we start at the point $(3, -3)$ and move in the positive y-direction, we see that both b and c decrease, while a increases. Both b and c have a low point at about $(3, -1.5)$, while a is 0 at this point. So a is definitely the graph of f_y, and one of b and c is the graph of f. To see which is which, we start at the point $(-3, -1.5)$ and move in the positive x-direction. b traces out a line with negative slope, while c traces out a parabola opening downward. This tells us that b is the x-derivative of c. So c is the graph of f, b is the graph of f_x, and a is the graph of f_y.

7. $f(x, y) = 16 - 4x^2 - y^2 \;\Rightarrow\; f_x(x, y) = -8x$ and $f_y(x, y) = -2y \;\Rightarrow\; f_x(1, 2) = -8$ and $f_y(1, 2) = -4$. The graph of f is the paraboloid $z = 16 - 4x^2 - y^2$ and the vertical plane $y = 2$ intersects it in the parabola $z = 12 - 4x^2$, $y = 2$ (the curve C_1 in the first figure).
The slope of the tangent line to this parabola at $(1, 2, 8)$ is $f_x(1, 2) = -8$. Similarly the plane $x = 1$ intersects the paraboloid in the parabola $z = 12 - y^2$, $x = 1$ (the curve C_2 in the second figure) and the slope of the tangent line at $(1, 2, 8)$ is $f_y(1, 2) = -4$.

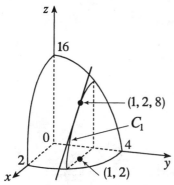 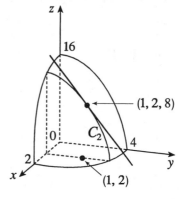

9. $f(x, y) = x^2 + y^2 + x^2 y \Rightarrow f_x = 2x + 2xy, f_y = 2y + x^2$

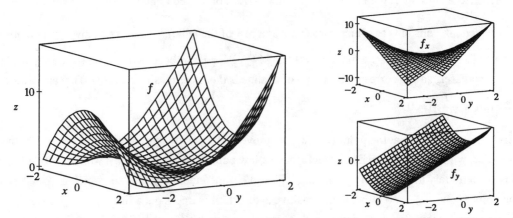

Note that the traces of f in planes parallel to the xz-plane are parabolas which open downward for $y < -1$ and upward for $y > -1$, and the traces of f_x in these planes are straight lines, which have negative slopes for $y < -1$ and positive slopes for $y > -1$. The traces of f in planes parallel to the yz-plane are parabolas which always open upward, and the traces of f_y in these planes are straight lines with positive slopes.

11. $f(x, y) = \dfrac{x - y}{x + y} \Rightarrow f_x(x, y) = \dfrac{(1)(x + y) - (x - y)(1)}{(x + y)^2} = \dfrac{2y}{(x + y)^2}$,

$f_y(x, y) = \dfrac{(-1)(x + y) - (x - y)(1)}{(x + y)^2} = -\dfrac{2x}{(x + y)^2}$

13. $f(u, v) = \tan^{-1}\left(\dfrac{u}{v}\right) \Rightarrow f_u(u, v) = \dfrac{1}{1 + (u/v)^2}\left(\dfrac{1}{v}\right) = \dfrac{1}{v}\left(\dfrac{v^2}{u^2 + v^2}\right) = \dfrac{v}{u^2 + v^2}$,

$f_v(u, v) = \dfrac{1}{1 + (u/v)^2}\left(-\dfrac{u}{v^2}\right) = -\dfrac{u}{v^2}\left(\dfrac{v^2}{u^2 + v^2}\right) = -\dfrac{u}{u^2 + v^2}$

15. $z = \ln\left(x + \sqrt{x^2 + y^2}\right) \Rightarrow$

$\dfrac{\partial z}{\partial x} = \dfrac{1}{x + \sqrt{x^2 + y^2}}\left[1 + \tfrac{1}{2}\left(x^2 + y^2\right)^{-1/2}(2x)\right] = \dfrac{\left(\sqrt{x^2 + y^2} + x\right)\Big/\sqrt{x^2 + y^2}}{\left(x + \sqrt{x^2 + y^2}\right)}$

$= \dfrac{1}{\sqrt{x^2 + y^2}}$

$\dfrac{\partial z}{\partial y} = \dfrac{1}{x + \sqrt{x^2 + y^2}}\left(\dfrac{1}{2}\right)\left(x^2 + y^2\right)^{-1/2}(2y) = \dfrac{y}{x\sqrt{x^2 + y^2} + x^2 + y^2}$

17. $f(x, y, z) = x^2 y z^3 + xy - z \Rightarrow f_x(x, y, z) = 2xyz^3 + y, f_y(x, y, z) = x^2 z^3 + x,$

$f_z(x, y, z) = 3x^2 y z^2 - 1$

19. $u = z \sin \left(\dfrac{y}{x+z} \right) \Rightarrow u_x = z \cos \left(\dfrac{y}{x+z} \right) \left[-y(x+z)^{-2} \right] = \dfrac{-yz}{(x+z)^2} \cos \left(\dfrac{y}{x+z} \right),$

$u_y = z \cos \left(\dfrac{y}{x+z} \right) \left(\dfrac{1}{x+z} \right) = \dfrac{z}{x+z} \cos \left(\dfrac{y}{x+z} \right),$

$u_z = \sin \left(\dfrac{y}{x+z} \right) + z \cos \left(\dfrac{y}{x+z} \right) \left[-y(x+z)^{-2} \right] = \sin \left(\dfrac{y}{x+z} \right) - \dfrac{yz}{(x+z)^2} \cos \left(\dfrac{y}{x+z} \right)$

21. $f(x,y,z,t) = \dfrac{x-y}{z-t} \Rightarrow f_x(x,y,z,t) = \dfrac{1}{z-t}, \ f_y(x,y,z,t) = -\dfrac{1}{z-t},$

$f_z(x,y,z,t) = (x-y)(-1)(z-t)^{-2} = \dfrac{y-x}{(z-t)^2}, \text{ and}$

$f_t(x,y,z,t) = (x-y)(-1)(z-t)^{-2}(-1) = \dfrac{x-y}{(z-t)^2}.$

23. $u = \sqrt{x_1^2 + x_2^2 + \cdots + x_n^2}.$ For each $i = 1, \ldots, n,$

$u_{x_i} = \frac{1}{2} \left(x_1^2 + x_2^2 + \cdots + x_n^2 \right)^{-1/2} (2x_i) = \dfrac{x_i}{\sqrt{x_1^2 + x_2^2 + \cdots + x_n^2}}.$

25. $f(x,y) = xe^{-y} + 3y \Rightarrow \dfrac{\partial f}{\partial y} = x(-1)e^{-y} + 3, \ \dfrac{\partial f}{\partial y}(1,0) = -1 + 3 = 2$

27. $f(x,y) = \sin(y-x) \Rightarrow \dfrac{\partial f}{\partial x} = -\cos(y-x), \ \dfrac{\partial f}{\partial x}(3,3) = -\cos(0) = -1$

29. $xy + yz = xz \Rightarrow \dfrac{\partial}{\partial x}(xy + yz) = \dfrac{\partial}{\partial x}(xz) \Leftrightarrow y + y\dfrac{\partial z}{\partial x} = z + x\dfrac{\partial z}{\partial x} \Leftrightarrow$

$(y-x)\dfrac{\partial z}{\partial x} = z - y, \text{ so } \dfrac{\partial z}{\partial x} = \dfrac{z-y}{y-x}. \ \dfrac{\partial}{\partial y}(xy + yz) = \dfrac{\partial}{\partial y}(xz) \Leftrightarrow x + z + y\dfrac{\partial z}{\partial y} = x\dfrac{\partial z}{\partial y} \Leftrightarrow$

$(y-x)\dfrac{\partial z}{\partial y} = -(x+z), \text{ so } \dfrac{\partial z}{\partial y} = \dfrac{x+z}{x-y}.$

31. $x^2 + y^2 - z^2 = 2x(y+z) \Leftrightarrow \dfrac{\partial}{\partial x}(x^2 + y^2 - z^2) = \dfrac{\partial}{\partial x}[2x(y+z)] \Leftrightarrow$

$2x - 2z\dfrac{\partial z}{\partial x} = 2(y+z) + 2x\dfrac{\partial z}{\partial x} \Leftrightarrow 2(x+z)\dfrac{\partial z}{\partial x} = 2(x-y-z), \text{ so } \dfrac{\partial z}{\partial x} = \dfrac{x-y-z}{x+z}.$

$\dfrac{\partial}{\partial y}(x^2 + y^2 - z^2) = \dfrac{\partial}{\partial y}[2x(y+z)] \Leftrightarrow 2y - 2z\dfrac{\partial z}{\partial y} = 2x\left(1 + \dfrac{\partial z}{\partial y}\right) \Leftrightarrow$

$2(x+z)\dfrac{\partial z}{\partial y} = 2(y-x), \text{ so } \dfrac{\partial z}{\partial y} = \dfrac{y-x}{x+z}.$

33. $f(x,y,z) = xyz \Rightarrow f_y(x,y,z) = xz, \text{ so } f_y(0,1,2) = 0.$

35. $f(x, y) = x^2 - xy + 2y^2 \Rightarrow$

$$f_x(x, y) = \lim_{h \to 0} \frac{f(x+h, y) - f(x, y)}{h} = \lim_{h \to 0} \frac{(x+h)^2 - (x+h)y + 2y^2 - (x^2 - xy + 2y^2)}{h}$$

$$= \lim_{h \to 0} \frac{h(2x - y + h)}{h} = \lim_{h \to 0}(2x - y + h) = 2x - y$$

$$f_y(x, y) = \lim_{h \to 0} \frac{f(x, y+h) - f(x, y)}{h} = \lim_{h \to 0} \frac{x^2 - x(y+h) + 2(y+h)^2 - (x^2 - xy + 2y^2)}{h}$$

$$= \lim_{h \to 0} \frac{h(4y - x + 2h)}{h} = \lim_{h \to 0}(4y - x + 2h) = 4y - x$$

37. (a) $z = f(x) + g(y) \Rightarrow \dfrac{\partial z}{\partial x} = f'(x), \dfrac{\partial z}{\partial y} = g'(y)$

(b) $z = f(x + y)$. Let $u = x + y$. Then $\dfrac{\partial z}{\partial x} = \dfrac{df}{du}\dfrac{\partial u}{\partial x} = \dfrac{df}{d(x+y)} = f'(x+y)$,

$$\frac{\partial z}{\partial y} = \frac{df}{du}\frac{\partial u}{\partial y} = \frac{df}{d(x+y)} = f'(x+y).$$

39. $f(x, y) = x^2 y + x\sqrt{y} \Rightarrow f_x = 2xy + \sqrt{y}, f_y = x^2 + \dfrac{x}{2\sqrt{y}}$. Thus $f_{xx} = 2y, f_{xy} = 2x + \dfrac{1}{2\sqrt{y}}$,

$f_{yx} = 2x + \dfrac{1}{2\sqrt{y}}$ and $f_{yy} = -\dfrac{x}{4y^{3/2}}$.

41. $z = (x^2 + y^2)^{3/2} \Rightarrow z_x = \frac{3}{2}(x^2 + y^2)^{1/2}(2x) = 3x(x^2 + y^2)^{1/2}$ and $z_y = 3y(x^2 + y^2)^{1/2}$.

Thus $z_{xx} = 3(x^2 + y^2)^{1/2} + 3x(x^2 + y^2)^{-1/2}(\frac{1}{2})(2x) = \dfrac{3(x^2 + y^2) + 3x^2}{\sqrt{x^2 + y^2}} = \dfrac{3(2x^2 + y^2)}{\sqrt{x^2 + y^2}}$ and

$z_{xy} = 3x(\frac{1}{2})(x^2 + y^2)^{-1/2}(2y) = \dfrac{3xy}{\sqrt{x^2 + y^2}}$.

By symmetry $z_{yx} = \dfrac{3xy}{\sqrt{x^2 + y^2}}$ and $z_{yy} = \dfrac{3(x^2 + 2y^2)}{\sqrt{x^2 + y^2}}$.

43. $u = x^5 y^4 - 3x^2 y^3 + 2x^2 \Rightarrow u_x = 5x^4 y^4 - 6xy^3 + 4x, u_{xy} = 20x^4 y^3 - 18xy^2$ and
$u_y = 4x^5 y^3 - 9x^2 y^2, u_{yx} = 20x^4 y^3 - 18xy^2$. Thus $u_{xy} = u_{yx}$.

45. $f(x, y) = x^2 y^3 - 2x^4 y \Rightarrow f_x = 2xy^3 - 8x^3 y, f_{xx} = 2y^3 - 24x^2 y, f_{xxx} = -48xy$

47. $f(x, y, z) = x^5 + x^4 y^4 z^3 + yz^2 \Rightarrow f_x = 5x^4 + 4x^3 y^4 z^3, f_{xy} = 16x^3 y^3 z^3$, and $f_{xyz} = 48x^3 y^3 z^2$

49. $u = \ln(x + 2y^2 + 3z^3) \Rightarrow \dfrac{\partial u}{\partial z} = \dfrac{1}{x + 2y^2 + 3z^3}(9z^2) = \dfrac{9z^2}{x + 2y^2 + 3z^3}$,

$\dfrac{\partial^2 u}{\partial y\, \partial z} = -9z^2(x + 2y^2 + 3z^3)^{-2}(4y) = -\dfrac{36yz^2}{(x + 2y^2 + 3z^3)^2}$, and $\dfrac{\partial^3 u}{\partial x\, \partial y\, \partial z} = \dfrac{72yz^2}{(x + 2y^2 + 3z^3)^3}$.

51. By Definition 4, $f_x(3,2) = \lim\limits_{h \to 0} \dfrac{f(3+h,2) - f(3,2)}{h}$ which we can approximate by considering

$h = 0.5$ and $h = -0.5$: $f_x(3,2) \approx \dfrac{f(3.5,2) - f(3,2)}{0.5} = \dfrac{22.4 - 17.5}{0.5} = 9.8$,

$f_x(3,2) \approx \dfrac{f(2.5,2) - f(3,2)}{-0.5} = \dfrac{10.2 - 17.5}{-0.5} = 14.6$. Averaging these values, we estimate $f_x(3,2)$

to be approximately 12.2. Similarly, $f_x(3,2.2) = \lim\limits_{h \to 0} \dfrac{f(3+h,2.2) - f(3,2.2)}{h}$

which we can approximate by considering $h = 0.5$ and

$h = -0.5$: $f_x(3,2.2) \approx \dfrac{f(3.5,2.2) - f(3,2.2)}{0.5} = \dfrac{26.1 - 15.9}{0.5} = 20.4$,

$f_x(3,2.2) \approx \dfrac{f(2.5,2.2) - f(3,2.2)}{-0.5} = \dfrac{9.3 - 15.9}{-0.5} = 13.2$. Averaging these values, we have

$f_x(3,2.2) \approx 16.8$.

To estimate $f_{xy}(3,2)$, we first need an estimate for $f_x(3,1.8)$:

$f_x(3,1.8) \approx \dfrac{f(3.5,1.8) - f(3,1.8)}{0.5} = \dfrac{20.0 - 18.1}{0.5} = 3.8$,

$f_x(3,1.8) \approx \dfrac{f(2.5,1.8) - f(3,1.8)}{-0.5} = \dfrac{12.5 - 18.1}{-0.5} = 11.2$. Averaging these values, we get

$f_x(3,1.8) \approx 7.5$. Now $f_{xy}(x,y) = \dfrac{\partial}{\partial y}[f_x(x,y)]$ and $f_x(x,y)$ is itself a function of 2 variables, so

Definition 4 says that $f_{xy}(x,y) = \dfrac{\partial}{\partial y}[f_x(x,y)] = \lim\limits_{h \to 0} \dfrac{f_x(x,y+h) - f_x(x,y)}{h} \quad \Rightarrow$

$f_{xy}(3,2) = \lim\limits_{h \to 0} \dfrac{f_x(3,2+h) - f_x(3,2)}{h}$. We can estimate this value using our previous work with

$h = 0.2$ and $h = -0.2$: $f_{xy}(3,2) \approx \dfrac{f_x(3,2.2) - f_x(3,2)}{0.2} = \dfrac{16.8 - 12.2}{0.2} = 23$,

$f_{xy}(3,2) \approx \dfrac{f_x(3,1.8) - f_x(3,2)}{-0.2} = \dfrac{7.5 - 12.2}{-0.2} = 23.5$. Averaging these values, we estimate

$f_{xy}(3,2)$ to be approximately 23.25.

53. $u = e^{-\alpha^2 k^2 t} \sin kx \quad \Rightarrow \quad u_x = k e^{-\alpha^2 k^2 t} \cos kx, \ u_{xx} = -k^2 e^{-\alpha^2 k^2 t} \sin kx$, and

$u_t = -\alpha^2 k^2 e^{-\alpha^2 k^2 t} \sin kx$. Thus $\alpha^2 u_{xx} = u_t$.

55. $u = \dfrac{1}{\sqrt{x^2 + y^2 + z^2}} \quad \Rightarrow \quad u_x = \left(-\tfrac{1}{2}\right)\left(x^2 + y^2 + z^2\right)^{-3/2}(2x) = -x\left(x^2 + y^2 + z^2\right)^{-3/2}$ and

$u_{xx} = -\left(x^2 + y^2 + z^2\right)^{-3/2} - x\left(-\tfrac{3}{2}\right)\left(x^2 + y^2 + z^2\right)^{-5/2}(2x) = \dfrac{2x^2 - y^2 - z^2}{\left(x^2 + y^2 + z^2\right)^{5/2}}.$

By symmetry, $u_{yy} = \dfrac{2y^2 - x^2 - z^2}{\left(x^2 + y^2 + z^2\right)^{5/2}}$ and $u_{zz} = \dfrac{2z^2 - x^2 - y^2}{\left(x^2 + y^2 + z^2\right)^{5/2}}$. Thus

$u_{xx} + u_{yy} + u_{zz} = \dfrac{2x^2 - y^2 - z^2 + 2y^2 - x^2 - z^2 + 2z^2 - x^2 - y^2}{\left(x^2 + y^2 + z^2\right)^{5/2}} = 0.$

57. Let $v = x + at$, $w = x - at$. Then

$$u_t = \frac{\partial \left[f(v) + g(w) \right]}{\partial t} = \frac{df(v)}{dv} \frac{\partial v}{\partial t} + \frac{dg(w)}{dw} \frac{\partial w}{\partial t} = af'(v) - ag'(w) \text{ and}$$

$$u_{tt} = \frac{\partial \left[af'(v) - ag'(w) \right]}{\partial t} = a \left[af''(v) + ag''(w) \right] = a^2 \left[f''(v) + g''(w) \right]. \text{ Similarly, by using the}$$

Chain Rule we have $u_x = f'(v) + g'(w)$ and $u_{xx} = f''(v) + g''(w)$. Thus $u_{tt} = a^2 u_{xx}$.

59. If we fix $K = K_0$, $P(L, K_0)$ is a function of a single variable L, and $\dfrac{dP}{dL} = \alpha \dfrac{P}{L}$ is a separable

differential equation. Then $\dfrac{dP}{P} = \alpha \dfrac{dL}{L} \Rightarrow \displaystyle\int \dfrac{dP}{P} = \int \alpha \dfrac{dL}{L} \Rightarrow \ln |P| = \alpha \ln |L| + C(K_0)$,

where $C(K_0)$ can depend on K_0. Then $|P| = e^{\alpha \ln |L| + C(K_0)}$, and since $P > 0$ and $L > 0$, we have

$P = e^{\alpha \ln L} e^{C(K_0)} = e^{C(K_0)} e^{\ln L^\alpha} = C_1(K_0) L^\alpha$ where $C_1(K_0) = e^{C(K_0)}$.

61. By the Chain Rule, taking the partial derivative of both sides with respect to R_1 gives

$$\frac{\partial R^{-1}}{\partial R} \frac{\partial R}{\partial R_1} = \frac{\partial \left[(1/R_1) + (1/R_2) + (1/R_3) \right]}{\partial R_1} \text{ or } -R^{-2} \frac{\partial R}{\partial R_1} = -R_1^{-2}. \text{ Thus } \frac{\partial R}{\partial R_1} = \frac{R^2}{R_1^2}.$$

63. $\dfrac{\partial K}{\partial m} = \frac{1}{2} V^2$, $\dfrac{\partial K}{\partial V} = mV$, $\dfrac{\partial^2 K}{\partial V^2} = m$. Thus $\dfrac{\partial K}{\partial m} \cdot \dfrac{\partial^2 K}{\partial V^2} = \frac{1}{2} V^2 m = K$.

65. $f_x(x, y) = x + 4y \Rightarrow f_{xy}(x, y) = 4$ and $f_y(x, y) = 3x - y \Rightarrow f_{yx}(x, y) = 3$. Since f_{xy} and f_{yx} are continuous everywhere but $f_{xy}(x, y) \neq f_{yx}(x, y)$, Clairaut's Theorem implies that such a function $f(x, y)$ does not exist.

67. By the geometry of partial derivatives, the slope of the tangent line is $f_x(1, 2)$. By implicit differentiation of $4x^2 + 2y^2 + z^2 = 16$, we get $8x + 2z(\partial z/\partial x) = 0 \Rightarrow \partial z/\partial x = -4x/z$, so when $x = 1$ and $z = 2$ we have $\partial z/\partial x = -2$. So the slope is $f_x(1, 2) = -2$. Thus the tangent line is given by $z - 2 = -2(x - 1)$, $y = 2$. Taking the parameter to be $t = x - 1$, we can write parametric equations for this line: $x = 1 + t$, $y = 2$, $z = 2 - 2t$.

69. Let $g(x) = f(x, 0) = x \left(x^2 \right)^{-3/2} e^0 = x |x|^{-3}$. But we are using the point $(1, 0)$, so near $(1, 0)$, $g(x) = x^{-2}$. Then $g'(x) = -2x^{-3}$ and $g'(1) = -2$, so using (1) we have $f_x(1, 0) = g'(1) = -2$.

71. (a)

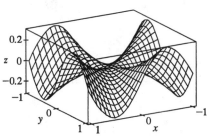

(b) For $(x, y) \neq (0, 0)$, $f_x(x, y) = \dfrac{(3x^2y - y^3)(x^2 + y^2) - (x^3y - xy^3)(2x)}{(x^2 + y^2)^2} = \dfrac{x^4y + 4x^2y^3 - y^5}{(x^2 + y^2)^2}$,

and by symmetry $f_y(x, y) = \dfrac{x^5 - 4x^3y^2 - xy^4}{(x^2 + y^2)^2}$.

(c) $f_x(0, 0) = \lim\limits_{h \to 0} \dfrac{f(h, 0) - f(0, 0)}{h} = \lim\limits_{h \to 0} \dfrac{(0/h^2) - 0}{h} = 0$ and

$f_y(0, 0) = \lim\limits_{h \to 0} \dfrac{f(0, h) - f(0, 0)}{h} = 0.$

(d) By (3), $f_{xy}(0, 0) = \dfrac{\partial f_x}{\partial y} = \lim\limits_{h \to 0} \dfrac{f_x(0, h) - f_x(0, 0)}{h} = \lim\limits_{h \to 0} \dfrac{(-h^5 - 0)/h^4}{h} = -1$ while by (2),

$f_{yx}(0, 0) = \dfrac{\partial f_y}{\partial x} = \lim\limits_{h \to 0} \dfrac{f_y(h, 0) - f_y(0, 0)}{h} = \lim\limits_{h \to 0} \dfrac{h^5/h^4}{h} = 1.$

(e) For $(x, y) \neq (0, 0)$, we use a CAS to compute $f_{xy}(x, y) = \dfrac{x^6 + 9x^4y^2 - 4x^2y^4 + 4y^6}{(x^2 + y^2)^3}$. Now as

$(x, y) \to (0, 0)$ along the x-axis, $f_{xy}(x, y) \to 1$ while as $(x, y) \to (0, 0)$ along the y-axis,

$f_{xy}(x, y) \to 4$. Thus f_{xy} isn't continuous at $(0, 0)$ and Clairaut's Theorem doesn't apply, so there is

no contradiction. The graphs of f_{xy} and f_{yx} are identical except at the origin, where we observe the

discontinuity.

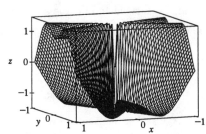

Section 11.4   Tangent Planes and Linear Approximations

1. $z = f(x,y) = y^2 - x^2$ $\Rightarrow$ $f_x(x,y) = -2x$, $f_y(x,y) = 2y$, $f_x(-4,5) = 8$,
$f_y(-4,5) = 10$. Thus an equation of the tangent plane is
$z - 9 = f_x(-4,5)(x-(-4)) + f_y(-4,5)(y-5) \Rightarrow z - 9 = 8(x+4) + 10(y-5)$ or
$z = 8x + 10y - 9$.

3. $z = f(x,y) = \ln(2x+y)$ $\Rightarrow$ $f_x(x,y) = \dfrac{2}{2x+y}$, $f_y(x,y) = \dfrac{1}{2x+y}$, $f_x(-1,3) = 2$,
$f_y(-1,3) = 1$. Thus an equation of the tangent plane is $z = 2(x+1) + (y-3)$ or $z = 2x + y - 1$.

5. $z = f(x,y) = xy$, so $f_x(x,y) = y$ $\Rightarrow$ $f_x(-1,2) = 2$, $f_y(x,y) = x$ $\Rightarrow$ $f_y(-1,2) = -1$ and an
equation of the tangent plane is $z + 2 = 2(x+1) + (-1)(y-2)$ or $2x - y - z = -2$. After zooming
in, the surface and the tangent plane become almost indistinguishable. (Here, the tangent plane is shown
with fewer traces than the surface.) If we zoom in farther, the surface and the tangent plane will appear
to coincide.

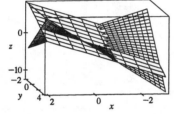
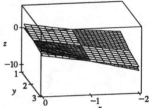

7. $f(x,y) = e^{-(x^2+y^2)/15}\left(\sin^2 x + \cos^2 y\right)$. A CAS gives
$f_x = -\frac{2}{15}e^{-(x^2+y^2)/15}\left(x\sin^2 x + x\cos^2 y - 15\sin x\cos x\right)$ and
$f_y = -\frac{2}{15}e^{-(x^2+y^2)/15}\left(y\sin^2 x + y\cos^2 y + 15\sin y\cos y\right)$.
We use the CAS to evaluate these at $(2,3)$, and then substitute the results into Equation 2 in order to plot
the tangent plane. After zooming in, the surface and the tangent plane become almost indistinguishable.
(Here, the tangent plane is above the surface.) If we zoom in farther, the surface and the tangent plane
will appear to coincide.

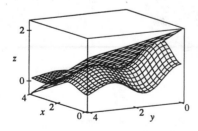
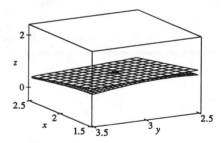

9. $f(x,y) = x\sqrt{y}$. The partial derivatives are $f_x(x,y) = \sqrt{y}$ and $f_y(x,y) = \dfrac{x}{2\sqrt{y}}$, so $f_x(1,4) = 2$ and

$f_y(1,4) = \frac{1}{4}$. Both f_x and f_y are continuous functions for $y > 0$, so by Theorem 8, f is

differentiable at $(1,4)$. By Equation 3, the linearization of f at $(1,4)$ is given by

$$L(x,y) = f(1,4) + f_x(1,4)(x-1) + f_y(1,4)(y-4) = 2 + 2(x-1) + \tfrac{1}{4}(y-4) = 2x + \tfrac{1}{4}y - 1.$$

11. $f(x,y) = e^x \cos xy$. The partial derivatives are $f_x(x,y) = e^x(\cos xy - y\sin xy)$ and

$f_y(x,y) = -xe^x \sin xy$, so $f_x(0,0) = 1$ and $f_y(0,0) = 0$. Both f_x and f_y are continuous functions, so

f is differentiable at $(0,0)$ by Theorem 8. The linearization of f at $(0,0)$ is given by

$$L(x,y) = f(0,0) + f_x(0,0)(x-0) + f_y(0,0)(y-0) = 1 + 1(x-0) + 0(y-0) = x + 1.$$

13. $f(x,y) = \sqrt{20 - x^2 - 7y^2} \quad \Rightarrow \quad f_x(x,y) = -\dfrac{x}{\sqrt{20 - x^2 - 7y^2}}$ and

$f_y(x,y) = -\dfrac{7y}{\sqrt{20 - x^2 - 7y^2}}$, so $f_x(2,1) = -\frac{2}{3}$ and $f_y(2,1) = -\frac{7}{3}$. Then the linear approximation

of f at $(2,1)$ is given by

$$f(x,y) \approx f(2,1) + f_x(2,1)(x-2) + f_y(2,1)(y-1) = 3 - \tfrac{2}{3}(x-2) - \tfrac{7}{3}(y-1)$$

$$= -\tfrac{2}{3}x - \tfrac{7}{3}y + \tfrac{20}{3}$$

Thus $f(1.95, 1.08) \approx -\frac{2}{3}(1.95) - \frac{7}{3}(1.08) + \frac{20}{3} = 2.84\overline{6}$.

15. $f(x,y,z) = \sqrt{x^2 + y^2 + z^2} \quad \Rightarrow \quad f_x(x,y,z) = \dfrac{x}{\sqrt{x^2 + y^2 + z^2}}, \; f_y(x,y,z) = \dfrac{y}{\sqrt{x^2 + y^2 + z^2}},$

and $f_z(x,y,z) = \dfrac{z}{\sqrt{x^2 + y^2 + z^2}}$, so $f_x(3,2,6) = \frac{3}{7}$, $f_y(3,2,6) = \frac{2}{7}$, and $f_z(3,2,6) = \frac{6}{7}$. Then the

linear approximation of f at $(3,2,6)$ is given by

$$f(x,y,z) \approx f(3,2,6) + f_x(3,2,6)(x-3) + f_y(3,2,6)(y-2) + f_z(3,2,6)(z-6)$$

$$= 7 + \tfrac{3}{7}(x-3) + \tfrac{2}{7}(y-2) + \tfrac{6}{7}(z-6) = \tfrac{3}{7}x + \tfrac{2}{7}y + \tfrac{6}{7}z$$

Then $\sqrt{(3.02)^2 + (1.97)^2 + (5.99)^2} = f(3.02, 1.97, 5.99) \approx \frac{3}{7}(3.02) + \frac{2}{7}(1.97) + \frac{6}{7}(5.99) \approx 6.9914$.

17. From the table, $f(94, 80) = 127$. To estimate $f_T(94, 80)$ and $f_H(94, 80)$ we follow the procedure used in Section 11.3. Since $f_T(94, 80) = \lim\limits_{h \to 0} \dfrac{f(94 + h, 80) - f(94, 80)}{h}$, we approximate this quantity with $h = \pm 2$ and use the values given in the table:

$$f_T(94, 80) \approx \frac{f(96, 80) - f(94, 80)}{2} = \frac{135 - 127}{2} = 4,$$

$$f_T(94, 80) \approx \frac{f(92, 80) - f(94, 80)}{-2} = \frac{119 - 127}{-2} = 4.$$

Averaging these values gives $f_T(94, 80) \approx 4$. Similarly,

$$f_H(94, 80) = \lim\limits_{h \to 0} \frac{f(94, 80 + h) - f(94, 80)}{h}, \text{ so we}$$

use $h = \pm 5$: $f_H(94, 80) \approx \dfrac{f(94, 85) - f(94, 80)}{5} = \dfrac{132 - 127}{5} = 1,$

$$f_H(94, 80) \approx \frac{f(94, 75) - f(94, 80)}{-5} = \frac{122 - 127}{-5} = 1. \text{ Averaging these values gives}$$

$f_H(94, 80) \approx 1$. The linear approximation, then, is

$$f(T, H) \approx f(94, 80) + f_T(94, 80)(T - 94) + f_H(94, 80)(H - 80)$$
$$\approx 127 + 4(T - 94) + 1(H - 80)$$

Thus when $T = 95$ and $H = 78$, $f(95, 78) \approx 127 + 4(95 - 94) + 1(78 - 80) = 129$, so we estimate the heat index to be approximately $129\,°F$.

19. $z = x^2 y^3 \quad \Rightarrow \quad dz = \dfrac{\partial z}{\partial x}\, dx + \dfrac{\partial z}{\partial y}\, dy = 2xy^3\, dx + 3x^2 y^2\, dy$

21. $w = \ln \sqrt{x^2 + y^2 + z^2} \quad \Rightarrow$

$$dw = \left(\frac{1}{2}\right) \frac{2x\left(x^2 + y^2 + z^2\right)^{-1/2} dx + 2y\left(x^2 + y^2 + z^2\right)^{-1/2} dy + 2z\left(x^2 + y^2 + z^2\right)^{-1/2} dz}{\left(x^2 + y^2 + z^2\right)^{1/2}}$$

$$= \frac{x\, dx + y\, dy + z\, dz}{x^2 + y^2 + z^2}$$

23. $\Delta x = 0.05$, $\Delta y = 0.1$, $z = 5x^2 + y^2$, $z_x = 10x$, $z_y = 2y$. Thus when
$x = 1$, $y = 2$, $dz = (10)(0.05) + (4)(0.1) = 0.9$, while
$\Delta z = f(1.05, 2.1) - f(1, 2) = 5(1.05)^2 + (2.1)^2 - 5 - 4 = 0.9225$.

25. $dA = \dfrac{\partial A}{\partial x}\, dx + \dfrac{\partial A}{\partial y}\, dy = y\, dx + x\, dy$ and $|\Delta x| \le 0.1$, $|\Delta y| \le 0.1$. Thus the maximum error in the area is about $dA = 24(0.1) + 30(0.1) = 5.4 \text{ cm}^2$.

27. The volume of a can is $V = \pi r^2 h$ and $\Delta V \approx dV$ is an estimate of the amount of tin. Here $dV = 2\pi r h\, dr + \pi r^2\, dh$, so $\Delta V \approx dV = 2\pi(48)(-0.04) + \pi(16)(-0.08) \approx -16.08 \text{ cm}^3$. Thus the amount of tin is about 16 cm^3.

29. The area of the rectangle is $A = xy$, and $\Delta A \approx dA$ is an estimate of the area of paint in the stripe. Here $dA = y\, dx + x\, dy$, so $\Delta A \approx dA = (100)\left(\frac{1}{2}\right) + (200)\left(\frac{1}{2}\right) = 150 \text{ ft}^2$. Thus there are approximately 150 ft^2 of paint in the stripe.

31. By the Chain Rule, taking partial derivatives of both sides with respect to R_1 gives

$$\frac{\partial R^{-1}}{\partial R}\frac{\partial R}{\partial R_1} = \frac{\partial\left[(1/R_1)+(1/R_2)+(1/R_3)\right]}{\partial R_1} \Rightarrow -R^{-2}\frac{\partial R}{\partial R_1} = -R_1^{-2} \Rightarrow \frac{\partial R}{\partial R_1} = \frac{R^2}{R_1^2}, \text{ and by}$$

symmetry, $\dfrac{\partial R}{\partial R_2} = \dfrac{R^2}{R_2^2}$, $\dfrac{\partial R}{\partial R_3} = \dfrac{R^2}{R_3^2}$. When $R_1 = 25$, $R_2 = 40$ and $R_3 = 50$, $\dfrac{1}{R} = \dfrac{17}{200} \Leftrightarrow$

$R = \frac{200}{17}$ ohms. Since the possible error for each R_i is 0.5%, the maximum error of R is attained by
setting $\Delta R_i = 0.005 R_i$. So

$$\Delta R \approx dR = \frac{\partial R}{\partial R_1}\Delta R_1 + \frac{\partial R}{\partial R_2}\Delta R_2 + \frac{\partial R}{\partial R_3}\Delta R_3 = (0.005)\,R^2\left[\frac{1}{R_1} + \frac{1}{R_2} + \frac{1}{R_3}\right]$$

$$= (0.005)\,R = \frac{1}{17} \approx 0.059 \text{ ohms}$$

33. $r\,(u, v) = (u + v)\,i + 3u^2\,j + (u - v)\,k$

$r_u = i + 6u\,j + k$ and $r_v = i - k$, so
$r_u \times r_v = -6u\,i + 2\,j - 6u\,k$. Since the point $(2, 3, 0)$
corresponds to $u = 1$, $v = 1$, a normal vector to the surface
at $(2, 3, 0)$ is $-6\,i + 2\,j - 6\,k$, and an equation of the tangent
plane is $-6x + 2y - 6z = -6$ or $3x - y + 3z = 3$.

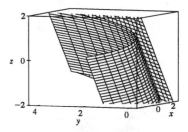

35. $r\,(u, v) = uv\,i + ue^v\,j + ve^u\,k$

$r_u = \langle v, e^v, ve^u \rangle$, $r_v = \langle u, ue^v, e^u \rangle$, and
$r_u \times r_v = e^{u+v}\,(1 - uv)\,i + e^u\,(uv - v)\,j + e^v\,(uv - u)\,k$.
The point $(0, 0, 0)$ corresponds to $u = 0$, $v = 0$. Thus a
normal vector to the surface at $(0, 0, 0)$ is i, and an equation
of the tangent plane is $x = 0$.

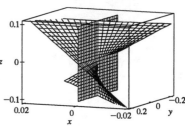

37. $\Delta z = f\,(a + \Delta x, b + \Delta y) - f\,(a, b) = (a + \Delta x)^2 + (b + \Delta y)^2 - (a^2 + b^2)$
$= a^2 + 2a\,\Delta x + (\Delta x)^2 + b^2 + 2b\,\Delta y + (\Delta y)^2 - a^2 - b^2 = 2a\,\Delta x + (\Delta x)^2 + 2b\,\Delta y + (\Delta y)^2$
But $f_x\,(a, b) = 2a$ and $f_y\,(a, b) = 2b$ and so $\Delta z = f_x\,(a, b)\,\Delta x + f_y\,(a, b)\,\Delta y + \Delta x\Delta x + \Delta y\Delta y$,
which is Definition 7 with $\epsilon_1 = \Delta x$ and $\epsilon_2 = \Delta y$. Hence f is differentiable.

39. To show that f is continuous at (a, b) we need to show that $\displaystyle\lim_{(x,y)\to(a,b)} f\,(x, y) = f\,(a, b)$ or equivalently

$\displaystyle\lim_{(\Delta x,\Delta y)\to(0,0)} f\,(a + \Delta x, b + \Delta y) = f\,(a, b)$. Since f is differentiable at (a, b),
$f\,(a + \Delta x, b + \Delta y) - f\,(a, b) = \Delta z = f_x\,(a, b)\,\Delta x + f_y\,(a, b)\,\Delta y + \epsilon_1\,\Delta x + \epsilon_2\,\Delta y$,
where ϵ_1 and $\epsilon_2 \to 0$ as $(\Delta x, \Delta y) \to (0, 0)$. Thus
$f\,(a + \Delta x, b + \Delta y) = f\,(a, b) + f_x\,(a, b)\,\Delta x + f_y\,(a, b)\,\Delta y + \epsilon_1\,\Delta x + \epsilon_2\,\Delta y$. Taking the limit of
both sides as $(\Delta x, \Delta y) \to (0, 0)$ gives $\displaystyle\lim_{(\Delta x,\Delta y)\to(0,0)} f\,(a + \Delta x, b + \Delta y) = f\,(a, b)$. Thus f is

continuous at (a, b).

Section 11.5   The Chain Rule

1. $z = x^2 + y^2$, $x = t^3$, $y = 1 + t^2$ $\Rightarrow$

$$\frac{dz}{dt} = 2x\frac{dx}{dt} + 2y\frac{dy}{dt} = (2t^3)(3t^2) + 2(1 + t^2)(2t) = 6t^5 + 4t^3 + 4t$$

3. $w = xy^2z^3$, $x = \sin t$, $y = \cos t$, $z = 1 + e^{2t}$ $\Rightarrow$

$$\frac{dw}{dt} = y^2z^3(\cos t) + 2xyz^3(-\sin t) + 3xy^2z^2(2e^{2t})$$

5. $z = x^2 \sin y$, $x = s^2 + t^2$, $y = 2st$ $\Rightarrow$

$$\frac{\partial z}{\partial s} = (2x \sin y)(2s) + (x^2 \cos y)(2t) = 4sx \sin y + 2tx^2 \cos y,$$

$$\frac{\partial z}{\partial t} = (2x \sin y)(2t) + (x^2 \cos y)(2s) = 4xt \sin y + 2sx^2 \cos y$$

7. $z = x^2 - 3x^2y^3$, $x = se^t$, $y = se^{-t}$ $\Rightarrow$

$\partial z/\partial s = (2x - 6xy^3)(e^t) + (-9x^2y^2)(e^{-t}) = (2x - 6xy^3)e^t - 9x^2y^2e^{-t}$,

$\partial z/\partial t = (2x - 6xy^3)(se^t) + (-9x^2y^2)(-se^{-t}) = (2x - 6xy^3)se^t + 9x^2y^2se^{-t}$

9. When $t = 3$, $x = g(3) = 2$ and $y = h(3) = 7$. By the Chain Rule (2),

$$\frac{dz}{dt} = \frac{\partial f}{\partial x}\frac{dx}{dt} + \frac{\partial f}{\partial y}\frac{dy}{dt} = f_x(2,7)\,g'(3) + f_y(2,7)\,h'(3) = (6)(5) + (-8)(-4) = 62.$$

11.

$u = f(x, y)$, $x = x(r, s, t)$, $y = y(r, s, t)$ $\Rightarrow$

$$\frac{\partial u}{\partial r} = \frac{\partial u}{\partial x}\frac{\partial x}{\partial r} + \frac{\partial u}{\partial y}\frac{\partial y}{\partial r}, \quad \frac{\partial u}{\partial s} = \frac{\partial u}{\partial x}\frac{\partial x}{\partial s} + \frac{\partial u}{\partial y}\frac{\partial y}{\partial s},$$

$$\frac{\partial u}{\partial t} = \frac{\partial u}{\partial x}\frac{\partial x}{\partial t} + \frac{\partial u}{\partial y}\frac{\partial y}{\partial t}$$

13.

$v = f(p, q, r)$, $p = p(x, y, z)$, $q = q(x, y, z)$, $r = r(x, y, z)$ $\Rightarrow$

$$\frac{\partial v}{\partial x} = \frac{\partial v}{\partial p}\frac{\partial p}{\partial x} + \frac{\partial v}{\partial q}\frac{\partial q}{\partial x} + \frac{\partial v}{\partial r}\frac{\partial r}{\partial x},$$

$$\frac{\partial v}{\partial y} = \frac{\partial v}{\partial p}\frac{\partial p}{\partial y} + \frac{\partial v}{\partial q}\frac{\partial q}{\partial y} + \frac{\partial v}{\partial r}\frac{\partial r}{\partial y}, \quad \frac{\partial v}{\partial z} = \frac{\partial v}{\partial p}\frac{\partial p}{\partial z} + \frac{\partial v}{\partial q}\frac{\partial q}{\partial z} + \frac{\partial v}{\partial r}\frac{\partial r}{\partial z}$$

15. $w = x^2 + y^2 + z^2$, $x = st$, $y = s \cos t$, $z = s \sin t$ $\Rightarrow$

$$\frac{\partial w}{\partial s} = \frac{\partial w}{\partial x}\frac{\partial x}{\partial s} + \frac{\partial w}{\partial y}\frac{\partial y}{\partial s} + \frac{\partial w}{\partial z}\frac{\partial z}{\partial s} = 2xt + 2y \cos t + 2z \sin t. \text{ When } s = 1, t = 0,$$

we have $x = 0$, $y = 1$ and $z = 0$, so $\partial w/\partial s = 2 \cos 0 = 2$. Similarly

$\partial w/\partial t = 2xs + 2y(-s \sin t) + 2z(s \cos t) = 0 + (-2) \sin 0 + 0 = 0$ when $s = 1$ and $t = 0$.

17. $z = y^2 \tan x$, $x = t^2 uv$, $y = u + tv^2$ $\Rightarrow$
$\partial z/\partial t = (y^2 \sec^2 x) 2tuv + (2y \tan x) v^2$, $\partial z/\partial u = (y^2 \sec^2 x) t^2 v + 2y \tan x$,
$\partial z/\partial v = (y^2 \sec^2 x) t^2 u + (2y \tan x) 2tv$. When $t = 2$, $u = 1$ and $v = 0$, we have $x = 0$, $y = 1$, so
$\partial z/\partial t = 0$, $\partial z/\partial u = 0$, $\partial z/\partial v = 4$.

19. $u = \dfrac{x + y}{y + z}$, $x = p + r + t$, $y = p - r + t$, $z = p + r - t$ $\Rightarrow$

$$\frac{\partial u}{\partial p} = \frac{1}{y + z} + \frac{(y + z) - (x + y)}{(y + z)^2} - \frac{x + y}{(y + z)^2} = \frac{(y + z) + (z - x) - (x + y)}{(y + z)^2} = 2\frac{z - x}{(y + z)^2}$$

$$= 2\frac{-2t}{4p^2} = -\frac{t}{p^2},$$

$$\frac{\partial u}{\partial r} = \frac{1}{y + z} + \frac{z - x}{(y + z)^2}(-1) - \frac{x + y}{(y + z)^2} = 0, \text{ and}$$

$$\frac{\partial u}{\partial t} = \frac{1}{y + z} + \frac{z - x}{(y + z)^2} + \frac{x + y}{(y + z)^2} = 2\frac{y + z}{(y + z)^2} = \frac{2}{2p} = \frac{1}{p}.$$

21. $x^2 - xy + y^3 = 8$, so let $F(x, y) = x^2 - xy + y^3 - 8 = 0$. Then
$$\frac{dy}{dx} = -\frac{F_x}{F_y} = -\frac{(2x - y)}{-x + 3y^2} = \frac{y - 2x}{3y^2 - x}.$$

23. Let $F(x, y, z) = xy + yz - xz = 0$. Then $\dfrac{\partial z}{\partial x} = -\dfrac{F_x}{F_z} = -\dfrac{y - z}{y - x} = \dfrac{z - y}{y - x}$,
$$\frac{\partial z}{\partial y} = -\frac{F_y}{F_z} = -\frac{x + z}{y - x} = \frac{x + z}{x - y}.$$

25. Let $F(x, y, z) = xe^y + yz + ze^x = 0$. Then $\dfrac{\partial z}{\partial x} = -\dfrac{F_x}{F_z} = -\dfrac{e^y + ze^x}{y + e^x}$, $\dfrac{\partial z}{\partial y} = -\dfrac{F_y}{F_z} = -\dfrac{xe^y + z}{y + e^x}$.

27. Since x and y are each functions of t, $T(x, y)$ is a function of t, so by the Chain Rule,
$\dfrac{dT}{dt} = \dfrac{\partial T}{\partial x}\dfrac{dx}{dt} + \dfrac{\partial T}{\partial y}\dfrac{dy}{dt}$. After 3 seconds, $x = \sqrt{1 + t} = \sqrt{1 + 3} = 2$,

$y = 2 + \frac{1}{3}t = 2 + \frac{1}{3}(3) = 3$, $\dfrac{dx}{dt} = \dfrac{1}{2\sqrt{1 + t}} = \dfrac{1}{2\sqrt{1 + 3}} = \dfrac{1}{4}$, and $\dfrac{dy}{dt} = \dfrac{1}{3}$. Then

$\dfrac{dT}{dt} = T_x(2, 3)\dfrac{dx}{dt} + T_y(2, 3)\dfrac{dy}{dt} = 4\left(\frac{1}{4}\right) + 3\left(\frac{1}{3}\right) = 2$. Thus the temperature is rising at a rate of
2 degrees Celsius per second.

29. $C = 1449.2 + 4.6T - 0.055T^2 + 0.00029T^3 + 0.016D$, so $\dfrac{\partial C}{\partial T} = 4.6 - 0.11T + 0.00087T^2$ and

$\dfrac{\partial C}{\partial D} = 0.016$. From the graph, the diver is experiencing a temperature of approximately $12.5\,°C$ at

$t = 20$ minutes, so $\dfrac{\partial C}{\partial T} = 4.6 - 0.11(12.5) + 0.00087(12.5)^2 \approx 3.36$. By sketching tangent lines at

$t = 20$ to the graphs given, we estimate $\dfrac{dD}{dt} \approx \dfrac{1}{2}$ and $\dfrac{dT}{dt} \approx -\dfrac{1}{10}$. Then, by the Chain Rule,

$\dfrac{dC}{dt} = \dfrac{\partial C}{\partial T}\dfrac{dT}{dt} + \dfrac{\partial C}{\partial D}\dfrac{dD}{dt} \approx (3.36)\left(-\frac{1}{10}\right) + (0.016)\left(\frac{1}{2}\right) \approx -0.33$. Thus the speed of sound
experienced by the diver is decreasing at a rate of approximately 0.33 m/s per minute.

31. (a) $V = \ell wh$, so by the Chain Rule,

$$\frac{dV}{dt} = \frac{\partial V}{\partial \ell}\frac{d\ell}{dt} + \frac{\partial V}{\partial w}\frac{dw}{dt} + \frac{\partial V}{\partial h}\frac{dh}{dt} = wh\frac{d\ell}{dt} + \ell h\frac{dw}{dt} + \ell w\frac{dh}{dt}$$

$$= 2 \cdot 2 \cdot 2 + 1 \cdot 2 \cdot 2 + 1 \cdot 2 \cdot (-3) = 6 \text{ m}^3/\text{s}$$

(b) $S = 2\left(\ell w + \ell h + wh\right)$, so by the Chain Rule,

$$\frac{dS}{dt} = \frac{\partial S}{\partial \ell}\frac{d\ell}{dt} + \frac{\partial S}{\partial w}\frac{dw}{dt} + \frac{\partial S}{\partial h}\frac{dh}{dt} = 2\left(w+h\right)\frac{d\ell}{dt} + 2\left(\ell+h\right)\frac{dw}{dt} + 2\left(\ell+w\right)\frac{dh}{dt}$$

$$= 2\left(2+2\right)2 + 2\left(1+2\right)2 + 2\left(1+2\right)\left(-3\right) = 10 \text{ m}^2/\text{s}$$

(c) $L^2 = \ell^2 + w^2 + h^2 \;\Rightarrow\; 2L\dfrac{dL}{dt} = 2\ell\dfrac{d\ell}{dt} + 2w\dfrac{dw}{dt} + 2h\dfrac{dh}{dt} = 2\left(1\right)\left(2\right) + 2\left(2\right)\left(2\right) + 2\left(2\right)\left(-3\right) = 0$

$\;\Rightarrow\; dL/dt = 0 \text{ m/s}.$

33. $\dfrac{dP}{dt} = 0.05$, $\dfrac{dT}{dt} = 0.15$, $V = 8.31\dfrac{T}{P}$ and $\dfrac{dV}{dt} = \dfrac{8.31}{P}\dfrac{dT}{dt} - 8.31\dfrac{T}{P^2}\dfrac{dP}{dt}$. Thus when $P = 20$ and

$T = 320$, $\dfrac{dV}{dt} = 8.31\left[\dfrac{0.15}{20} - \dfrac{\left(0.05\right)\left(320\right)}{400}\right] \approx -0.27 \text{ L/s}.$

35. (a) Using the Chain Rule, $\dfrac{\partial z}{\partial r} = \dfrac{\partial z}{\partial x}\cos\theta + \dfrac{\partial z}{\partial y}\sin\theta$, $\dfrac{\partial z}{\partial \theta} = \dfrac{\partial z}{\partial x}\left(-r\sin\theta\right) + \dfrac{\partial z}{\partial y}r\cos\theta.$

(b) $\left(\dfrac{\partial z}{\partial r}\right)^2 = \left(\dfrac{\partial z}{\partial x}\right)^2 \cos^2\theta + 2\dfrac{\partial z}{\partial x}\dfrac{\partial z}{\partial y}\cos\theta\sin\theta + \left(\dfrac{\partial z}{\partial y}\right)^2 \sin^2\theta,$

$\left(\dfrac{\partial z}{\partial \theta}\right)^2 = \left(\dfrac{\partial z}{\partial x}\right)^2 r^2\sin^2\theta - 2\dfrac{\partial z}{\partial x}\dfrac{\partial z}{\partial y}r^2\cos\theta\sin\theta + \left(\dfrac{\partial z}{\partial y}\right)^2 r^2\cos^2\theta.$ Thus

$\left(\dfrac{\partial z}{\partial r}\right)^2 + \dfrac{1}{r^2}\left(\dfrac{\partial z}{\partial \theta}\right)^2 = \left[\left(\dfrac{\partial z}{\partial x}\right)^2 + \left(\dfrac{\partial z}{\partial y}\right)^2\right]\left(\cos^2\theta + \sin^2\theta\right) = \left(\dfrac{\partial z}{\partial x}\right)^2 + \left(\dfrac{\partial z}{\partial y}\right)^2.$

37. Let $u = x - y$. Then $\dfrac{\partial z}{\partial x} = \dfrac{dz}{du}\dfrac{\partial u}{\partial x} = \dfrac{dz}{du}$ and $\dfrac{\partial z}{\partial y} = \dfrac{dz}{du}\left(-1\right)$. Thus $\dfrac{\partial z}{\partial x} + \dfrac{\partial z}{\partial y} = 0.$

39. Let $u = x + at$, $v = x - at$. Then $z = f\left(u\right) + g\left(v\right)$, so $\partial z/\partial u = f'\left(u\right)$ and

$\partial z/\partial v = g'\left(v\right)$. Thus $\dfrac{\partial z}{\partial t} = \dfrac{\partial z}{\partial u}\dfrac{\partial u}{\partial t} + \dfrac{\partial z}{\partial v}\dfrac{\partial v}{\partial t} = af'\left(u\right) - ag'\left(v\right)$ and

$\dfrac{\partial^2 z}{\partial t^2} = a\dfrac{\partial}{\partial t}\left[f'\left(u\right) - g'\left(v\right)\right] = a\left(\dfrac{df'\left(u\right)}{du}\dfrac{\partial u}{\partial t} - \dfrac{dg'\left(v\right)}{dv}\dfrac{\partial v}{\partial t}\right) = a^2 f''\left(u\right) + a^2 g''\left(v\right).$ Similarly

$\dfrac{\partial z}{\partial x} = f'\left(u\right) + g'\left(v\right)$ and $\dfrac{\partial^2 z}{\partial x^2} = f''\left(u\right) + g''\left(v\right).$ Thus $\dfrac{\partial^2 z}{\partial t^2} = a^2\dfrac{\partial^2 z}{\partial x^2}.$

41. $\dfrac{\partial z}{\partial s} = \dfrac{\partial z}{\partial x} 2s + \dfrac{\partial z}{\partial y} 2r$. Then

$$\dfrac{\partial^2 z}{\partial r \partial s} = \dfrac{\partial}{\partial r}\left(\dfrac{\partial z}{\partial x} 2s\right) + \dfrac{\partial}{\partial r}\left(\dfrac{\partial z}{\partial y} 2r\right)$$

$$= \dfrac{\partial^2 z}{\partial x^2}\dfrac{\partial x}{\partial r} 2s + \dfrac{\partial}{\partial y}\left(\dfrac{\partial z}{\partial x}\right)\dfrac{\partial y}{\partial r} 2s + \dfrac{\partial z}{\partial x}\dfrac{\partial}{\partial r}(2s) + \dfrac{\partial^2 z}{\partial y^2}\dfrac{\partial y}{\partial r} 2r + \dfrac{\partial}{\partial x}\left(\dfrac{\partial z}{\partial y}\right)\dfrac{\partial x}{\partial r} 2r + \dfrac{\partial z}{\partial y} 2$$

$$= 4rs\dfrac{\partial^2 z}{\partial x^2} + \dfrac{\partial^2 z}{\partial y \partial x} 4s^2 + 0 + 4rs\dfrac{\partial^2 z}{\partial y^2} + \dfrac{\partial^2 z}{\partial x \partial y} 4r^2 + 2\dfrac{\partial z}{\partial y}$$

By the continuity of the partials, $\dfrac{\partial^2 z}{\partial r \partial s} = 4rs\dfrac{\partial^2 z}{\partial x^2} + 4rs\dfrac{\partial^2 z}{\partial y^2} + \left(4r^2 + 4s^2\right)\dfrac{\partial^2 z}{\partial x \partial y} + 2\dfrac{\partial z}{\partial y}$.

43. $\dfrac{\partial z}{\partial r} = \dfrac{\partial z}{\partial x}\cos\theta + \dfrac{\partial z}{\partial y}\sin\theta$ and $\dfrac{\partial z}{\partial\theta} = -\dfrac{\partial z}{\partial x} r\sin\theta + \dfrac{\partial z}{\partial y} r\cos\theta$. Then

$$\dfrac{\partial^2 z}{\partial r^2} = \cos\theta\left(\dfrac{\partial^2 z}{\partial x^2}\cos\theta + \dfrac{\partial^2 z}{\partial y \partial x}\sin\theta\right) + \sin\theta\left(\dfrac{\partial^2 z}{\partial y^2}\sin\theta + \dfrac{\partial^2 z}{\partial x \partial y}\cos\theta\right)$$

$$= \cos^2\theta\dfrac{\partial^2 z}{\partial x^2} + 2\cos\theta\sin\theta\dfrac{\partial^2 z}{\partial x \partial y} + \sin^2\theta\dfrac{\partial^2 z}{\partial y^2}$$

and

$$\dfrac{\partial^2 z}{\partial\theta^2} = -r\cos\theta\dfrac{\partial z}{\partial x} + (-r\sin\theta)\left(\dfrac{\partial^2 z}{\partial x^2}(-r\sin\theta) + \dfrac{\partial^2 z}{\partial y \partial x} r\cos\theta\right)$$

$$-r\sin\theta\dfrac{\partial z}{\partial y} + r\cos\theta\left(\dfrac{\partial^2 z}{\partial y^2} r\cos\theta + \dfrac{\partial^2 z}{\partial x \partial y}(-r\sin\theta)\right)$$

$$= -r\cos\theta\dfrac{\partial z}{\partial x} - r\sin\theta\dfrac{\partial z}{\partial y} + r^2\sin^2\theta\dfrac{\partial^2 z}{\partial x^2} - 2r^2\cos\theta\sin\theta\dfrac{\partial^2 z}{\partial x \partial y} + r^2\cos^2\theta\dfrac{\partial^2 z}{\partial y^2}$$

Thus

$$\dfrac{\partial^2 z}{\partial r^2} + \dfrac{1}{r^2}\dfrac{\partial^2 z}{\partial\theta^2} + \dfrac{1}{r}\dfrac{\partial z}{\partial r} = \left(\cos^2\theta + \sin^2\theta\right)\dfrac{\partial^2 z}{\partial x^2} + \left(\sin^2\theta + \cos^2\theta\right)\dfrac{\partial^2 z}{\partial y^2} - \dfrac{1}{r}\cos\theta\dfrac{\partial z}{\partial x}$$

$$-\dfrac{1}{r}\sin\theta\dfrac{\partial z}{\partial y} + \dfrac{1}{r}\left(\cos\theta\dfrac{\partial z}{\partial x} + \sin\theta\dfrac{\partial z}{\partial y}\right)$$

$$= \dfrac{\partial^2 z}{\partial x^2} + \dfrac{\partial^2 z}{\partial y^2}\ \text{as desired.}$$

45. $F(x, y, z) = 0$ is assumed to define z as a function of x and y, that is, $z = f(x, y)$. So by (6),

$\dfrac{\partial z}{\partial x} = -\dfrac{F_x}{F_z}$ since $F_z \neq 0$. Similarly, it is assumed that $F(x, y, z) = 0$ defines x as a function of y and

z, that is $x = h(x, z)$. Then $F(h(y, z), y, z) = 0$ and by the Chain Rule, $F_x\dfrac{\partial x}{\partial y} + F_y\dfrac{\partial y}{\partial y} + F_z\dfrac{\partial z}{\partial y} = 0$.

But $\dfrac{\partial z}{\partial y} = 0$ and $\dfrac{\partial y}{\partial y} = 1$, so $F_x\dfrac{\partial x}{\partial y} + F_y = 0 \Rightarrow \dfrac{\partial x}{\partial y} = -\dfrac{F_y}{F_x}$. A similar calculation shows that

$\dfrac{\partial y}{\partial z} = -\dfrac{F_z}{F_y}$. Thus $\dfrac{\partial z}{\partial x}\dfrac{\partial x}{\partial y}\dfrac{\partial y}{\partial z} = \left(-\dfrac{F_x}{F_z}\right)\left(-\dfrac{F_y}{F_x}\right)\left(-\dfrac{F_z}{F_y}\right) = -1$.

Section 11.6 Directional Derivatives and the Gradient Vector

1. First we draw a line passing through Raleigh and the eye of the hurricane. We can approximate the directional derivative at Raleigh in the direction of the eye of the hurricane by the average rate of change of pressure between the points where this line intersects the contour lines closest to Raleigh. In the direction of the eye of the hurricane, the pressure changes from 996 millibars to 992 millibars. We estimate the distance between these two points to be approximately 40 miles, so the rate of change of pressure in the direction given is approximately $\frac{992 - 996}{40} = -0.1$ millibars/mi.

3. $f(x, y) = x^2y^3 + 2x^4y \Rightarrow f_x(x, y) = 2xy^3 + 8x^3y$ and $f_y(x, y) = 3x^2y^2 + 2x^4$. If $\mathbf{u}$ is a unit vector in the direction of $\theta = \frac{\pi}{3}$, then from Equation 6,

$$D_\mathbf{u} f(1, -2) = f_x(1, -2) \cos \frac{\pi}{3} + f_y(1, -2) \sin \frac{\pi}{3} = (-32)\left(\frac{1}{2}\right) + (14)\left(\frac{\sqrt{3}}{2}\right) = 7\sqrt{3} - 16.$$

5. $f(x, y) = x^3 - 4x^2y + y^2$

   (a) $\nabla f(x, y) = f_x\mathbf{i} + f_y\mathbf{j} = \left(3x^2 - 8xy\right)\mathbf{i} + \left(2y - 4x^2\right)\mathbf{j}$

   (b) $\nabla f(0, -1) = -2\mathbf{j}$

   (c) By Equation 9, $D_\mathbf{u} f(0, -1) = \nabla f(0, -1) \cdot \mathbf{u} = \langle 0, -2\rangle \cdot \langle \frac{3}{5}, \frac{4}{5}\rangle = -\frac{8}{5}$.

7. $f(x, y, z) = xy^2z^3$

   (a) $\nabla f(x, y, z) = f_x\mathbf{i} + f_y\mathbf{j} + f_z\mathbf{k} = y^2z^3\mathbf{i} + 2xyz^3\mathbf{j} + 3xy^2z^2\mathbf{k}$

   (b) $\nabla f(1, -2, 1) = 4\mathbf{i} - 4\mathbf{j} + 12\mathbf{k}$

   (c) $\nabla f(1, -2, 1) \cdot \mathbf{u} = \frac{4}{\sqrt{3}} + \frac{4}{\sqrt{3}} + \frac{12}{\sqrt{3}} = \frac{20}{\sqrt{3}}$

9. $f(x, y) = \sqrt{x - y} \Rightarrow \nabla f(x, y) = \left\langle \frac{1}{2}(x - y)^{-1/2}, -\frac{1}{2}(x - y)^{-1/2}\right\rangle$, $\nabla f(5, 1) = \langle \frac{1}{4}, -\frac{1}{4}\rangle$, and a unit vector in the direction of $\mathbf{v}$ is $\mathbf{u} = \langle \frac{12}{13}, \frac{5}{13}\rangle$, so $D_\mathbf{u} f(5, 1) = \nabla f(5, 1) \cdot \mathbf{u} = \frac{12}{52} - \frac{5}{52} = \frac{7}{52}$.

11. $g(x, y, z) = x \tan^{-1}(y/z) \Rightarrow \nabla g(x, y, z) = \left\langle \tan^{-1}(y/z), xz/\left(y^2 + z^2\right), -xy/\left(y^2 + z^2\right)\right\rangle$, $\nabla g(1, 2, -2) = \langle -\frac{\pi}{4}, -\frac{1}{4}, -\frac{1}{4}\rangle$, $\mathbf{u} = \frac{1}{\sqrt{3}}\langle 1, 1, -1\rangle$ and

$$D_\mathbf{u} g(1, 2, -2) = \frac{(-\pi)(1)}{4\sqrt{3}} + \frac{(-1)(1)}{4\sqrt{3}} + \frac{(-1)(-1)}{4\sqrt{3}} = -\frac{\pi}{4\sqrt{3}}.$$

13. $f(x, y) = xe^{-y} + 3y \Rightarrow \nabla f(x, y) = \langle e^{-y}, 3 - xe^{-y}\rangle$, $\nabla f(1, 0) = \langle 1, 2\rangle$ is the direction and the maximum rate is $|\nabla f(1, 0)| = \sqrt{5}$.

15. $f(x, y, z) = x + y/z \Rightarrow \nabla f(x, y, z) = \left\langle 1, \frac{1}{z}, -\frac{y}{z^2}\right\rangle$, so the maximum rate of change is $|\nabla f(4, 3, -1)| = \sqrt{11}$ in the direction $\langle 1, -1, -3\rangle$.

17. (a) As in the proof of Theorem 15, $D_{\mathbf{u}}f = |\nabla f|\cos\theta$. Since the minimum value of $\cos\theta$ is -1 occurring when $\theta = \pi$, the minimum value of $D_{\mathbf{u}}f$ is $-|\nabla f|$ occurring when $\theta = \pi$, that is when $\mathbf{u}$ is in the opposite direction of ∇f (assuming $\nabla f \neq \mathbf{0}$).

(b) $f(x,y) = x^4 y - x^2 y^3 \Rightarrow \nabla f(x,y) = \langle 4x^3 y - 2xy^3, x^4 - 3x^2 y^2 \rangle$, so f decreases fastest at the point $(2,-3)$ in the direction $-\nabla f(2,-3) = -\langle 12, -92 \rangle = \langle -12, 92 \rangle$.

19. $T = k/\sqrt{x^2 + y^2 + z^2}$ and $120 = T(1,2,2) = k/3$ so $k = 360$.

(a) $\mathbf{u} = \dfrac{\langle 1, -1, 1 \rangle}{\sqrt{3}}$,

$$D_{\mathbf{u}}T(1,2,2) = \nabla T(1,2,2) \cdot \mathbf{u} = \left[-360 \left(x^2 + y^2 + z^2\right)^{-3/2} \langle x, y, z \rangle \right]_{(1,2,2)} \cdot \mathbf{u}$$

$$= -\tfrac{40}{3}\langle 1,2,2 \rangle \cdot \tfrac{1}{\sqrt{3}}\langle 1,-1,1 \rangle = -\tfrac{40}{3\sqrt{3}}$$

(b) From (a), $\nabla T = -360\left(x^2 + y^2 + z^2\right)^{-3/2}\langle x, y, z \rangle$, and since $\langle x, y, z \rangle$ is the position vector of the point (x, y, z), the vector $-\langle x, y, z \rangle$, and thus ∇T, always points toward the origin.

21. $\nabla V(x, y, z) = \langle 10x - 3y + yz, xz - 3x, xy \rangle$, $\nabla V(3,4,5) = \langle 38, 6, 12 \rangle$

(a) $D_{\mathbf{u}}V(3,4,5) = \langle 38, 6, 12 \rangle \cdot \tfrac{1}{\sqrt{3}}\langle 1, 1, -1 \rangle = \tfrac{32}{\sqrt{3}}$

(b) $\langle 38, 6, 12 \rangle$ or $\langle 19, 3, 6 \rangle$.

(c) $|\nabla V(x, y, z)| = \sqrt{(10x - 3y + yz)^2 + (xz - 3x)^2 + x^2 y^2}$. In particular, $|\nabla V(3,4,5)| = \sqrt{1624} = 2\sqrt{406}$.

23. A unit vector in the direction of $\overrightarrow{AB}$ is $\mathbf{i}$ and a unit vector in the direction of $\overrightarrow{AC}$ is $\mathbf{j}$. Thus $D_{\overrightarrow{AB}}f(1,3) = f_x(1,3) = 3$ and $D_{\overrightarrow{AC}}f(1,3) = f_y(1,3) = 26$. Therefore $\nabla f(1,3) = \langle f_x(1,3), f_y(1,3)\rangle = \langle 3, 26 \rangle$, and by definition, $D_{\overrightarrow{AD}}f(1,3) = \nabla f \cdot \mathbf{u}$ where $\mathbf{u}$ is a unit vector in the direction of $\overrightarrow{AD}$, which is $\langle \tfrac{5}{13}, \tfrac{12}{13} \rangle$. Therefore, $D_{\overrightarrow{AD}}f(1,3) = \langle 3, 26 \rangle \cdot \langle \tfrac{5}{13}, \tfrac{12}{13} \rangle = 3 \cdot \tfrac{5}{13} + 26 \cdot \tfrac{12}{13} = \tfrac{327}{13}$.

25. (a) $\nabla(au + bv) = \left\langle \dfrac{\partial(au + bv)}{\partial x}, \dfrac{\partial(au + bv)}{\partial y} \right\rangle = \left\langle a\dfrac{\partial u}{\partial x} + b\dfrac{\partial v}{\partial x}, a\dfrac{\partial u}{\partial y} + b\dfrac{\partial v}{\partial y} \right\rangle$

$$= a\left\langle \dfrac{\partial u}{\partial x}, \dfrac{\partial u}{\partial y} \right\rangle + b\left\langle \dfrac{\partial v}{\partial x}, \dfrac{\partial v}{\partial y} \right\rangle = a\nabla u + b\nabla v$$

(b) $\nabla(uv) = \left\langle v\dfrac{\partial u}{\partial x} + u\dfrac{\partial v}{\partial x}, v\dfrac{\partial u}{\partial y} + u\dfrac{\partial v}{\partial y} \right\rangle = v\left\langle \dfrac{\partial u}{\partial x}, \dfrac{\partial u}{\partial y} \right\rangle + u\left\langle \dfrac{\partial v}{\partial x}, \dfrac{\partial v}{\partial y} \right\rangle = v\nabla u + u\nabla v$

(c) $\nabla\left(\dfrac{u}{v}\right) = \left\langle \dfrac{v\dfrac{\partial u}{\partial x} - u\dfrac{\partial v}{\partial x}}{v^2}, \dfrac{v\dfrac{\partial u}{\partial y} - u\dfrac{\partial v}{\partial y}}{v^2} \right\rangle = \dfrac{v\left\langle \dfrac{\partial u}{\partial x}, \dfrac{\partial u}{\partial y} \right\rangle - u\left\langle \dfrac{\partial v}{\partial x}, \dfrac{\partial v}{\partial y} \right\rangle}{v^2} = \dfrac{v\nabla u - u\nabla v}{v^2}$

(d) $\nabla u^n = \left\langle \dfrac{\partial(u^n)}{\partial x}, \dfrac{\partial(u^n)}{\partial y} \right\rangle = \left\langle nu^{n-1}\dfrac{\partial u}{\partial x}, nu^{n-1}\dfrac{\partial u}{\partial y} \right\rangle = nu^{n-1}\nabla u$

27. $\nabla F\left(x,y,z\right) = \langle 2x - 2y + 4z, 2y - 2x, -2z + 4x\rangle$, $\nabla F\left(1,0,1\right) = \langle 6, -2, 2\rangle$

    (a) $6x - 2y + 2z = 8$ or $3x - y + z = 4$ **(b)** $\dfrac{x-1}{3} = -y = z - 1$

29. $F\left(x,y,z\right) = -z + xe^y \cos z$, $\nabla F\left(x,y,z\right) = \langle e^y \cos z, xe^y \cos z, -1 - xe^y \sin z\rangle$,

    $\nabla F\left(1,0,0\right) = \langle 1, 1, -1\rangle$

    (a) $x + y - z = 1$ **(b)** $x - 1 = y = -z$

31. $\nabla F\left(x,y,z\right) = \langle y + z, x + z, y + x\rangle$,

    $\nabla F\left(1,1,1\right) = \langle 2, 2, 2\rangle$, so the equation of

    the tangent plane is $2x + 2y + 2z = 6$ or

    $x + y + z = 3$, and the normal line is given

    by $x - 1 = y - 1 = z - 1$ or $x = y = z$.

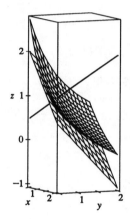

33. $\nabla f\left(x,y\right) = \langle 2x, 8y\rangle$, $\nabla f\left(2,1\right) = \langle 4, 8\rangle$. The tangent line

    has equation $\nabla f\left(2,1\right) \cdot \langle x - 2, y - 1\rangle = 0 \;\Rightarrow$

    $4\left(x - 2\right) + 8\left(y - 1\right) = 0$, which simplifies to $x + 2y = 4$.

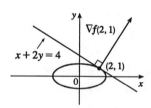

35. $\nabla F\left(x_0, y_0, z_0\right) = \left\langle \dfrac{2x_0}{a^2}, \dfrac{2y_0}{b^2}, \dfrac{2z_0}{c^2} \right\rangle$. Thus an equation of the tangent plane at $\left(x_0, y_0, z_0\right)$ is

    $\dfrac{2x_0}{a^2}x + \dfrac{2y_0}{b^2}y + \dfrac{2z_0}{c^2}z = 2\left(\dfrac{x_0^2}{a^2} + \dfrac{y_0^2}{b^2} + \dfrac{z_0^2}{c^2}\right) = 2\left(1\right) = 2$ since $\left(x_0, y_0, z_0\right)$ is a point on the ellipsoid.

    Hence $\dfrac{x_0}{a^2}x + \dfrac{y_0}{b^2}y + \dfrac{z_0}{c^2}z = 1$ is an equation of the tangent plane.

37. $\nabla f\left(x_0, y_0, z_0\right) = \langle 2x_0, -2y_0, 4z_0\rangle$ and the given line has direction numbers $2, 4, 6$, so

    $\langle 2x_0, -2y_0, 4z_0\rangle = k\langle 2, 4, 6\rangle$ or $x_0 = k$, $y_0 = -2k$ and $z_0 = \frac{3}{2}k$. But $x_0^2 - y_0^2 + 2z_0^2 = 1$ or

    $\left(1 - 4 + \frac{9}{2}\right)k^2 = 1$, so $k = \pm\sqrt{\frac{2}{3}} = \pm\frac{\sqrt{6}}{3}$ and there are two such points: $\left(\pm\frac{\sqrt{6}}{3}, \mp\frac{2\sqrt{6}}{3}, \pm\frac{\sqrt{6}}{2}\right)$.

39. Let (x_0, y_0, z_0) be a point on the surface. Then the equation of the tangent plane at the point is

$$\frac{x}{2\sqrt{x_0}} + \frac{y}{2\sqrt{y_0}} + \frac{z}{2\sqrt{z_0}} = \frac{\sqrt{x_0} + \sqrt{y_0} + \sqrt{z_0}}{2}. \text{ But } \sqrt{x_0} + \sqrt{y_0} + \sqrt{z_0} = \sqrt{c}, \text{ so the equation is}$$

$\dfrac{x}{\sqrt{x_0}} + \dfrac{y}{\sqrt{y_0}} + \dfrac{z}{\sqrt{z_0}} = \sqrt{c}$. The x-, y-, and z-intercepts are $\sqrt{cx_0}$, $\sqrt{cy_0}$ and $\sqrt{cz_0}$ respectively. (The x-intercept is found by setting $y = z = 0$ and solving the resulting equation for x, and the y- and z-intercepts are found similarly.) So the sum of the intercepts is $\sqrt{c}\left(\sqrt{x_0} + \sqrt{y_0} + \sqrt{z_0}\right) = c$, a constant.

41. If $f(x, y, z) = z - x^2 - y^2$ and $g(x, y, z) = 4x^2 + y^2 + z^2$, then the tangent line is perpendicular to both ∇f and ∇g at $(-1, 1, 2)$. The vector $\mathbf{v} = \nabla f \times \nabla g$ will therefore be parallel to the tangent line. We have: $\nabla f(x, y, z) = \langle -2x, -2y, 1 \rangle \;\Rightarrow\; \nabla f(-1, 1, 2) = \langle 2, -2, 1 \rangle$, and $\nabla g(x, y, z) = \langle 8x, 2y, 2z \rangle \;\Rightarrow\; \nabla g(-1, 1, 2) = \langle -8, 2, 4 \rangle$. Hence

$$\mathbf{v} = \nabla f \times \nabla g = \begin{vmatrix} \mathbf{i} & \mathbf{j} & \mathbf{k} \\ 2 & -2 & 1 \\ -8 & 2 & 4 \end{vmatrix} = -10\,\mathbf{i} - 16\,\mathbf{j} - 12\,\mathbf{k}. \text{ Parametric equations are: } x = -1 - 10t,$$

$y = 1 - 16t$, $z = 2 - 12t$.

43. (a) The direction of the normal line of F is given by ∇F, and that of G by ∇G. Assuming that $\nabla F \neq 0 \neq \nabla G$, the two normal lines are be perpendicular at P if $\nabla F \cdot \nabla G = 0$ at P $\;\Leftrightarrow\;$ $\langle \partial F/\partial x, \partial F/\partial y, \partial F/\partial z \rangle \cdot \langle \partial G/\partial x, \partial G/\partial y, \partial G/\partial z \rangle = 0$ at P $\;\Leftrightarrow\;$ $F_x G_x + F_y G_y + F_z G_z = 0$ at P.

(b) Here $F = x^2 + y^2 - z^2$ and $G = x^2 + y^2 + z^2 - r^2$, so $\nabla F \cdot \nabla G = \langle 2x, 2y, -2z \rangle \cdot \langle 2x, 2y, 2z \rangle = 4x^2 + 4y^2 - 4z^2 = 4F = 0$, since the point $\langle x, y, z \rangle$ lies on the graph of $F = 0$. To see that this is true without using calculus, note that $G = 0$ is the equation of a sphere centered at the origin and $F = 0$ is the equation of a right circular cone with vertex at the origin (which is generated by lines through the origin). At any point of intersection, the sphere's normal line (which passes through the origin) lies on the cone, and thus is perpendicular to the cone's normal line. So the surfaces with equations $F = 0$ and $G = 0$ are everywhere orthogonal.

45. Let $\mathbf{u} = \langle a, b \rangle$ and $\mathbf{v} = \langle c, d \rangle$. Then we know that at the given point, $D_{\mathbf{u}}f = \nabla f \cdot \mathbf{u} = af_x + bf_y$ and $D_{\mathbf{v}}f = \nabla f \cdot \mathbf{v} = cf_x + df_y$. But these are just two linear equations in the two unknowns f_x and f_y, and since $\mathbf{u}$ and $\mathbf{v}$ are not parallel, we can solve the equations to find $\nabla f = \langle f_x, f_y \rangle$ at the given point. In fact, $\nabla f = \left\langle \dfrac{dD_{\mathbf{u}}f - bD_{\mathbf{v}}f}{ad - bc}, \dfrac{aD_{\mathbf{v}}f - cD_{\mathbf{u}}f}{ad - bc} \right\rangle$.

Section 11.7   Maximum and Minimum Values

1. (a) First we compute $D(1,1) = f_{xx}(1,1) f_{yy}(1,1) - [f_{xy}(1,1)]^2 = (4)(2) - (1)^2 = 7$. Since $D(1,1) > 0$ and $f_{xx}(1,1) > 0$, f has a local minimum at $(1,1)$ by the Second Derivatives Test.

(b) $D(1,1) = f_{xx}(1,1) f_{yy}(1,1) - [f_{xy}(1,1)]^2 = (4)(2) - (3)^2 = -1$. Since $D(1,1) < 0$, f has a saddle point at $(1,1)$ by the Second Derivatives Test.

3. In the figure, a point at approximately $(1,1)$ is enclosed by level curves which are oval in shape and indicate that as we move away from the point in any direction the values of f are increasing. Hence we would expect a local minimum at or near $(1,1)$. The level curves near $(0,0)$ resemble hyperbolas, and as we move away from the origin, the values of f increase in some directions and decrease in others, so we would expect to find a saddle point there.

To verify our predictions, we have $f(x,y) = 4 + x^3 + y^3 - 3xy \Rightarrow f_x(x,y) = 3x^2 - 3y$, $f_y(x,y) = 3y^2 - 3x$. We have critical points where these partial derivatives are equal to 0: $3x^2 - 3y = 0$, $3y^2 - 3x = 0$. Substituting $y = x^2$ from the first equation into the second equation gives $3(x^2)^2 - 3x = 0 \Rightarrow 3x(x^3 - 1) = 0 \Rightarrow x = 0$ or $x = 1$. Then we have two critical points, $(0,0)$ and $(1,1)$. The second partial derivatives are $f_{xx}(x,y) = 6x$, $f_{xy}(x,y) = -3$, and $f_{yy}(x,y) = 6y$, so $D(x,y) = f_{xx}(x,y) f_{yy}(x,y) - [f_{xy}(x,y)]^2 = (6x)(6y) - (-3)^2 = 36xy - 9$. Then $D(0,0) = 36(0)(0) - 9 = -9$, and $D(1,1) = 36(1)(1) - 9 = 27$. Since $D(0,0) < 0$, f has a saddle point at $(0,0)$ by the Second Derivatives Test. Since $D(1,1) > 0$ and $f_{xx}(1,1) > 0$, f has a local minimum at $(1,1)$.

5. $f(x,y) = x^2 + y^2 + 4x - 6y \Rightarrow f_x = 2x + 4$, $f_y = 2y - 6$, $f_{xx} = f_{yy} = 2$, $f_{xy} = 0$. Then $f_x = 0$ and $f_y = 0$ implies $(x,y) = (-2,3)$ and $D(-2,3) = 4 > 0$, so $f(-2,3) = -13$ is a local minimum.

7. $f(x,y) = x^2 + y^2 + x^2y + 4 \Rightarrow f_x = 2x + 2xy$, $f_y = 2y + x^2$, $f_{xx} = 2 + 2y$, $f_{yy} = 2$, $f_{xy} = 2x$. Then $f_y = 0$ implies $y = -\frac{1}{2}x^2$, substituting into $f_x = 0$ gives $2x - x^3 = 0$ so $x = 0$ or $x = \pm\sqrt{2}$. Thus the critical points are $(0,0)$, $(\sqrt{2}, -1)$ and $(-\sqrt{2}, -1)$. Now $D(0,0) = 4$, $D(\sqrt{2}, -1) = -8 = D(-\sqrt{2}, -1)$, $f_{xx}(0,0) = 2$, $f_{xx}(\pm\sqrt{2}, -1) = 0$.

Thus $f(0,0) = 4$ is a local minimum and $(\pm\sqrt{2}, -1)$ are saddle points.

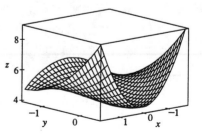

9. $f(x, y) = xy - 2x - y \Rightarrow f_x = y - 2, f_y = x - 1,$
$f_{xx} = f_{yy} = 0, f_{xy} = 1$ and the only critical point is $(1, 2)$.
Now $D(1, 2) = -1$, so $(1, 2)$ is a saddle point and f has no
local maximum or minimum.

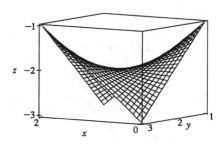

11. $f(x, y) = e^x \cos y \Rightarrow f_x = e^x \cos y, f_y = -e^x \sin y.$ Now
$f_x = 0$ implies $\cos y = 0$ or $y = \frac{\pi}{2} + n\pi$ for n an integer. But
$\sin\left(\frac{\pi}{2} + n\pi\right) \neq 0$, so there are no critical points.

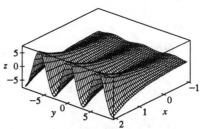

13. $f(x, y) = x \sin y \Rightarrow f_x = \sin y, f_y = x \cos y, f_{xx} = 0,$
$f_{yy} = -x \sin y$ and $f_{xy} = \cos y.$ Then $f_x = 0$ if and only if
$y = n\pi, n$ an integer, and substituting into $f_y = 0$ requires
$x = 0$ for each of these y-values. Thus the critical points are
$(0, n\pi), n$ an integer. But $D(0, n\pi) = -\cos^2(n\pi) < 0$ so
each critical point is a saddle point.

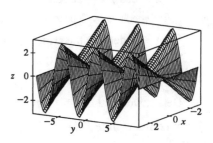

15. $f(x, y) = 3x^2 y + y^3 - 3x^2 - 3y^2 + 2$

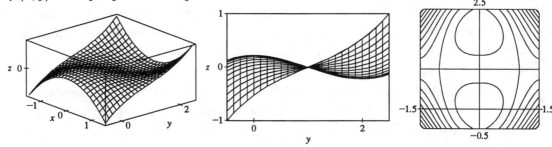

From the graphs, it appears that f has a local maximum $f(0, 0) \approx 2$ and a local minimum
$f(0, 2) \approx -2$. There appear to be saddle points near $(\pm 1, 1)$.
$f_x = 6xy - 6x, f_y = 3x^2 + 3y^2 - 6y.$ Then $f_x = 0$ implies $x = 0$ or $y = 1$ and when $x = 0, f_y = 0$
implies $y = 0$ or $y = 2$; when $y = 1, f_y = 0$ implies $x^2 = 1$ or $x = \pm 1$. Thus the critical points are
$(0, 0), (0, 2), (\pm 1, 1).$ Now $f_{xx} = 6y - 6, f_{yy} = 6y - 6$ and $f_{xy} = 6x,$ so $D(0, 0) = D(0, 2) = 36 > 0$
while $D(\pm 1, 1) = -36 < 0$ and $f_{xx}(0, 0) = -6, f_{xx}(0, 2) = 6.$ Hence $(\pm 1, 1)$ are saddle points while
$f(0, 0) = 2$ is a local maximum and $f(0, 2) = -2$ is a local minimum.

17. $f(x, y) = \sin x + \sin y + \sin(x + y), 0 \le x \le 2\pi, 0 \le y \le 2\pi$

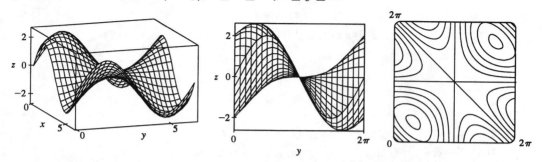

From the graphs it appears that f has a local maximum at about $(1, 1)$ with value approximately 2.6, a local minimum at about $(5, 5)$ with value approximately -2.6, and a saddle point at about $(3, 3)$.
$f_x = \cos x + \cos(x + y)$, $f_y = \cos y + \cos(x + y)$, $f_{xx} = -\sin x - \sin(x + y)$,
$f_{yy} = -\sin y - \sin(x + y)$, $f_{xy} = -\sin(x + y)$. Setting $f_x = 0$ and $f_y = 0$ and subtracting gives
$\cos x - \cos y = 0$ or $\cos x = \cos y$. Thus $x = y$ or $x = 2\pi - y$. If $x = y$, $f_x = 0$ becomes
$\cos x + \cos 2x = 0$ or $2\cos^2 x + \cos x - 1 = 0$, a quadratic in $\cos x$. Thus $\cos x = -1$ or $\frac{1}{2}$ and $x = \pi$,
$\frac{\pi}{3}$, or $\frac{5\pi}{3}$, yielding the critical points (π, π), $\left(\frac{\pi}{3}, \frac{\pi}{3}\right)$ and $\left(\frac{5\pi}{3}, \frac{5\pi}{3}\right)$. Similarly if $x = 2\pi - y$, $f_x = 0$
becomes $(\cos x) + 1 = 0$ and the resulting critical point is (π, π). Now
$D(x, y) = \sin x \sin y + \sin x \sin(x + y) + \sin y \sin(x + y)$. So $D(\pi, \pi) = 0$ and the Second
Derivatives Test doesn't apply. $D\left(\frac{\pi}{3}, \frac{\pi}{3}\right) = \frac{9}{4} > 0$ and $f_{xx}\left(\frac{\pi}{3}, \frac{\pi}{3}\right) < 0$ so $f\left(\frac{\pi}{3}, \frac{\pi}{3}\right) = \frac{3\sqrt{3}}{2}$ is a local
maximum while $D\left(\frac{5\pi}{3}, \frac{5\pi}{3}\right) = \frac{9}{4} > 0$ and $f_{xx}\left(\frac{5\pi}{3}, \frac{5\pi}{3}\right) > 0$, so $f\left(\frac{5\pi}{3}, \frac{5\pi}{3}\right) = -\frac{3\sqrt{3}}{2}$ is a local
minimum.

19. $f(x, y) = x^4 - 5x^2 + y^2 + 3x + 2 \Rightarrow f_x(x, y) = 4x^3 - 10x + 3$ and $f_y(x, y) = 2y$. $f_y = 0 \Rightarrow$
$y = 0$, and the graph of f_x shows that the roots of $f_x = 0$ are approximately $x = -1.714, 0.312$ and
1.402. (Alternatively, we could have used a calculator or a CAS to find these roots.) So to three
decimal places, the critical points are $(-1.714, 0)$, $(1.402, 0)$, and $(0.312, 0)$. Now since
$f_{xx} = 12x^2 - 10$, $f_{xy} = 0$, $f_{yy} = 2$, and $D = 24x^2 - 20$, we have $D(-1.714, 0) > 0$,
$f_{xx}(-1.714, 0) > 0$, $D(1.402, 0) > 0$, $f_{xx}(1.402, 0) > 0$, and $D(0.312, 0) < 0$. Therefore
$f(-1.714, 0) \approx -9.200$ and $f(1.402, 0) \approx 0.242$ are local minima, and $(0.312, 0)$ is a saddle point.
The lowest point on the graph is approximately $(-1.714, 0, -9.200)$.

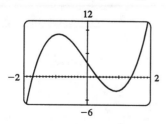

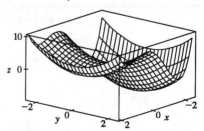

21. $f(x,y) = 2x + 4x^2 - y^2 + 2xy^2 - x^4 - y^4$ $\Rightarrow$ $f_x(x,y) = 2 + 8x + 2y^2 - 4x^3$,
$f_y(x,y) = -2y + 4xy - 4y^3$. Now $f_y = 0$ $\Leftrightarrow$ $2y(2y^2 - 2x + 1) = 0$ $\Leftrightarrow$ $y = 0$ or $y^2 = x - \frac{1}{2}$.
The first of these implies that $f_x = -4x^3 + 8x + 2$, and the second implies that
$f_x = 2 + 8x + 2\left(x - \frac{1}{2}\right) - 4x^3 = -4x^3 + 10x + 1$. From the graphs, we see that the first possibility
for f_x has roots at approximately -1.267, -0.259, and 1.526, and the second has a root at
approximately 1.629 (the negative roots do not give critical points, since $y^2 = x - \frac{1}{2}$ must be positive).
So to three decimal places, f has critical points at $(-1.267, 0)$, $(-0.259, 0)$, $(1.526, 0)$, and
$(1.629, \pm 1.063)$. Now since $f_{xx} = 8 - 12x^2$, $f_{xy} = 4y$, $f_{yy} = 4x - 12y^2$, and
$D = (8 - 12x^2)(4x - 12y^2) - 16y^2$, we have $D(-1.267, 0) > 0$, $f_{xx}(-1.267, 0) > 0$,
$D(-0.259, 0) < 0$, $D(1.526, 0) < 0$, $D(1.629, \pm 1.063) > 0$, and $f_{xx}(1.629, \pm 1.063) < 0$. Therefore,
to three decimal places, $f(-1.267, 0) \approx 1.310$ and $f(1.629, \pm 1.063) \approx 8.105$ are local maxima, and
$(-0.259, 0)$ and $(1.526, 0)$ are saddle points. The highest points on the graph are approximately
$(1.629, \pm 1.063, 8.105)$.

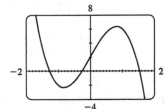
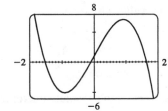
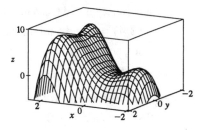

23. Since f is a polynomial it is continuous on D, so an absolute maximum
and minimum exist. Here $f_x = -3$, $f_y = 4$ so there are no critical
points inside D. Thus the absolute extrema must both occur on the
boundary. Along L_1, $y = 0$ and $f(x, 0) = 5 - 3x$, a decreasing
function in x, so the maximum value is $f(0, 0) = 5$ and the minimum
value is $f(4, 0) = -7$. Along L_2, $x = 4$ and $f(4, y) = -7 + 4y$, an
increasing function in y, so the minimum value is $f(4, 0) = -7$ and
the maximum value is $f(4, 5) = 13$. Along L_3, $y = \frac{5}{4}x$ and $f\left(x, \frac{5}{4}x\right) = 5 + 2x$, an increasing function
in x, so the minimum value is $f(0, 0) = 5$ and the maximum value is $f(4, 5) = 13$. Thus the absolute
minimum of f on D is $f(4, 0) = -7$ and the absolute maximum is $f(4, 5) = 13$.

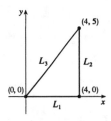

25. In Exercise 7, we found the critical points of f; only $(0,0)$ with

$f(0,0) = 4$ is in D. On L_1: $y = -1$, $f(x,-1) = 5$, a constant. On L_2:

$x = 1$, $f(1,y) = y^2 + y + 5$, a quadratic in y which attains its

maximum at $(1,1)$, $f(1,1) = 7$ and its minimum at $\left(1,-\frac{1}{2}\right)$,

$f\left(1,-\frac{1}{2}\right) = \frac{17}{4}$. On L_3: $f(x,1) = 2x^2 + 5$ which attains its maximum

at $(-1,1)$ and $(1,1)$ with $f(\pm 1,1) = 7$ and its minimum at $(0,1)$,

$f(0,1) = 5$. On L_4: $f(-1,y) = y^2 + y + 5$ with maximum

at $(-1,1)$, $f(-1,1) = 7$ and minimum at $\left(-1,-\frac{1}{2}\right)$, $f\left(-1,-\frac{1}{2}\right) = \frac{17}{4}$. Thus the absolute maximum is

attained at both $(\pm 1,1)$ with $f(\pm 1,1) = 7$ and the absolute minimum on D is attained at $(0,0)$ with

$f(0,0) = 4$.

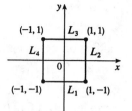

27. $f_x(x,y) = y - 1$ and $f_y(x,y) = x - 1$ and so the critical

point is $(1,1)$ (in D), where $f(1,1) = 0$. Along L_1: $y = 4$, so

$f(x,4) = 1 + 4x - x - 4 = 3x - 3$, $-2 \le x \le 2$, which is an

increasing function and has a maximum value when $x = 2$

where $f(2,4) = 3$ and a minimum of $f(-2,4) = -9$. Along

L_2: $y = x^2$, so let $g(x) = f(x,x^2) = x^3 - x^2 - x + 1$. Then

$g'(x) = 3x^2 - 2x - 1 = 0 \iff x = -\frac{1}{3}$ or $x = 1$.

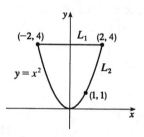

$f\left(-\frac{1}{3},\frac{1}{9}\right) = \frac{32}{27}$ and $f(1,1) = 0$. As a result, the absolute maximum and minimum values of f on D

are $f(2,4) = 3$ and $f(-2,4) = -9$.

29. $f(x,y) = -\left(x^2 - 1\right)^2 - \left(x^2 y - x - 1\right)^2 \Rightarrow$

$f_x(x,y) = -2\left(x^2 - 1\right)(2x) - 2\left(x^2 y - x - 1\right)(2xy - 1)$ and $f_y(x,y) = -2\left(x^2 y - x - 1\right)x^2$.

Setting $f_y(x,y) = 0$ gives either $x = 0$ or $x^2 y - x - 1 = 0$. There are no critical points for $x = 0$, since

$f_x(0,y) = -2$, so we set $x^2 y - x - 1 = 0 \iff y = \dfrac{x+1}{x^2}$ $(x \ne 0)$, so

$f_x\left(x,\dfrac{x+1}{x^2}\right) = -2\left(x^2 - 1\right)(2x) - 2\left(x^2\dfrac{x+1}{x^2} - x - 1\right)\left(2x\dfrac{x+1}{x^2} - 1\right) = -4x\left(x^2 - 1\right)$.

Therefore $f_x(x,y) = f_y(x,y) = 0$ at the points $(1,2)$ and $(-1,0)$.

To classify these critical points, we calculate

$f_{xx}(x,y) = -12x^2 - 12x^2 y^2 + 12xy + 4y + 2$,

$f_{yy}(x,y) = -2x^4$, and $f_{xy}(x,y) = -8x^3 y + 6x^2 + 4x$. In

order to use the Second Derivatives Test we calculate

$D(-1,0) = f_{xx}(-1,0)\, f_{yy}(-1,0) - [f_{xy}(-1,0)]^2$

$\qquad\quad = 16 > 0$,

$f_{xx}(-1,0) = -10 < 0$, $D(1,2) = 16 > 0$, and

$f_{xx}(1,2) = -26 < 0$, so both $(-1,0)$ and $(1,2)$ give local

maxima.

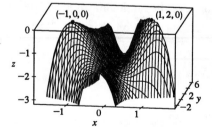

31. $d = \sqrt{(x-2)^2 + (y+2)^2 + (z-3)^2}$, where $z = \frac{1}{3}(6x + 4y - 2)$, so we minimize

$d^2 = f(x, y) = (x-2)^2 + (y+2)^2 + \left(2x + \frac{4}{3}y - \frac{11}{3}\right)^2$. Then $f_x = 10x + \frac{16}{3}y - \frac{56}{3}$ and

$f_y = \frac{50}{9}y + \frac{16}{3}x - \frac{52}{9}$. Solving $50y + 48x = 52$ and $16y + 30x = 56$ simultaneously gives $x = \frac{164}{61}$,

$y = -\frac{94}{61}$. The absolute minimum must occur at a critical point. Thus $d^2 = \left(\frac{42}{61}\right)^2 + \left(\frac{28}{61}\right)^2 + \left(-\frac{21}{61}\right)^2$ or

$d = \frac{7}{\sqrt{61}}$.

33. Minimize $d^2 = x^2 + y^2 + z^2 = x^2 + y^2 + xy + 1$. Then $f_x = 2x + y$, $f_y = 2y + x$ so the critical point is $(0, 0)$ and $D(0, 0) = 4 - 1 > 0$ with $f_{xx}(0, 0) = 2$ so this is a minimum. Thus $z^2 = 1$ or $z = \pm 1$ and the points on the surface are $(0, 0, \pm 1)$.

35. $x + y + z = 100$, so maximize $f(x, y) = xy(100 - x - y)$. $f_x = 100y - 2xy - y^2$,

$f_y = 100x - x^2 - 2xy$, $f_{xx} = -2y$, $f_{yy} = -2x$, $f_{xy} = 100 - 2x - 2y$. Then $f_x = 0$ implies $y = 0$ or

$y = 100 - 2x$. Substituting $y = 0$ into $f_y = 0$ gives $x = 0$ or $x = 100$ and substituting $y = 100 - 2x$

into $f_y = 0$ gives $3x^2 - 100x = 0$ so $x = 0$ or $\frac{100}{3}$. Thus the critical points are $(0, 0)$, $(100, 0)$, $(0, 100)$

and $\left(\frac{100}{3}, \frac{100}{3}\right)$. $D(0, 0) = D(100, 0) = D(0, 100) = -10{,}000$ while $D\left(\frac{100}{3}, \frac{100}{3}\right) = \frac{10{,}000}{3}$ and

$f_{xx}\left(\frac{100}{3}, \frac{100}{3}\right) = -\frac{200}{3} < 0$. Thus $(0, 0)$, $(100, 0)$ and $(0, 100)$ are saddle points whereas $f\left(\frac{100}{3}, \frac{100}{3}\right)$ is

a local maximum. Thus the numbers are $x = y = z = \frac{100}{3}$.

37. Maximize $f(x, y) = xy\left(36 - 9x^2 - 36y^2\right)^{1/2}/2$ with (x, y, z) in first octant. Then

$f_x = \dfrac{y\left(36 - 9x^2 - 36y^2\right)^{1/2}}{2} + \dfrac{-9x^2y\left(36 - 9x^2 - 36y^2\right)^{-1/2}}{2} = \dfrac{\left(36y - 18x^2y - 36y^3\right)}{2\left(36 - 9x^2 - 36y^2\right)^{1/2}}$ and

$f_y = \dfrac{36x - 9x^3 - 72xy^2}{2\left(36 - 9x^2 - 36y^2\right)^{1/2}}$. Setting $f_x = 0$ gives $y = 0$ or $y^2 = \dfrac{2 - x^2}{2}$ but $y > 0$, so only the latter

solution applies. Substituting this y into $f_y = 0$ gives $x^2 = \frac{4}{3}$ or $x = \frac{2}{\sqrt{3}}$, $y = \frac{1}{\sqrt{3}}$ and then

$z^2 = (36 - 12 - 12)/4 = 3$. The fact that this gives a maximum volume follows from the geometry.

This maximum volume is $V = (2x)(2y)(2z) = 8\left(\frac{2}{\sqrt{3}}\right)\left(\frac{1}{\sqrt{3}}\right)\left(\sqrt{3}\right) = \frac{16}{\sqrt{3}}$.

39. Maximize $f(x, y) = \dfrac{xy}{3}(6 - x - 2y)$, then the maximum volume is $V = xyz$.

$f_x = \frac{1}{3}\left(6y - 2xy - y^2\right) = \frac{1}{3}y(6 - 2x - 2y)$ and $f_y = \frac{1}{3}x(6 - x - 4y)$. Setting $f_x = 0$ and $f_y = 0$

gives the critical point $(2, 1)$ which geometrically must yield a maximum. Thus the volume of the largest

such box is $V = (2)(1)\left(\frac{2}{3}\right) = \frac{4}{3}$.

41. Let the dimensions be x, y, and z; then $4x + 4y + 4z = c$ and the volume is

$V = xyz = xy\left(\frac{1}{4}c - x - y\right) = \frac{1}{4}cxy - x^2y - xy^2$, $x > 0$, $y > 0$. Then $V_x = \frac{1}{4}cy - 2xy - y^2$ and

$V_y = \frac{1}{4}cx - x^2 - 2xy$, so $V_x = 0 = V_y$ when $2x + y = \frac{1}{4}c$ and $x + 2y = \frac{1}{4}c$. Solving, we get $x = \frac{1}{12}c$,

$y = \frac{1}{12}c$ and $z = \frac{1}{4}c - x - y = \frac{1}{12}c$. From the geometrical nature of the problem, this critical point must

give an absolute maximum. Thus the box is a cube with edge length $\frac{1}{12}c$.

43. Let the dimensions be x, y and z, then minimize $xy + 2(xz + yz)$ if $xyz = 32{,}000$ m^3. Then
$f(x, y) = xy + [64{,}000\,(x + y)\,/xy] = xy + 64{,}000\,(x^{-1} + y^{-1})$, $f_x = y - 64{,}000x^{-2}$,
$f_y = x - 64{,}000y^{-2}$. And $f_x = 0$ implies $y = 64{,}000/x^2$; substituting into $f_y = 0$ implies $x^3 = 64{,}000$
or $x = 40$ and then $y = 40$. Now $D(x, y) = [(2)(64{,}000)]^2\, x^{-3}y^{-3} - 1 > 0$ for $(40, 40)$ and
$f_{xx}(40, 40) > 0$ so this is indeed a minimum. Thus the dimensions of the box are $x = y = 40$ cm,
$z = 20$ cm.

45. Note that here the variables are m and b, and $f(m, b) = \sum\limits_{i=1}^{n} [y_i - (mx_i + b)]^2$. Then

$$f_m = \sum_{i=1}^{n} -2x_i\,[y_i - (mx_i + b)] = 0 \text{ implies } \sum_{i=1}^{n} (x_i y_i - mx_i^2 - bx_i) = 0 \text{ or}$$

$$\sum_{i=1}^{n} x_i y_i = m \sum_{i=1}^{n} x_i^2 + b \sum_{i=1}^{n} x_i \text{ and } f_b = \sum_{i=1}^{n} -2\,[y_i - (mx_i + b)] = 0 \text{ implies}$$

$$\sum_{i=1}^{n} y_i = m \sum_{i=1}^{n} x_i + \sum_{i=1}^{n} b = m \left(\sum_{i=1}^{n} x_i \right) + nb. \text{ Thus we have the two desired equations. Now}$$

$$f_{mm} = \sum_{i=1}^{n} 2x_i^2, \ f_{bb} = \sum_{i=1}^{n} 2 = 2n \text{ and } f_{mb} = \sum_{i=1}^{n} 2x_i. \text{ And } f_{mm}(m, b) > 0 \text{ always and}$$

$$D(m, b) = 4n \left(\sum_{i=1}^{n} x_i^2 \right) - 4 \left(\sum_{i=1}^{n} x_i \right)^2 = 4 \left[n \left(\sum_{i=1}^{n} x_i^2 \right) - \left(\sum_{i=1}^{n} x_i \right)^2 \right] > 0 \text{ always so the}$$

solutions of these two equations do indeed minimize $\sum\limits_{i=1}^{n} d_i^2$.

Section 11.8 Lagrange Multipliers

1. At the extreme values of f, the level curves of f just touch the curve $g(x, y) = 8$ with a common tangent line. (See Figure 1 and the accompanying discussion.) We can observe several such occurrences on the contour map, but the level curve $f(x, y) = c$ with the largest value of c which still intersects the curve $g(x, y) = 8$ is approximately $c = 59$, and the smallest value of c corresponding to a level curve which intersects $g(x, y) = 8$ appears to be $c = 30$. Thus we estimate the maximum value of f subject to the constraint $g(x, y) = 8$ to be about 59 and the minimum to be 30.

3. $f(x, y) = x^2 - y^2$, $g(x, y) = x^2 + y^2 = 1$ $\Rightarrow$ $\nabla f = \langle 2x, -2y \rangle$, $\lambda \nabla g = \langle 2\lambda x, 2\lambda y \rangle$. Then $2x = 2\lambda x$ implies $x = 0$ or $\lambda = 1$. If $x = 0$, then $x^2 + y^2 = 1$ implies $y = \pm 1$ and if $\lambda = 1$, then $-2y = 2\lambda y$ implies $y = 0$ and thus $x = \pm 1$. Thus the possible points for the extrema of f are $(\pm 1, 0)$, $(0, \pm 1)$. But $f(\pm 1, 0) = 1$ while $f(0, \pm 1) = -1$ so the maximum value of f on $x^2 + y^2 = 1$ is $f(\pm 1, 0) = 1$ and the minimum value is $f(0, \pm 1) = -1$.

5. $f(x, y) = xy$, $g(x, y) = 9x^2 + y^2 = 4$ $\Rightarrow$ $\nabla f = \langle y, x \rangle$, $\lambda \nabla g = \langle 18\lambda x, 2\lambda y \rangle$. Then $y = 18\lambda x$ implies $(x, y) = (0, 0)$ or $\lambda = \dfrac{y}{18x}$ and $x = 2\lambda y$ implies $(x, y) = (0, 0)$ or $\lambda = \dfrac{x}{2y}$. Thus $(x, y) = (0, 0)$ or $\dfrac{y}{18x} = \dfrac{x}{2y}$ implies $y^2 = 9x^2$. Now $(x, y) = (0, 0)$ doesn't satisfy $g(x, y) = 4$, and when $y^2 = 9x^2$, $g(x, y) = 4$ implies $x^2 = \frac{2}{9}$ or $x = \pm\frac{\sqrt{2}}{3}$. Hence the possible points are $\left(\pm\frac{\sqrt{2}}{3}, \sqrt{2}\right)$, $\left(\pm\frac{\sqrt{2}}{3}, -\sqrt{2}\right)$ and the maximum value of f on the ellipse is $f\left(\frac{\sqrt{2}}{3}, \sqrt{2}\right) = f\left(-\frac{\sqrt{2}}{3}, -\sqrt{2}\right) = \frac{2}{3}$ while the minimum value is $f\left(-\frac{\sqrt{2}}{3}, \sqrt{2}\right) = f\left(\frac{\sqrt{2}}{3}, -\sqrt{2}\right) = -\frac{2}{3}$.

7. $f(x, y, z) = x + 3y + 5z$, $g(x, y, z) = x^2 + y^2 + z^2 = 1$ $\Rightarrow$ $\nabla f = \langle 1, 3, 5 \rangle$, $\lambda \nabla g = \langle 2\lambda x, 2\lambda y, 2\lambda z \rangle$. Then $\nabla f = \lambda \nabla g$ implies $\lambda = \dfrac{1}{2x} = \dfrac{3}{2y} = \dfrac{5}{2z}$ so $x = \frac{1}{5}z$, $y = \frac{3}{5}z$. Then $x^2 + y^2 + z^2 = 1$ implies $\frac{1}{25}z^2 + \frac{9}{25}z^2 + z^2 = 1$ or $z = \pm\sqrt{\frac{5}{7}}$. Thus the possible points are $\left(\pm\frac{1}{\sqrt{35}}, \pm\frac{3}{\sqrt{35}}, \pm\frac{5}{\sqrt{35}}\right)$ with the maximum being $f\left(\frac{1}{\sqrt{35}}, \frac{3}{\sqrt{35}}, \frac{5}{\sqrt{35}}\right) = \sqrt{35}$ and the minimum being $f\left(-\frac{1}{\sqrt{35}}, -\frac{3}{\sqrt{35}}, -\frac{5}{\sqrt{35}}\right) = -\sqrt{35}$.

9. $f(x, y, z) = xyz$, $g(x, y, z) = x^2 + 2y^2 + 3z^2 = 6$ $\Rightarrow$ $\nabla f = \langle yz, xz, xy \rangle$, $\lambda \nabla g = \langle 2\lambda x, 4\lambda y, 6\lambda z \rangle$. Then $\nabla f = \lambda \nabla g$ implies $\lambda = (yz)/(2x) = (xz)/(4y) = (xy)/(6z)$ or $x^2 = 2y^2$ and $z^2 = \frac{2}{3}y^2$. Thus $x^2 + 2y^2 + 3z^2 = 6$ implies $6y^2 = 6$ or $y = \pm 1$. Then the possible points are $\left(\sqrt{2}, \pm 1, \sqrt{\frac{2}{3}}\right)$, $\left(\sqrt{2}, \pm 1, -\sqrt{\frac{2}{3}}\right)$, $\left(-\sqrt{2}, \pm 1, \sqrt{\frac{2}{3}}\right)$, $\left(-\sqrt{2}, \pm 1, -\sqrt{\frac{2}{3}}\right)$. And the maximum value of f on the ellipsoid is $\frac{2}{\sqrt{3}}$, occurring when all coordinates are positive or exactly two are negative and the minimum is $-\frac{2}{\sqrt{3}}$ occurring when 1 or 3 of the coordinates are negative.

11. $f(x, y, z) = x^2 + y^2 + z^2$, $g(x, y, z) = x^4 + y^4 + z^4 = 1$ $\Rightarrow$ $\nabla f = \langle 2x, 2y, 2z \rangle$,
$\lambda \nabla g = \langle 4\lambda x^3, 4\lambda y^3, 4\lambda z^3 \rangle$.

Case 1: If $x \neq 0$, $y \neq 0$ and $z \neq 0$, then $\nabla f = \lambda \nabla g$ implies $\lambda = 1/(2x^2) = 1/(2y^2) = 1/(2z^2)$ or
$x^2 = y^2 = z^2$ and $3x^4 = 1$ or $x = \pm\frac{1}{\sqrt[4]{3}}$ giving the points $\left(\pm\frac{1}{\sqrt[4]{3}}, \frac{1}{\sqrt[4]{3}}, \frac{1}{\sqrt[4]{3}}\right)$, $\left(\pm\frac{1}{\sqrt[4]{3}}, -\frac{1}{\sqrt[4]{3}}, \frac{1}{\sqrt[4]{3}}\right)$,
$\left(\pm\frac{1}{\sqrt[4]{3}}, \frac{1}{\sqrt[4]{3}}, -\frac{1}{\sqrt[4]{3}}\right)$, $\left(\pm\frac{1}{\sqrt[4]{3}}, -\frac{1}{\sqrt[4]{3}}, -\frac{1}{\sqrt[4]{3}}\right)$ all with an f-value of $\sqrt{3}$.

Case 2: If one of the variables equals zero and the other two are not zero, then the squares of the two
nonzero coordinates are equal with common value $\frac{1}{\sqrt{2}}$ and corresponding f value of $\sqrt{2}$.

Case 3: If exactly two of the variables are zero, then the third variable has value ± 1 with the
corresponding f value of 1. Thus on $x^4 + y^4 + z^4 = 1$, the maximum value of f is $\sqrt{3}$ and the
minimum value is 1.

13. $f(x, y, z, t) = x + y + z + t$, $g(x, y, z, t) = x^2 + y^2 + z^2 + t^2 = 1$ $\Rightarrow$
$\langle 1, 1, 1, 1 \rangle = \langle 2\lambda x, 2\lambda y, 2\lambda z, 2\lambda t \rangle$, so $\lambda = 1/(2x) = 1/(2y) = 1/(2z) = 1/(2t)$ and $x = y = z = t$.
But $x^2 + y^2 + z^2 + t^2 = 1$, so the possible points are $\left(\pm\frac{1}{2}, \pm\frac{1}{2}, \pm\frac{1}{2}, \pm\frac{1}{2}\right)$. Thus the maximum value of
f is $f\left(\frac{1}{2}, \frac{1}{2}, \frac{1}{2}, \frac{1}{2}\right) = 2$ and the minimum value is $f\left(-\frac{1}{2}, -\frac{1}{2}, -\frac{1}{2}, -\frac{1}{2}\right) = -2$.

15. $f(x, y, z) = x + 2y$, $g(x, y, z) = x + y + z = 1$, $h(x, y, z) = y^2 + z^2 = 4$ $\Rightarrow$ $\nabla f = \langle 1, 2, 0 \rangle$,
$\lambda \nabla g = \langle \lambda, \lambda, \lambda \rangle$ and $\mu \nabla h = \langle 0, 2\mu y, 2\mu z \rangle$. Then $1 = \lambda$, $2 = \lambda + 2\mu y$ and $0 = \lambda + 2\mu z$ so
$\mu y = \frac{1}{2} = -\mu z$ or $y = 1/(2\mu)$, $z = -1/(2\mu)$. Thus $x + y + z = 1$ implies $x = 1$ and $y^2 + z^2 = 4$
implies $\mu = \pm\frac{1}{2\sqrt{2}}$. Then the possible points are $\left(1, \pm\sqrt{2}, \mp\sqrt{2}\right)$ and the maximum value is
$f\left(1, \sqrt{2}, -\sqrt{2}\right) = 1 + 2\sqrt{2}$ and the minimum value is $f\left(1, -\sqrt{2}, \sqrt{2}\right) = 1 - 2\sqrt{2}$.

17. $f(x, y, z) = yz + xy$, $g(x, y, z) = xy = 1$, $h(x, y, z) = y^2 + z^2 = 1$ $\Rightarrow$ $\nabla f = \langle y, x + z, y \rangle$,
$\lambda \nabla g = \langle \lambda y, \lambda x, 0 \rangle$, $\mu \nabla h = \langle 0, 2\mu y, 2\mu z \rangle$. Then $y = \lambda y$ implies $\lambda = 1$ [$y \neq 0$ since $g(x, y, z) = 1$],
$x + z = \lambda x + 2\mu y$ and $y = 2\mu z$. Thus $\mu = z/(2y) = y/(2z)$ or $y^2 = z^2$, and so $y^2 + z^2 = 1$ implies
$y = \pm\frac{1}{\sqrt{2}}$, $z = \pm\frac{1}{\sqrt{2}}$. Then $xy = 1$ implies $x = \pm\sqrt{2}$ and the possible points are
$\left(\pm\sqrt{2}, \pm\frac{1}{\sqrt{2}}, \frac{1}{\sqrt{2}}\right)$, $\left(\pm\sqrt{2}, \pm\frac{1}{\sqrt{2}}, -\frac{1}{\sqrt{2}}\right)$. Hence the maximum of f subject to the constraints is
$f\left(\pm\sqrt{2}, \pm\frac{1}{\sqrt{2}}, \pm\frac{1}{\sqrt{2}}\right) = \frac{3}{2}$ and the minimum is $f\left(\pm\sqrt{2}, \pm\frac{1}{\sqrt{2}}, \mp\frac{1}{\sqrt{2}}\right) = \frac{1}{2}$.

Note: Since $xy = 1$ is one of the constraints we could have solved the problem by solving
$f(y, z) = yz + 1$ subject to $y^2 + z^2 = 1$.

19. $f(x, y) = e^{-xy}$. For the interior of the region, we find the critical points: $f_x = -ye^{-xy}$, $f_y = -xe^{-xy}$, so the only critical point is $(0, 0)$, and $f(0, 0) = 1$. For the boundary, we use Lagrange multipliers. $g(x, y) = x^2 + 4y^2 = 1 \Rightarrow \lambda \nabla g = \langle 2\lambda x, 8\lambda y \rangle$, so setting $\nabla f = \lambda \nabla g$ we get $-ye^{-xy} = 2\lambda x$ and $-xe^{-xy} = 8\lambda y$. The first of these gives $e^{-xy} = -2\lambda x/y$, and then the second gives $-x(-2\lambda x/y) = 8\lambda y \Rightarrow x^2 = 4y^2$. Solving this last equation with the constraint $x^2 + 4y^2 = 1$ gives $x = \pm\frac{1}{\sqrt{2}}$ and $y = \pm\frac{1}{2\sqrt{2}}$. Now $f\left(\pm\frac{1}{\sqrt{2}}, \mp\frac{1}{2\sqrt{2}}\right) = e^{1/4} \approx 1.284$ and $f\left(\pm\frac{1}{\sqrt{2}}, \pm\frac{1}{2\sqrt{2}}\right) = e^{-1/4} \approx 0.779$. The former are the maxima on the region and the latter are the minima.

21. $P(L, K) = bL^\alpha K^{1-\alpha}$, $g(L, K) = mL + nK = p \Rightarrow \nabla P = \langle \alpha bL^{\alpha-1}K^{1-\alpha}, (1-\alpha)bL^\alpha K^{-\alpha} \rangle$, $\lambda \nabla g = \langle \lambda m, \lambda n \rangle$. Then $\alpha b(K/L)^{1-\alpha} = \lambda m$ and $(1-\alpha)b(L/K)^\alpha = \lambda n$ and $mL + nK = p$, so $\alpha b(K/L)^{1-\alpha}/m = (1-\alpha)b(L/K)^\alpha/n$ or $n\alpha/[m(1-\alpha)] = (L/K)^\alpha(L/K)^{1-\alpha}$ or $L = Kn\alpha/[m(1-\alpha)]$. Substituting into $mL + nK = p$ gives $K = (1-\alpha)p/n$ and $L = \alpha p/m$ for the maximum production.

23. Let the sides of the rectangle be x and y. Then $f(x, y) = xy$, $g(x, y) = 2x + 2y = p \Rightarrow \nabla f(x, y) = \langle y, x \rangle$, $\lambda \nabla g = \langle 2\lambda, 2\lambda \rangle$. Then $\lambda = \frac{1}{2}y = \frac{1}{2}x$ implies $x = y$ and the rectangle with maximum area is a square with side length $\frac{1}{4}p$.

25. $f(x, y, z) = (x - 2)^2 + (y + 2)^2 + (z - 3)^2$, $g(x, y, z) = 6x + 4y - 3z = 2 \Rightarrow \nabla f = \langle 2(x - 2), 2(y + 2), 2(z - 3) \rangle = \lambda \nabla g = \langle 6\lambda, 4\lambda, -3\lambda \rangle$, so $x = 3\lambda + 2$, $y = 2\lambda - 2$, $z = -\frac{3}{2}\lambda + 3$ and $(18\lambda + 12) + (8\lambda - 8) + \frac{9}{2}\lambda - 9 = 2$ implies $\lambda = \frac{14}{61}$. Thus the shortest distance is $\sqrt{\left(\frac{42}{61}\right)^2 + \left(\frac{28}{61}\right)^2 + \left(-\frac{21}{61}\right)^2} = \frac{7}{\sqrt{61}}$.

27. $f(x, y, z) = x^2 + y^2 + z^2$, $g(x, y, z) = z^2 - xy - 1 = 0 \Rightarrow \nabla f = \langle 2x, 2y, 2z \rangle = \lambda \nabla g = \langle -\lambda y, -\lambda x, 2\lambda z \rangle$. Then $2z = 2\lambda z$ implies $z = 0$ or $\lambda = 1$. If $z = 0$ then $g(x, y, z) = 1$ implies $xy = -1$ or $x = -1/y$. Thus $2x = -\lambda y$ and $2y = -\lambda x$ imply $\lambda = 2/y^2 = 2y^2$ or $y = \pm 1$, $x = \pm 1$. If $\lambda = 1$, then $2x = -y$ and $2y = -x$ imply $x = y = 0$, so $z = \pm 1$. Hence the possible points are $(\pm 1, \mp 1, 0)$, $(0, 0, \pm 1)$ and the minimum value of f is $f(0, 0, \pm 1) = 1$, so the points closest to the origin are $(0, 0, \pm 1)$.

29. $f(x, y, z) = xyz$, $g(x, y, z) = x + y + z = 100 \Rightarrow \nabla f = \langle yz, xz, xy \rangle = \lambda \nabla g = \langle \lambda, \lambda, \lambda \rangle$. Then $\lambda = yz = xz = xy$ implies $x = y = z = \frac{100}{3}$.

31. If the dimensions are $2x$, $2y$ and $2z$, then $f(x, y, z) = 8xyz$ and $g(x, y, z) = 9x^2 + 36y^2 + 4z^2 = 36 \Rightarrow \nabla f = \langle 8yz, 8xz, 8xy \rangle = \lambda \nabla g = \langle 18\lambda x, 72\lambda y, 8\lambda z \rangle$. Thus $18\lambda x = 8yz$, $72\lambda y = 8xz$, $8\lambda z = 8xy$ so $x^2 = 4y^2$, $z^2 = 9y^2$ and $36y^2 + 36y^2 + 36y^2 = 36$ or $y = \frac{1}{\sqrt{3}}$ $(y > 0)$. Thus the volume of the largest such rectangle is $8\left(\frac{1}{\sqrt{3}}\right)\left(\frac{2}{\sqrt{3}}\right)\left(\frac{3}{\sqrt{3}}\right) = 16\sqrt{3}$.

33. $f(x, y, z) = xyz$, $g(x, y, z) = x + 2y + 3z = 6$ $\Rightarrow$ $\nabla f = \langle yz, xz, xy \rangle = \lambda \nabla g = \langle \lambda, 2\lambda, 3\lambda \rangle$.
Then $\lambda = yz = \frac{1}{2}xz = \frac{1}{3}xy$ implies $x = 2y$, $z = \frac{2}{3}y$. But $2y + 2y + 2y = 6$ so $y = 1$, $x = 2$, $z = \frac{2}{3}$
and the volume is $V = \frac{4}{3}$.

35. $f(x, y, z) = xyz$, $g(x, y, z) = 4(x + y + z) = c$ $\Rightarrow$ $\nabla f = \langle yz, xz, xy \rangle$, $\lambda \nabla g = \langle 4\lambda, 4\lambda, 4\lambda \rangle$.
Thus $4\lambda = yz = xz = xy$ or $x = y = z = \frac{1}{12}c$ are the dimensions giving the maximum volume.

37. We need to find the extrema of $f(x, y, z) = x^2 + y^2 + z^2$ subject to the two constraints
$g(x, y, z) = x + y + 2z = 2$ and $h(x, y, z) = x^2 + y^2 - z = 0$. $\nabla f = \langle 2x, 2y, 2z \rangle$, $\lambda \nabla g = \langle \lambda, \lambda, 2\lambda \rangle$
and $\mu \nabla h = \langle 2\mu x, 2\mu y, -\mu \rangle$. Thus we need (1) $2x = \lambda + 2\mu x$, (2) $2y = \lambda + 2\mu y$, (3) $2z = 2\lambda - \mu$,
(4) $x + y + 2z = 2$, and (5) $x^2 + y^2 - z = 0$. From (1) and (2), $2(x - y) = 2\mu(x - y)$, so if $x \neq y$,
$\mu = 1$. Putting this in (3) gives $2z = 2\lambda - 1$ or $\lambda = z + \frac{1}{2}$, but putting $\mu = 1$ into (1) says $\lambda = 0$. Hence
$z + \frac{1}{2} = 0$ or $z = -\frac{1}{2}$. Then (4) and (5) become $x + y - 3 = 0$ and $x^2 + y^2 + \frac{1}{2} = 0$. The last equation
cannot be true, so this case gives no solution. So we must have $x = y$. Then (4) and (5) become
$2x + 2z = 2$ and $2x^2 - z = 0$ which imply $z = 1 - x$ and $z = 2x^2$. Thus $2x^2 = 1 - x$ or
$2x^2 + x - 1 = (2x - 1)(x + 1) = 0$ so $x = \frac{1}{2}$ or $x = -1$. The two points to check are $\left(\frac{1}{2}, \frac{1}{2}, \frac{1}{2}\right)$ and
$(-1, -1, 2)$: $f\left(\frac{1}{2}, \frac{1}{2}, \frac{1}{2}\right) = \frac{3}{4}$ and $f(-1, -1, 2) = 6$. Thus $\left(\frac{1}{2}, \frac{1}{2}, \frac{1}{2}\right)$ is the point on the ellipse nearest
the origin and $(-1, -1, 2)$ is the one farthest from the origin.

Chapter 11 Review

Concept Check

1. (a) A function f of two variables is a rule that assigns to each ordered pair (x, y) of real numbers in its domain a unique real number denoted by $f(x, y)$.

(b) One way to visualize a function of two variables is by graphing it, resulting in the surface $z = f(x, y)$. Another method for visualizing a function of two variables is a contour map. The contour map consists of level curves of the function which are horizontal traces of the graph of the function projected onto the xy-plane.

2. A function f of three variables is a rule that assigns to each ordered triple (x, y, z) in its domain a unique real number $f(x, y, z)$. We can visualize a function of three variables by examining its level surfaces $f(x, y, z) = k$, where k is a constant.

3. See Definition 1 and the preceding discussion in Section 11.2. We can show that a limit at a point does not exist by finding two different paths approaching the point along which $f(x, y)$ has different limits.

4. (a) See Definition 3 in Section 11.2.

(b) If f is continuous on $\mathbb{R}^2$, its graph will appear as a surface without holes or breaks.

5. (a) See (2) and (3) in Section 11.3.

(b) See the discussion preceding Example 2 on page 769.

(c) To find f_x, regard y as a constant and differentiate $f(x, y)$ with respect to x. To find f_y, regard x as a constant and differentiate $f(x, y)$ with respect to y.

6. See the statement of Clairaut's Theorem on page 773.

7. (a) See (2) in Section 11.4.

(b) See (19) and the preceding discussion in Section 11.6.

(c) See the discussion following Example 6 on page 786.

8. See (3) and (4) and the accompanying discussion in Section 11.4. We can interpret the linearization of f at (a, b) geometrically as the linear function whose graph is the tangent plane to the graph of f at (a, b). Thus it is the linear function which best approximates f near (a, b).

9. (a) See Definition 7 in Section 11.4.

(b) Use Theorem 8 in Section 11.4.

10. See (10) and the associated discussion in Section 11.4.

11. See (2) and (3) in Section 11.5.

12. See (7) and the preceding discussion in Section 11.5.

13. (a) See Definition 2 in Section 11.6. We can interpret it as the rate of change of f at (x_0, y_0) in the direction of $\mathbf{u}$. Geometrically, if P is the point $(x_0, y_0, f(x_0, y_0))$ on the graph of f and C is the curve of intersection of the graph of f with the vertical plane that passes through P in the direction $\mathbf{u}$, the directional derivative of f at (x_0, y_0) in the direction of $\mathbf{u}$ is the slope of the tangent line to C at P. (See Figure 5 on page 801.)

(b) See Theorem 3 in Section 11.6.

14. (a) See (8) and (13) in Section 11.6.

(b) $D_{\mathbf{u}} f(x, y) = \nabla f(x, y) \cdot \mathbf{u}$

(c) The gradient vector of a function points in the direction of maximum rate of increase of the function. On a graph of the function, the gradient points in the direction of steepest ascent.

15. (a) f has a local maximum at (a, b) if $f(x, y) \leq f(a, b)$ when (x, y) is near (a, b).

(b) f has an absolute maximum at (a, b) if $f(x, y) \leq f(a, b)$ for all points (x, y) in the domain of f.

(c) f has a local minimum at (a, b) if $f(x, y) \geq f(a, b)$ when (x, y) is near (a, b).

(d) f has an absolute minimum at (a, b) if $f(x, y) \geq f(a, b)$ for all points (x, y) in the domain of f.

(e) f has a saddle point at (a, b) if $f(a, b)$ is a local maximum in one direction but a local minimum in another.

16. (a) By Theorem 2 in Section 11.7, if f has a local maximum at (a, b) and the first-order partial derivatives of f exist there, then $f_x(a, b) = 0$ and $f_y(a, b) = 0$.

(b) A critical point of f is a point (a, b) such that $f_x(a, b) = 0$ and $f_y(a, b) = 0$ or one of these partial derivatives does not exist.

17. See (3) in Section 11.7.

18. (a) See Figure 11 and the accompanying discussion on page 817.

(b) See Theorem 8 in Section 11.7.

(c) See the procedure outlined in (9) in Section 11.7.

19. See the discussion beginning on page 822; see the discussion preceding Example 5 on page 827.

Review

True-False Quiz

1. True. $f_y(a,b) = \lim\limits_{h \to 0} \dfrac{f(a, b+h) - f(a,b)}{h}$ from Equation 11.3.3. Let $h = y - b$. As $h \to 0$, $y \to b$.

Then by substituting, we get $f_y(a,b) = \lim\limits_{y \to b} \dfrac{f(a,y) - f(a,b)}{y - b}$.

3. False. $f_{xy} = \dfrac{\partial^2 f}{\partial y \partial x}$

5. False. See Example 3 in Section 11.2.

7. True. If f has a local minimum and f is differentiable at (a,b) then by Theorem 11.7.2, $f_x(a,b) = 0$ and $f_y(a,b) = 0$, so $\nabla f(a,b) = \langle f_x(a,b), f_y(a,b) \rangle = \langle 0,0 \rangle = \mathbf{0}$.

9. False. $\nabla f(x,y) = \langle 0, 1/y \rangle$.

11. True. $\nabla f = \langle \cos x, \cos y \rangle$, so $|\nabla f| = \sqrt{\cos^2 x + \cos^2 y}$. But $|\cos \theta| \leq 1$, so $|\nabla f| \leq \sqrt{2}$. Now $D_{\mathbf{u}} f(x,y) = \nabla f \cdot \mathbf{u} = |\nabla f| |\mathbf{u}| \cos \theta$, but $\mathbf{u}$ is a unit vector, so $|D_{\mathbf{u}} f(x,y)| \leq \sqrt{2} \cdot 1 \cdot 1 = \sqrt{2}$.

Exercises

1. The domain of $\sin^{-1} x$ is $-1 \leq x \leq 1$ while the domain of $\tan^{-1} y$ is all real numbers, so the domain of $f(x,y) = \sin^{-1} x + \tan^{-1} y$ is $\{(x,y) \mid -1 \leq x \leq 1\}$.

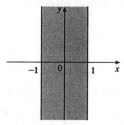

3. $z = f(x,y) = 1 - x^2 - y^2$, a paraboloid with vertex $(0,0,1)$.

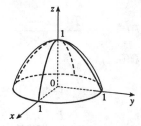

5. Let $k = e^{-c} = e^{-(x^2 + y^2)}$ be the level curves. Then $-\ln k = c = x^2 + y^2$, so we have a family of concentric circles.

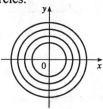

7. f is a rational function, so it is continuous on its domain. Since f is defined at $(1,1)$, we use direct substitution to evaluate the limit:

$$\lim_{(x,y) \to (1,1)} \frac{2xy}{x^2 + 2y^2} = \frac{2(1)(1)}{1^2 + 2(1)^2} = \frac{2}{3}.$$

9. (a) $T_x(6,4) = \lim\limits_{h \to 0} \dfrac{T(6+h,4) - T(6,4)}{h}$, so we can approximate $T_x(6,4)$ by considering $h = \pm 2$

and using the values given in the table: $T_x(6,4) \approx \dfrac{T(8,4) - T(6,4)}{2} = \dfrac{86 - 80}{2} = 3$,

$T_x(6,4) \approx \dfrac{T(4,4) - T(6,4)}{-2} = \dfrac{72 - 80}{-2} = 4$. Averaging these values, we estimate $T_x(6,4)$ to be

approximately $3.5\,°C/m$. Similarly, $T_y(6,4) = \lim\limits_{h \to 0} \dfrac{T(6,4+h) - T(6,4)}{h}$, which we can

approximate with $h = \pm 2$: $T_y(6,4) \approx \dfrac{T(6,6) - T(6,4)}{2} = \dfrac{75 - 80}{2} = -2.5$,

$T_y(6,4) \approx \dfrac{T(6,2) - T(6,4)}{-2} = \dfrac{87 - 80}{-2} = -3.5$. Averaging these values, we estimate $T_y(6,4)$ to

be approximately $-3.0\,°C/m$.

(b) Here $\mathbf{u} = \left\langle \frac{1}{\sqrt{2}}, \frac{1}{\sqrt{2}} \right\rangle$, so by Equation 11.6.9,

$D_{\mathbf{u}}T(6,4) = \nabla T(6,4) \cdot \mathbf{u} = T_x(6,4)\frac{1}{\sqrt{2}} + T_y(6,4)\frac{1}{\sqrt{2}}$. Using our estimates from part (a), we

have $D_{\mathbf{u}}T(6,4) \approx (3.5)\frac{1}{\sqrt{2}} + (-3.0)\frac{1}{\sqrt{2}} = \frac{1}{2\sqrt{2}} \approx 0.35$. This means that as we move through the

point $(6,4)$ in the direction of $\mathbf{u}$, the temperature increases at a rate of approximately $0.35\,°C/m$.

Alternatively, we can use Definition 11.6.2: $D_{\mathbf{u}}T(6,4) = \lim\limits_{h \to 0} \dfrac{T\left(6 + h\frac{1}{\sqrt{2}}, 4 + h\frac{1}{\sqrt{2}}\right) - T(6,4)}{h}$,

which we can estimate with $h = \pm 2\sqrt{2}$. Then $D_{\mathbf{u}}T(6,4) \approx \dfrac{T(8,6) - T(6,4)}{2\sqrt{2}} = \dfrac{80 - 80}{2\sqrt{2}} = 0$,

$D_{\mathbf{u}}T(6,4) \approx \dfrac{T(4,2) - T(6,4)}{-2\sqrt{2}} = \dfrac{74 - 80}{-2\sqrt{2}} = \dfrac{3}{\sqrt{2}}$. Averaging these values, we have

$D_{\mathbf{u}}T(6,4) \approx \frac{3}{2\sqrt{2}} \approx 1.1\,°C/m$.

(c) $T_{xy}(x,y) = \dfrac{\partial}{\partial y}[T_x(x,y)] = \lim\limits_{h \to 0} \dfrac{T_x(x, y+h) - T_x(x,y)}{h}$, so

$T_{xy}(6,4) = \lim\limits_{h \to 0} \dfrac{T_x(6, 4+h) - T_x(6,4)}{h}$ which we can estimate with $h = \pm 2$. We have

$T_x(6,4) \approx 3.5$ from part (a), but we will also need values for $T_x(6,6)$ and $T_x(6,2)$. If we use

$h = \pm 2$ and the values given in the table, we have

$T_x(6,6) \approx \dfrac{T(8,6) - T(6,6)}{2} = \frac{80 - 75}{2} = 2.5$, $T_x(6,6) \approx \dfrac{T(4,6) - T(6,6)}{-2} = \frac{68 - 75}{-2} = 3.5$.

Averaging these values, we estimate $T_x(6,6) \approx 3.0$. Similarly,

$T_x(6,2) \approx \dfrac{T(8,2) - T(6,2)}{2} = \frac{90 - 87}{2} = 1.5$, $T_x(6,2) \approx \dfrac{T(4,2) - T(6,2)}{-2} = \frac{74 - 87}{-2} = 6.5$.

Averaging these values, we estimate $T_x(6,2) \approx 4.0$. Finally, we estimate

$T_{xy}(6,4)$: $T_{xy}(6,4) \approx \dfrac{T_x(6,6) - T_x(6,4)}{2} = \frac{3.0 - 3.5}{2} = -0.25$,

$T_{xy}(6,4) \approx \dfrac{T_x(6,2) - T_x(6,4)}{-2} = \frac{4.0 - 3.5}{-2} = -0.25$. Averaging these values, we have

$T_{xy}(6,4) \approx -0.25$.

11. $f(x,y) = 3x^4 - x\sqrt{y} \Rightarrow f_x = 12x^3 - \sqrt{y}, f_y = -\frac{1}{2}xy^{-1/2}$

13. $f(s,t) = e^{2s}\cos \pi t \Rightarrow f_s = 2e^{2s}\cos \pi t, f_t = -\pi e^{2s}\sin \pi t$

15. $f(x,y,z) = xy^z \Rightarrow f_x = y^z, f_y = xzy^{z-1}, f_z = xy^z \ln y$

17. $f(x,y) = x^2y^3 - 2x^4 + y^2 \Rightarrow f_x = 2xy^3 - 8x^3, f_y = 3x^2y^2 + 2y, f_{xx} = 2y^3 - 24x^2,$
$f_{yy} = 6x^2y + 2,$ and $f_{xy} = f_{yx} = 6xy^2.$

19. $f(x,y,z) = xy^2z^3 \Rightarrow f_x = y^2z^3, f_y = 2xyz^3, f_z = 3xy^2z^2, f_{xx} = 0, f_{yy} = 2xz^3, f_{zz} = 6xy^2z,$
$f_{xy} = f_{yx} = 2yz^3, f_{xz} = f_{zx} = 3y^2z^2,$ and $f_{yz} = f_{zy} = 6xyz^2.$

21. $u = x^y \Rightarrow u_x = yx^{y-1}, u_y = x^y \ln x$ and $(x/y)u_x + (\ln x)^{-1}u_y = x^y + x^y = 2u.$

23. $z_x(0,1) = 0, z_y(0,1) = 6$ and an equation of the tangent plane is $z - 5 = 6(y-1)$ or $z - 6y = -1.$

25. $F(x,y,z) = xy^2z^3, \nabla F = \langle y^2z^3, 2xyz^3, 3xy^2z^2 \rangle$ and $\nabla F(3,2,1) = \langle 4, 12, 36 \rangle.$ Thus an equation of
the tangent plane is $4x + 12y + 36z = 72$ or $x + 3y + 9z = 18.$

27. $\mathbf{r}_u = -v\mathbf{j} + 2u\mathbf{k}, \mathbf{r}_v = 2v\mathbf{i} - u\mathbf{j}$ and $\mathbf{r}_u \times \mathbf{r}_v = 2u^2\mathbf{i} + 4uv\mathbf{j} + 2v^2\mathbf{k}.$ Since the point $(4, -2, 1)$
corresponds to $u = 1, v = 2$ (or $u = -1, v = -2$ but $\mathbf{r}_u \times \mathbf{r}_v$ is the same for both), a normal vector to
the surface at $(4, -2, 1)$ is $2\mathbf{i} + 8\mathbf{j} + 8\mathbf{k}$ and an equation of the tangent plane is $2x + 8y + 8z = 0$ or
$x + 4y + 4z = 0.$

29. $F(x,y,z) = x^2 + y^2 + z^2, \nabla F(x_0, y_0, z_0) = \langle 2x_0, 2y_0, 2z_0 \rangle = k\langle 2, 1, -3 \rangle$ or $x_0 = k, y_0 = \frac{1}{2}k$ and
$z_0 = -\frac{3}{2}k.$ But $x_0^2 + y_0^2 + z_0^2 = 1,$ so $\frac{7}{2}k^2 = 1$ and $k = \pm\sqrt{\frac{2}{7}}.$ Hence there are two such points:
$\left(\pm\sqrt{\frac{2}{7}}, \pm\frac{1}{\sqrt{14}}, \mp\frac{3}{\sqrt{14}} \right).$

31. $f(x,y,z) = x^3\sqrt{y^2 + z^2} \Rightarrow f_x(x,y,z) = 3x^2\sqrt{y^2 + z^2}, f_y(x,y,z) = \dfrac{yx^3}{\sqrt{y^2 + z^2}},$ and
$f_z(x,y,z) = \dfrac{zx^3}{\sqrt{y^2 + z^2}},$ so $f(2,3,4) = 8(5) = 40, f_x(2,3,4) = 3(4)\sqrt{25} = 60,$
$f_y(2,3,4) = \frac{3(8)}{\sqrt{25}} = \frac{24}{5},$ and $f_z(2,3,4) = \frac{4(8)}{\sqrt{25}} = \frac{32}{5}.$ Then the linear approximation of f at $(2,3,4)$ is
$$f(x,y,z) \approx f(2,3,4) + f_x(2,3,4)(x-2) + f_y(2,3,4)(y-3) + f_z(2,3,4)(z-4)$$
$$= 40 + 60(x-2) + \tfrac{24}{5}(y-3) + \tfrac{32}{5}(z-4) = 60x + \tfrac{24}{5}y + \tfrac{32}{5}z - 120$$
Then
$$(1.98)^3\sqrt{(3.01)^2 + (3.97)^2} = f(1.98, 3.01, 3.97) \approx 60(1.98) + \tfrac{24}{5}(3.01) + \tfrac{32}{5}(3.97) - 120$$
$$= 38.656$$

33. $\dfrac{dw}{dt} = \dfrac{1}{2\sqrt{x}}(2e^{2t}) + \dfrac{2y}{z}(3t^2 + 4) + \dfrac{-y^2}{z^2}(2t) = e^t + \dfrac{2y}{z}(3t^2 + 4) - 2t\dfrac{y^2}{z^2}$

35. By the Chain Rule, $\dfrac{\partial z}{\partial s} = \dfrac{\partial z}{\partial x}\dfrac{\partial x}{\partial s} + \dfrac{\partial z}{\partial y}\dfrac{\partial y}{\partial s}$. When $s = 1$ and

$t = 2$, $x = g(1, 2) = 3$ and $y = h(1, 2) = 6$, so

$\dfrac{\partial z}{\partial s} = f_x(3, 6)\, g_s(1, 2) + f_y(3, 6)\, h_s(1, 2) = (7)(-1) + (8)(-5) = -47$. Similarly,

$\dfrac{\partial z}{\partial t} = \dfrac{\partial z}{\partial x}\dfrac{\partial x}{\partial t} + \dfrac{\partial z}{\partial y}\dfrac{\partial y}{\partial t}$, so $\dfrac{\partial z}{\partial t} = f_x(3, 6)\, g_t(1, 2) + f_y(3, 6)\, h_t(1, 2) = (7)(4) + (8)(10) = 108$.

37. $\dfrac{\partial z}{\partial x} = 2xf'\left(x^2 - y^2\right)$, $\dfrac{\partial z}{\partial y} = 1 - 2yf'\left(x^2 - y^2\right)$ $\left[\text{where } f' = \dfrac{df}{d\left(x^2 - y^2\right)}\right]$. Then

$y\dfrac{\partial z}{\partial x} + x\dfrac{\partial z}{\partial y} = 2xyf'\left(x^2 - y^2\right) + x - 2xyf'\left(x^2 - y^2\right) = x$.

39. $\dfrac{\partial z}{\partial x} = \dfrac{\partial z}{\partial u}y + \dfrac{\partial z}{\partial v}\dfrac{-y}{x^2}$ and

$$\frac{\partial^2 z}{\partial x^2} = y\frac{\partial}{\partial x}\left(\frac{\partial z}{\partial u}\right) + \frac{2y}{x^3}\frac{\partial z}{\partial v} + \frac{-y}{x^2}\frac{\partial}{\partial x}\left(\frac{\partial z}{\partial v}\right)$$

$$= \frac{2y}{x^3}\frac{\partial z}{\partial v} + y\left(\frac{\partial^2 z}{\partial u^2}y + \frac{\partial^2 z}{\partial v\partial u}\frac{-y}{x^2}\right) + \frac{-y}{x^2}\left(\frac{\partial^2 z}{\partial v^2}\frac{-y}{x^2} + \frac{\partial^2 z}{\partial u\partial v}y\right)$$

$$= \frac{2y}{x^3}\frac{\partial z}{\partial v} + y^2\frac{\partial^2 z}{\partial u^2} - \frac{2y^2}{x^2}\frac{\partial^2 z}{\partial u\partial v} + \frac{y^2}{x^4}\frac{\partial^2 z}{\partial v^2}$$

Also $\dfrac{\partial z}{\partial y} = x\dfrac{\partial z}{\partial u} + \dfrac{1}{x}\dfrac{\partial z}{\partial v}$ and

$$\frac{\partial^2 z}{\partial y^2} = x\frac{\partial}{\partial y}\left(\frac{\partial z}{\partial u}\right) + \frac{1}{x}\frac{\partial}{\partial y}\left(\frac{\partial z}{\partial v}\right) = x\left(\frac{\partial^2 z}{\partial u^2}x + \frac{\partial^2 z}{\partial v\partial u}\frac{1}{x}\right) + \frac{1}{x}\left(\frac{\partial^2 z}{\partial v^2}\frac{1}{x} + \frac{\partial^2 z}{\partial u\partial v}x\right)$$

$$= x^2\frac{\partial^2 z}{\partial u^2} + 2\frac{\partial^2 z}{\partial u\partial v} + \frac{1}{x^2}\frac{\partial^2 z}{\partial v^2}$$

Thus

$$x^2\frac{\partial^2 z}{\partial x^2} - y^2\frac{\partial^2 z}{\partial y^2} = \frac{2y}{x}\frac{\partial z}{\partial v} + x^2y^2\frac{\partial^2 z}{\partial u^2} - 2y^2\frac{\partial^2 z}{\partial u\partial v} + \frac{y^2}{x^2}\frac{\partial^2 z}{\partial v^2} - x^2y^2\frac{\partial^2 z}{\partial u^2} - 2y^2\frac{\partial^2 z}{\partial u\partial v} - \frac{y^2}{x^2}\frac{\partial^2 z}{\partial v^2}$$

$$= \frac{2y}{x}\frac{\partial z}{\partial v} - 4y^2\frac{\partial^2 z}{\partial u\partial v} = 2v\frac{\partial z}{\partial v} - 4uv\frac{\partial^2 z}{\partial u\partial v}$$

since $y = xv = \dfrac{uv}{y}$ or $y^2 = uv$.

41. $\nabla f = \left\langle z^2\sqrt{y}\,e^{x\sqrt{y}},\ \dfrac{xz^2 e^{x\sqrt{y}}}{2\sqrt{y}},\ 2ze^{x\sqrt{y}}\right\rangle = ze^{x\sqrt{y}}\left\langle z\sqrt{y},\ \dfrac{xz}{2\sqrt{y}},\ 2\right\rangle$

43. $\nabla f = \langle 1/\sqrt{x},\ -2y\rangle$, $\nabla f(1, 5) = \langle 1, -10\rangle$, $\mathbf{u} = \frac{1}{5}\langle 3, -4\rangle$. Then $D_{\mathbf{u}}f(1, 5) = \frac{43}{5}$.

45. $\nabla f = \left\langle 2xy,\ x^2 + 1/\left(2\sqrt{y}\right)\right\rangle$, $|\nabla f(2, 1)| = \left|\langle 4, \frac{9}{2}\rangle\right|$. Thus the maximum rate of change of f at $(2, 1)$ is $\dfrac{\sqrt{145}}{2}$ in the direction $\langle 4, \frac{9}{2}\rangle$.

47. First we draw a line passing through Homestead and the eye of the hurricane. We can approximate the directional derivative at Homestead in the direction of the eye of the hurricane by the average rate of change of wind speed between the points where this line intersects the contour lines closest to Homestead. In the direction of the eye of the hurricane, the wind speed changes from 45 to 50 knots. We estimate the distance between these two points to be approximately 8 miles, so the rate of change of wind speed in the direction given is approximately $\frac{50-45}{8} = \frac{5}{8} = 0.625$ knots/mi.

49. $f(x,y) = x^2 - xy + y^2 + 9x - 6y + 10 \Rightarrow$
$f_x = 2x - y + 9$, $f_y = -x + 2y - 6$, $f_{xx} = 2 = f_{yy}$,
$f_{xy} = -1$. Then $f_x = 0$ and $f_y = 0$ imply $y = 1$, $x = -4$.
Thus the only critical point is $(-4, 1)$ and $f_{xx}(-4, 1) > 0$,
$D(-4, 1) = 3 > 0$, so $f(-4, 1) = -11$ is a local minimum.

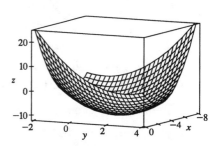

51. $f(x,y) = 3xy - x^2 y - xy^2 \Rightarrow f_x = 3y - 2xy - y^2$,
$f_y = 3x - x^2 - 2xy$, $f_{xx} = -2y$, $f_{yy} = -2x$, $f_{xy} = 3 - 2x - 2y$.
Then $f_x = 0$ implies $y(3 - 2x - y) = 0$ so $y = 0$ or $y = 3 - 2x$.
Substituting into $f_y = 0$ implies $x(3 - x) = 0$ or $3x(-1 + x) = 0$.
Hence the critical points are $(0,0)$, $(3,0)$, $(0,3)$ and $(1,1)$.
$D(0,0) = D(3,0) = D(0,3) = -9 < 0$ so $(0,0)$, $(3,0)$, and $(0,3)$
are saddle points. $D(1,1) = 3 > 0$ and $f_{xx}(1,1) = -2 < 0$, so
$f(1,1) = 1$ is a local maximum.

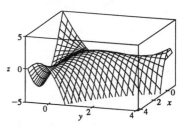

53. First solve inside D. Here $f_x = 4y^2 - 2xy^2 - y^3$,
$f_y = 8xy - 2x^2 y - 3xy^2$. Then $f_x = 0$ implies $y = 0$ or $y = 4 - 2x$,
but $y = 0$ isn't inside D. Substituting $y = 4 - 2x$ into $f_y = 0$ implies
$x = 0$, $x = 2$ or $x = 1$, but $x = 0$ isn't inside D, and when $x = 2$,
$y = 0$ but $(2,0)$ isn't inside D. Thus the only critical point inside D is
$(1,2)$ and $f(1,2) = 4$. Secondly we consider the boundary of D. On
L_1, $f(x,0) = 0$ and so $f = 0$ on L_1. On L_2, $x = -y + 6$ and
$f(-y+6, y) = y^2(6 - y)(-2) = -2(6y^2 - y^3)$ which has

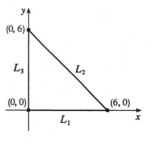

critical points at $y = 0$ and $y = 4$. Then $f(6,0) = 0$ while $f(2,4) = -64$. On L_3, $f(0,y) = 0$, so
$f = 0$ on L_3. Thus on D the absolute maximum of f is $f(1,2) = 4$ while the absolute minimum is
$f(2,4) = -64$.

55. $f(x, y) = x^3 - 3x + y^4 - 2y^2$

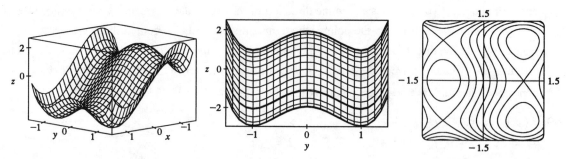

From the graphs, it appears that f has a local maximum $f(-1, 0) \approx 2$, local minima $f(1, \pm 1) \approx -3$, and saddle points at $(-1, \pm 1)$ and $(1, 0)$.

To find the exact quantities, we calculate $f_x = 3x^2 - 3 = 0 \iff x = \pm 1$ and $f_y = 4y^3 - 4y = 0 \iff y = 0, \pm 1$, giving the critical points estimated above. Also $f_{xx} = 6x$, $f_{xy} = 0$, $f_{yy} = 12y^2 - 4$, so using the Second Derivatives Test, $D(-1, 0) = 24 > 0$ and $f_{xx}(-1, 0) = -6 < 0$ indicating a local maximum $f(-1, 0) = 2$; $D(1, \pm 1) = 48 > 0$ and $f_{xx}(1, \pm 1) = 6 > 0$ indicating local minima $f(1, \pm 1) = -3$; and $D(-1, \pm 1) = -48$ and $D(1, 0) = -24$, indicating saddle points.

57. $f(x, y) = x^2 y$, $g(x, y) = x^2 + y^2 = 1 \implies \nabla f = \langle 2xy, x^2 \rangle = \lambda \nabla g = \langle 2\lambda x, 2\lambda y \rangle$. Then $2xy = 2\lambda x$ and $x^2 = 2\lambda y$ imply $\lambda = x^2/(2y)$ and $\lambda = y$ if $x \neq 0$ and $y \neq 0$. Hence $x^2 = 2y^2$. Then $x^2 + y^2 = 1$ implies $3y^2 = 1$ so $y = \pm \frac{1}{\sqrt{3}}$ and $x = \pm \sqrt{\frac{2}{3}}$. [Note if $x = 0$ then $x^2 = 2\lambda y$ implies $y = 0$ and $f(0, 0) = 0$.] Thus the possible points are $\left(\pm \sqrt{\frac{2}{3}}, \pm \frac{1}{\sqrt{3}} \right)$ and the absolute maxima are $f\left(\pm \sqrt{\frac{2}{3}}, \frac{1}{\sqrt{3}} \right) = \frac{2}{3\sqrt{3}}$ while the absolute minima are $f\left(\pm \sqrt{\frac{2}{3}}, -\frac{1}{\sqrt{3}} \right) = -\frac{2}{3\sqrt{3}}$.

59. $f(x, y, z) = x + y + z$, $g(x, y, z) = 1/x + 1/y + 1/z = 1 \implies$ $\nabla f = \langle 1, 1, 1 \rangle = \lambda \nabla g = \langle -\lambda x^{-2}, -\lambda y^{-2}, -\lambda z^{-2} \rangle$. Thus $\lambda = -x^2 = -y^2 = -z^2$ or $y = \pm x$, $z = \pm x$. Substituting into $1/x + 1/y + 1/z = 1$ gives (1) $3/x = 1$ so $x = 3$, or (2) $1/x = 1$ so $x = 1$, or (3) $-1/x = 1$ so $x = -1$ with the associated points (1) $(3, 3, 3)$, (2) $(1, 1, -1)$ or $(1, -1, 1)$, (3) $(-1, 1, 1)$. Thus the absolute maximum is $f(3, 3, 3) = 9$ and the absolute minimum is $f(1, 1, -1) = f(1, -1, 1) = f(-1, 1, 1) = 1$.

61. $f(x, y, z) = x^2 + y^2 + z^2$, $g(x, y, z) = xy^2z^3 = 2 \Rightarrow$

$\nabla f = \langle 2x, 2y, 2z \rangle = \lambda \nabla g = \langle \lambda y^2 z^3, 2\lambda xyz^3, 3\lambda xy^2 z^2 \rangle$. Since $xy^2z^3 = 2$, $x \neq 0$, $y \neq 0$ and $z \neq 0$,

so (1) $2x = \lambda y^2 z^3$, (2) $1 = \lambda xz^3$, (3) $2 = 3\lambda xy^2 z$. Then (2) and (3) imply $\dfrac{1}{xz^3} = \dfrac{2}{3xy^2z}$ or

$y^2 = \frac{2}{3}z^2$ so $y = \pm z\sqrt{\frac{2}{3}}$. Similarly (1) and (3) imply $\dfrac{2x}{y^2z^3} = \dfrac{2}{3xy^2z}$ or $3x^2 = z^2$ so $x = \pm\frac{1}{\sqrt{3}}z$. But

$xy^2z^3 = 2$ so x and z must have the same sign, that is, $x = \frac{1}{\sqrt{3}}z$. Thus $g(x, y, z) = 2$ implies

$\frac{1}{\sqrt{3}}z\left(\frac{2}{3}z^2\right)z^3 = 2$ or $z = \pm 3^{1/4}$ and the possible points are $\left(\pm 3^{-1/4}, 3^{-1/4}\sqrt{2}, \pm 3^{1/4}\right)$,

$\left(\pm 3^{-1/4}, -3^{-1/4}\sqrt{2}, \pm 3^{1/4}\right)$. However at each of these points f takes on the same value, $2\sqrt{3}$. But

$(2, 1, 1)$ also satisfies $g(x, y, z) = 2$ and $f(2, 1, 1) = 6 > 2\sqrt{3}$. Thus f has an absolute minimum value

of $2\sqrt{3}$ and no absolute maximum subject to the constraint $xy^2z^3 = 2$.

Alternate Solution: $g(x, y, z) = xy^2z^3 = 2$ implies $y^2 = \dfrac{2}{xz^3}$, so minimize $f(x, z) = x^2 + \dfrac{2}{xz^3} + z^2$.

Then $f_x = 2x - \dfrac{2}{x^2z^3}$, $f_z = -\dfrac{6}{xz^4} + 2z$, $f_{xx} = 2 + \dfrac{4}{x^3z^3}$, $f_{zz} = \dfrac{24}{xz^5} + 2$ and $f_{xz} = \dfrac{6}{x^2z^4}$. Now

$f_x = 0$ implies $2x^3z^3 - 2 = 0$ or $z = 1/x$. Substituting into $f_y = 0$ implies

$-6x^3 + 2x^{-1} = 0$ or $x = \frac{1}{\sqrt[4]{3}}$, so the two critical points are $\left(\pm\frac{1}{\sqrt[4]{3}}, \pm\sqrt[4]{3}\right)$. Then

$D\left(\pm\frac{1}{\sqrt[4]{3}}, \pm\sqrt[4]{3}\right) = (2 + 4)\left(2 + \frac{24}{3}\right) - \left(\frac{6}{\sqrt{3}}\right)^2 > 0$ and $f_{xx}\left(\pm\frac{1}{\sqrt[4]{3}}, \pm\sqrt[4]{3}\right) = 6 > 0$, so each point is a

minimum. Finally, $y^2 = \frac{2}{xz^3}$, so the four points closest to the origin are $\left(\pm\frac{1}{\sqrt[4]{3}}, \frac{\sqrt{2}}{\sqrt[4]{3}}, \pm\sqrt[4]{3}\right)$,

$\left(\pm\frac{1}{\sqrt[4]{3}}, -\frac{\sqrt{2}}{\sqrt[4]{3}}, \pm\sqrt[4]{3}\right)$.

63.

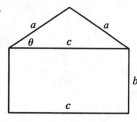

The area of the triangle is $\frac{1}{2}ca\sin\theta$ and the area of the rectangle is bc. Thus, the area of the whole object is $f(a,b,c) = \frac{1}{2}ca\sin\theta + bc$. The perimeter of the object is $g(a,b,c) = 2a + 2b + c = P$. To simplify $\sin\theta$ in terms of a, b, and c notice that $a^2\sin^2\theta + \left(\frac{1}{2}c\right)^2 = a^2 \Rightarrow$

$$\sin\theta = \frac{1}{2a}\sqrt{4a^2 - c^2}. \text{ Thus } f(a,b,c) = \frac{c}{4}\sqrt{4a^2 - c^2} + bc.$$

(Instead of using θ, we could just have used the Pythagorean Theorem.) As a result, by Lagrange's method, we must find a, b, c, and λ by solving $\nabla f = \lambda \nabla g$ which gives the following equations:

(1) $ca\left(4a^2 - c^2\right)^{-1/2} = 2\lambda$, (2) $c = 2\lambda$, (3) $\frac{1}{4}\left(4a^2 - c^2\right)^{1/2} - \frac{1}{4}c^2\left(4a^2 - c^2\right)^{-1/2} + b = \lambda$, and

(4) $2a + 2b + c = P$. From (2), $\lambda = \frac{1}{2}c$ and so (1) produces $ca\left(4a^2 - c^2\right)^{-1/2} = c \Rightarrow$

$\left(4a^2 - c^2\right)^{1/2} = a \Rightarrow 4a^2 - c^2 = a^2 \Rightarrow$ (5) $c = \sqrt{3}a$. Similarly, since $\left(4a^2 - c^2\right)^{1/2} = a$ and

$\lambda = \frac{1}{2}c$, (3) gives $\frac{a}{4} - \frac{c^2}{4a} + b = \frac{c}{2}$, so from (5), $\frac{a}{4} - \frac{3a}{4} + b = \frac{\sqrt{3}a}{2} \Rightarrow -\frac{a}{2} - \frac{\sqrt{3}a}{2} = -b \Rightarrow$

(6) $b = \frac{a}{2}\left(1 + \sqrt{3}\right)$. Substituting (5) and (6) into (4) we get: $2a + a\left(1 + \sqrt{3}\right) + \sqrt{3}a = P \Rightarrow$

$3a + 2\sqrt{3}a = P \Rightarrow a = \dfrac{P}{3 + 2\sqrt{3}} = \dfrac{2\sqrt{3} - 3}{3}P$ and thus $b = \dfrac{\left(2\sqrt{3} - 3\right)\left(1 + \sqrt{3}\right)}{6}P = \dfrac{3 - \sqrt{3}}{6}P$

and $c = \left(2 - \sqrt{3}\right)P$.

Focus on Problem Solving

1. The areas of the smaller rectangles are $A_1 = xy$,

$A_2 = (L - x)\,y$, $A_3 = (L - x)\,(W - y)$,

$A_4 = x\,(W - y)$. For $0 \le x \le L$, $0 \le y \le W$, let

$$f(x, y) = A_1^2 + A_2^2 + A_3^2 + A_4^2$$

$$= x^2 y^2 + (L - x)^2\, y^2 + (L - x)^2\,(W - y)^2$$

$$+ x^2\,(W - y)^2$$

$$= \left[x^2 + (L - x)^2\right]\left[y^2 + (W - y)^2\right]$$

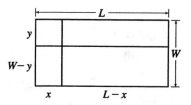

Then we need to find the maximum and minimum values of $f(x, y)$. Here

$f_x(x, y) = [2x - 2(L - x)]\left[y^2 + (W - y)^2\right] = 0 \quad \Rightarrow \quad 4x - 2L = 0$ or $x = \frac{1}{2}L$, and

$f_y(x, y) = \left[x^2 + (L - x)^2\right][2y - 2(W - y)] = 0 \quad \Rightarrow \quad 4y - 2W = 0$ or $y = W/2$. Also

$f_{xx} = 4\left[y^2 + (W - y)^2\right]$, $f_{yy} = 4\left[x^2 + (L - x)^2\right]$, and $f_{xy} = (4x - 2L)(4y - 2W)$. Then

$D = 16\left[y^2 + (W - y)^2\right]\left[x^2 + (L - x)^2\right] - (4x - 2L)^2\,(4y - 2W)^2$. Thus when $x = \frac{1}{2}L$ and

$y = \frac{1}{2}W$, $D > 0$ and $f_{xx} = 2W^2 > 0$. Thus a minimum of f occurs at $\left(\frac{1}{2}L, \frac{1}{2}W\right)$ and this minimum

value is $f\left(\frac{1}{2}L, \frac{1}{2}W\right) = \frac{1}{4}L^2 W^2$. There are no other critical points, so the maximum must occur on the

boundary. Now along the width of the rectangle let $g(y) = f(0, y) = f(L, y) = L^2\left[y^2 + (W - y)^2\right]$,

$0 \le y \le W$. Then $g'(y) = L^2\left[2y - 2(W - y)\right] = 0 \quad \Leftrightarrow \quad y = \frac{1}{2}W$. And $g\left(\frac{1}{2}\right) = \frac{1}{2}L^2 W^2$. Checking

the endpoints, we get $g(0) = g(W) = L^2 W^2$. Along the length of the rectangle let

$h(x) = f(x, 0) = f(x, W) = W^2\left[x^2 + (L - x)^2\right]$, $0 \le x \le L$. By symmetry $h'(x) = 0 \quad \Leftrightarrow$

$x = \frac{1}{2}L$ and $h\left(\frac{1}{2}L\right) = \frac{1}{2}L^2 W^2$. At the endpoints we have $h(0) = h(L) = L^2 W^2$. Therefore $L^2 W^2$ is

the maximum value of f. This maximum value of f occurs when the "cutting" lines correspond to sides

of the rectangle.

3. Let $g(x, y) = x f\left(\frac{y}{x}\right)$. Then $g_x(x, y) = f\left(\frac{y}{x}\right) + x f'\left(\frac{y}{x}\right)\left(-\frac{y}{x^2}\right) = f\left(\frac{y}{x}\right) - \frac{y}{x} f'\left(\frac{y}{x}\right)$ and

$g_y(x, y) = x f'\left(\frac{y}{x}\right)\left(\frac{1}{x}\right) = f'\left(\frac{y}{x}\right)$. Thus the tangent plane at (x_0, y_0, z_0) on the surface has equation

$$z - x_0 f\left(\frac{y_0}{x_0}\right) = \left[f\left(\frac{y_0}{x_0}\right) - y_0 x_0^{-1} f'\left(\frac{y_0}{x_0}\right)\right](x - x_0) + f'\left(\frac{y_0}{x_0}\right)(y - y_0) \quad \Rightarrow$$

$\left[f\left(\frac{y_0}{x_0}\right) - y_0 x_0^{-1} f'\left(\frac{y_0}{x_0}\right)\right] x + \left[f'\left(\frac{y_0}{x_0}\right)\right] y - z = 0$. But any plane whose equation is of the form

$ax + by + cz = 0$ passes through the origin. Thus the origin is the common point of intersection.

5. (a) The area of a trapezoid is $\frac{1}{2}h(b_1 + b_2)$, where h is the height (the distance between the two parallel sides) and b_1, b_2 are the lengths of the bases (the parallel sides). From the figure in the text, we see that $h = x \sin \theta$, $b_1 = w - 2x$, and $b_2 = w - 2x + 2x \cos \theta$. Therefore the cross-sectional area of the rain gutter is

$$A(x, \theta) = \tfrac{1}{2}x \sin \theta \left[(w - 2x) + (w - 2x + 2x \cos \theta)\right] = (x \sin \theta)(w - 2x + x \cos \theta)$$

$$= wx \sin \theta - 2x^2 \sin \theta + x^2 \sin \theta \cos \theta, \; 0 < x \le \tfrac{1}{2}w, 0 < \theta \le \tfrac{\pi}{2}$$

We look for the critical points of A: $\partial A/\partial x = w \sin \theta - 4x \sin \theta + 2x \sin \theta \cos \theta$ and $\partial A/\partial \theta = wx \cos \theta - 2x^2 \cos \theta + x^2 (\cos^2 \theta - \sin^2 \theta)$, so $\partial A/\partial x = 0 \; \Leftrightarrow$

$$\sin \theta (w - 4x + 2x \cos \theta) = 0 \; \Leftrightarrow \; \cos \theta = \frac{4x - w}{2x} = 2 - \frac{w}{2x} \; (0 < \theta \le \tfrac{\pi}{2} \; \Rightarrow \; \sin \theta > 0). \text{ If,}$$

in addition, $\partial A/\partial \theta = 0$, then

$$0 = wx \cos \theta - 2x^2 \cos \theta + x^2 \left(2 \cos^2 \theta - 1\right)$$

$$= wx \left(2 - \frac{w}{2x}\right) - 2x^2 \left(2 - \frac{w}{2x}\right) + x^2 \left[2 \left(2 - \frac{w}{2x}\right)^2 - 1\right]$$

$$= 2wx - \tfrac{1}{2}w^2 - 4x^2 + wx + x^2 \left[8 - \frac{4w}{x} + \frac{w^2}{2x^2} - 1\right] = -wx + 3x^2 = x(3x - w)$$

Since $x > 0$, we must have $x = \frac{1}{3}w$, in which case $\cos \theta = \frac{1}{2}$, so $\theta = \frac{\pi}{3}$, $\sin \theta = \frac{\sqrt{3}}{2}$, $k = \frac{\sqrt{3}}{6}w$, $b_1 = \frac{1}{3}w$, $b_2 = \frac{2}{3}w$, and $A = \frac{\sqrt{3}}{12}w^2$. As in Example 11.7.6, we can argue from the physical nature of this problem that we have found a relative maximum of A. Now checking the boundary of A, let $g(\theta) = A(w/2, \theta) = \frac{1}{2}w^2 \sin \theta - \frac{1}{2}w^2 \sin \theta + \frac{1}{4}w^2 \sin \theta \cos \theta = \frac{1}{8}w^2 \sin 2\theta, 0 < \theta \le \frac{\pi}{2}$. Clearly g is maximized when $\sin 2\theta = 1$ in which case $A = \frac{1}{8}w^2$. Also along the line $\theta = \frac{\pi}{2}$, let $h(x) = A\left(x, \frac{\pi}{2}\right) = wx - 2x^2, 0 < x < \frac{1}{2}w \; \Rightarrow \; h'(x) = w - 4x = 0 \; \Leftrightarrow \; x = \frac{1}{4}w$, and $h\left(\frac{1}{4}w\right) = w\left(\frac{1}{4}w\right) - 2\left(\frac{1}{4}w\right)^2 = \frac{1}{8}w^2$. Since $\frac{1}{8}w^2 < \frac{\sqrt{3}}{12}w^2$, we conclude that the relative maximum found earlier was an absolute maximum.

(b) If the metal were bent into a semi-circular gutter of radius r, we would have $w = \pi r$ and

$$A = \tfrac{1}{2}\pi r^2 = \tfrac{1}{2}\pi \left(\frac{w}{\pi}\right)^2 = \frac{w^2}{2\pi}. \text{ Since } \frac{w^2}{2\pi} > \frac{\sqrt{3}w^2}{12}, \text{ it would be better to bend the metal into a}$$

gutter with a semicircular cross-section.

Focus on Problem Solving

7. (a) $x = r\cos\theta$, $y = r\sin\theta$, $z = z$. Then $\dfrac{\partial u}{\partial r} = \dfrac{\partial u}{\partial x}\dfrac{\partial x}{\partial r} + \dfrac{\partial u}{\partial y}\dfrac{\partial y}{\partial r} + \dfrac{\partial u}{\partial z}\dfrac{\partial z}{\partial r} = \dfrac{\partial u}{\partial x}\cos\theta + \dfrac{\partial u}{\partial y}\sin\theta$ and

$$\frac{\partial^2 u}{\partial r^2} = \cos\theta\left[\frac{\partial^2 u}{\partial x^2}\frac{\partial x}{\partial r} + \frac{\partial^2 u}{\partial y\partial x}\frac{\partial y}{\partial r} + \frac{\partial^2 u}{\partial z\partial x}\frac{\partial z}{\partial r}\right] + \sin\theta\left[\frac{\partial^2 u}{\partial y^2}\frac{\partial y}{\partial r} + \frac{\partial^2 u}{\partial x\partial y}\frac{\partial x}{\partial r} + \frac{\partial^2 u}{\partial z\partial y}\frac{\partial z}{\partial r}\right]$$

$$= \frac{\partial^2 u}{\partial x^2}\cos^2\theta + \frac{\partial^2 u}{\partial y^2}\sin^2\theta + 2\frac{\partial^2 u}{\partial y\partial x}\cos\theta\sin\theta$$

Similarly $\dfrac{\partial u}{\partial\theta} = -\dfrac{\partial u}{\partial x}r\sin\theta + \dfrac{\partial u}{\partial y}r\cos\theta$ and

$$\frac{\partial^2 u}{\partial\theta^2} = \frac{\partial^2 u}{\partial x^2}r^2\sin^2\theta + \frac{\partial^2 u}{\partial y^2}r^2\cos^2\theta - 2\frac{\partial^2 u}{\partial y\partial x}r^2\sin\theta\cos\theta - \frac{\partial u}{\partial x}r\cos\theta - \frac{\partial u}{\partial y}r\sin\theta.\ \text{So}$$

$$\frac{\partial^2 u}{\partial r^2} + \frac{1}{r}\frac{\partial u}{\partial r} + \frac{1}{r^2}\frac{\partial^2 u}{\partial\theta^2} + \frac{\partial^2 u}{\partial z^2}$$

$$= \frac{\partial^2 u}{\partial x^2}\cos^2\theta + \frac{\partial^2 u}{\partial y^2}\sin^2\theta + 2\frac{\partial^2 u}{\partial y\partial x}\cos\theta\sin\theta + \frac{\partial u}{\partial x}\frac{\cos\theta}{r} + \frac{\partial u}{\partial y}\frac{\sin\theta}{r}$$

$$+ \frac{\partial^2 u}{\partial x^2}\sin^2\theta + \frac{\partial^2 u}{\partial y^2}\cos^2\theta - 2\frac{\partial^2 u}{\partial y\partial x}\sin\theta\cos\theta - \frac{\partial u}{\partial x}\frac{\cos\theta}{r} - \frac{\partial u}{\partial y}\frac{\sin\theta}{r} + \frac{\partial^2 u}{\partial z^2}$$

$$= \frac{\partial^2 u}{\partial x^2} + \frac{\partial^2 u}{\partial y^2} + \frac{\partial^2 u}{\partial z^2}$$

(b) $x = \rho\sin\phi\cos\theta$, $y = \rho\sin\phi\sin\theta$, $z = \rho\cos\phi$. Then

$$\frac{\partial u}{\partial\rho} = \frac{\partial u}{\partial x}\frac{\partial x}{\partial\rho} + \frac{\partial u}{\partial y}\frac{\partial y}{\partial\rho} + \frac{\partial u}{\partial z}\frac{\partial z}{\partial\rho} = \frac{\partial u}{\partial x}\sin\phi\cos\theta + \frac{\partial u}{\partial y}\sin\phi\sin\theta + \frac{\partial u}{\partial z}\cos\phi,\ \text{and}$$

$$\frac{\partial^2 u}{\partial\rho^2} = \sin\phi\cos\theta\left[\frac{\partial^2 u}{\partial x^2}\frac{\partial x}{\partial\rho} + \frac{\partial^2 u}{\partial y\partial x}\frac{\partial y}{\partial\rho} + \frac{\partial^2 u}{\partial z\partial x}\frac{\partial z}{\partial\rho}\right]$$

$$+ \sin\phi\sin\theta\left[\frac{\partial^2 u}{\partial y^2}\frac{\partial y}{\partial\rho} + \frac{\partial^2 u}{\partial x\partial y}\frac{\partial x}{\partial\rho} + \frac{\partial^2 u}{\partial z\partial y}\frac{\partial z}{\partial\rho}\right]$$

$$+ \cos\phi\left[\frac{\partial^2 u}{\partial z^2}\frac{\partial z}{\partial\rho} + \frac{\partial^2 u}{\partial x\partial z}\frac{\partial x}{\partial\rho} + \frac{\partial^2 u}{\partial y\partial z}\frac{\partial y}{\partial\rho}\right]$$

$$= 2\frac{\partial^2 u}{\partial y\partial x}\sin^2\phi\sin\theta\cos\theta + 2\frac{\partial^2 u}{\partial z\partial x}\sin\phi\cos\phi\cos\theta + 2\frac{\partial^2 u}{\partial y\partial z}\sin\phi\cos\phi\sin\theta$$

$$+ \frac{\partial^2 u}{\partial x^2}\sin^2\phi\cos^2\theta + \frac{\partial^2 u}{\partial y^2}\sin^2\phi\sin^2\theta + \frac{\partial^2 u}{\partial z^2}\cos^2\phi$$

Similarly $\dfrac{\partial u}{\partial\phi} = \dfrac{\partial u}{\partial x}\rho\cos\phi\cos\theta + \dfrac{\partial u}{\partial y}\rho\cos\phi\sin\theta - \dfrac{\partial u}{\partial z}\rho\sin\phi$, and

$$\frac{\partial^2 u}{\partial\phi^2} = 2\frac{\partial^2 u}{\partial y\partial x}\rho^2\cos^2\phi\sin\theta\cos\theta - 2\frac{\partial^2 u}{\partial x\partial z}\rho^2\sin\phi\cos\phi\cos\theta$$

$$- 2\frac{\partial^2 u}{\partial y\partial z}\rho^2\sin\phi\cos\phi\sin\theta + \frac{\partial^2 u}{\partial x^2}\rho^2\cos^2\phi\cos^2\theta + \frac{\partial^2 u}{\partial y^2}\rho^2\cos^2\phi\sin^2\theta$$

$$+ \frac{\partial^2 u}{\partial z^2}\rho^2\sin^2\phi - \frac{\partial u}{\partial x}\rho\sin\phi\cos\theta - \frac{\partial u}{\partial y}\rho\sin\phi\sin\theta - \frac{\partial u}{\partial z}\rho\cos\phi$$

165

Focus on Problem Solving

And $\dfrac{\partial u}{\partial \theta} = -\dfrac{\partial u}{\partial x}\rho \sin\phi \sin\theta + \dfrac{\partial u}{\partial y}\rho \sin\phi\cos\theta$, while

$$\dfrac{\partial^2 u}{\partial\theta^2} = -2\dfrac{\partial^2 u}{\partial y\partial x}\rho^2\sin^2\phi\cos\theta\sin\theta + \dfrac{\partial^2 u}{\partial x^2}\rho^2\sin^2\phi\sin^2\theta$$

$$+\dfrac{\partial^2 u}{\partial y^2}\rho^2\sin^2\phi\cos^2\theta - \dfrac{\partial u}{\partial x}\rho\sin\phi\cos\theta - \dfrac{\partial u}{\partial y}\rho\sin\phi\sin\theta$$

Therefore

$$\dfrac{\partial^2 u}{\partial\rho^2} + \dfrac{2}{\rho}\dfrac{\partial u}{\partial\rho} + \dfrac{\cot\phi}{\rho^2}\dfrac{\partial u}{\partial\phi} + \dfrac{1}{\rho^2}\dfrac{\partial^2 u}{\partial\phi^2} + \dfrac{1}{\rho^2\sin^2\phi}\dfrac{\partial^2 u}{\partial\theta^2}$$

$$= \dfrac{\partial^2 u}{\partial x^2}\left[\left(\sin^2\phi\cos^2\theta\right) + \left(\cos^2\phi\cos^2\theta\right) + \sin^2\theta\right]$$

$$+ \dfrac{\partial^2 u}{\partial y^2}\left[\left(\sin^2\phi\sin^2\theta\right) + \left(\cos^2\phi\sin^2\theta\right) + \cos^2\theta\right] + \dfrac{\partial^2 u}{\partial z^2}\left[\cos^2\phi + \sin^2\phi\right]$$

$$+ \dfrac{\partial u}{\partial x}\left[\dfrac{2\sin^2\phi\cos\theta + \cos^2\phi\cos\theta - \sin^2\phi\cos\theta - \cos\theta}{\rho\sin\phi}\right]$$

$$+ \dfrac{\partial u}{\partial y}\left[\dfrac{2\sin^2\phi\sin\theta + \cos^2\phi\sin\theta - \sin^2\phi\sin\theta - \sin\theta}{\rho\sin\phi}\right]$$

But $2\sin^2\phi\cos\theta + \cos^2\phi\cos\theta - \sin^2\phi\cos\theta - \cos\theta = \left(\sin^2\phi + \cos^2\phi - 1\right)\cos\theta = 0$ and similarly the coefficient of $\partial u/\partial y$ is 0. Also

$\sin^2\phi\cos^2\theta + \cos^2\phi\cos^2\theta + \sin^2\theta = \cos^2\theta\left(\sin^2\phi + \cos^2\phi\right) + \sin^2\theta = 1$, and similarly the coefficient of $\partial^2 u/\partial y^2$ is 1. So Laplace's Equation in spherical coordinates is as stated.

9. Since we are minimizing the area of the ellipse, and the circle lies above the x- axis, the ellipse will intersect the circle for only one value of y. This y- value must satisfy both the equation of the circle and the equation of the ellipse. Now $x^2/a^2 + y^2/b^2 = 1 \Rightarrow$

$x^2 = \left(a^2/b^2\right)\left(b^2 - y^2\right)$. Substituting into the equation of the

circle gives $\left(a^2/b^2\right)\left(b^2 - y^2\right) + y^2 - 2y = 0 \Rightarrow$

$\left(\dfrac{b^2 - a^2}{b^2}\right)y^2 - 2y + a^2 = 0$. In order for there to be only one solution to this quadratic equation, the

discriminant must be 0, so $4 - 4a^2\dfrac{b^2 - a^2}{b^2} = 0 \Rightarrow b^2 - a^2b^2 + a^4 = 0$. The area of the ellipse is

$A(a, b) = \pi ab$, and we minimize this function subject to the constraint $g(a, b) = b^2 - a^2b^2 + a^4 = 0$.

Now $\nabla A = \lambda\nabla g \Leftrightarrow \pi b = \lambda\left(4a^3 - 2ab^2\right), \pi a = \lambda\left(2b - 2ba^2\right) \Rightarrow$ (1) $\lambda = \dfrac{\pi b}{2a\left(2a^2 - b^2\right)}$,

(2) $\lambda = \dfrac{\pi a}{2b\left(1 - a^2\right)}$, (3) $b^2 - a^2b^2 + a^4 = 0$. Comparing (1) and (2) gives

$\dfrac{\pi b}{2a\left(2a^2 - b^2\right)} = \dfrac{\pi a}{2b\left(1 - a^2\right)} \Rightarrow 2\pi b^2 = 4\pi a^4 \Leftrightarrow a^2 = \tfrac{1}{\sqrt{2}}b$. Substitute this into (3) to get

$b = \tfrac{3}{\sqrt{2}} \Rightarrow a = \sqrt{\tfrac{3}{2}}$.

Chapter 12   Multiple Integrals

Section 12.1   Double Integrals over Rectangles

1. (a) $\sum\limits_{i=1}^{2}\sum\limits_{j=1}^{2} f\left(x_{ij}^*, y_{ij}^*\right)\Delta A = f\left(0, \frac{3}{2}\right)\Delta A + f\left(0,2\right)\Delta A + f\left(1, \frac{3}{2}\right)\Delta A + f\left(1,2\right)\Delta A$

$$= \left(-\tfrac{27}{4}\right)\tfrac{1}{2} + (-12)\tfrac{1}{2} + \left(1 - \tfrac{27}{4}\right)\tfrac{1}{2} + (1 - 12)\tfrac{1}{2} = -17.75$$

(b) $\frac{1}{2}\left[f\left(1, \frac{3}{2}\right) + f\left(1,2\right) + f\left(2, \frac{3}{2}\right) + f\left(2,2\right)\right] = \frac{1}{2}\left[-\frac{23}{4} + (-11) + \left(-\frac{19}{4}\right) + (-10)\right]$

$$= \tfrac{1}{2}\left(-\tfrac{63}{2}\right) = -15.75$$

(c) $\frac{1}{2}\left[f\left(0,1\right) + f\left(0, \frac{3}{2}\right) + f\left(1,1\right) + f\left(1, \frac{3}{2}\right)\right] = \frac{1}{2}\left[-3 - \frac{27}{4} - 2 - \frac{23}{4}\right] = -8.75$

(d) $\frac{1}{2}\left[f\left(1,1\right) + f\left(1, \frac{3}{2}\right) + f\left(2,1\right) + f\left(2, \frac{3}{2}\right)\right] = \frac{1}{2}\left[-2 - \frac{23}{4} - 1 - \frac{19}{4}\right] = -6.75$

3. (a) The subrectangles are shown in the figure.
The surface is the graph of $f\left(x,y\right) = x^2 + 4y$ and $\Delta A = 1$, so we
estimate

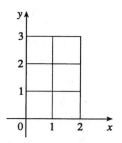

$$V \approx \sum_{i=1}^{2}\sum_{j=1}^{3} f\left(x_i, y_j\right)\Delta A$$

$$= f\left(1,1\right)\Delta A + f\left(1,2\right)\Delta A + f\left(1,3\right)\Delta A$$

$$+ f\left(2,1\right)\Delta A + f\left(2,2\right)\Delta A + f\left(2,3\right)\Delta A$$

$$= 5\left(1\right) + 9\left(1\right) + 13\left(1\right) + 8\left(1\right) + 12\left(1\right) + 16\left(1\right) = 63$$

(b) $V \approx \sum\limits_{i=1}^{2}\sum\limits_{j=1}^{3} f\left(\overline{x}_i, \overline{y}_j\right)\Delta A$

$$= f\left(\tfrac{1}{2}, \tfrac{1}{2}\right)\Delta A + f\left(\tfrac{1}{2}, \tfrac{3}{2}\right)\Delta A + f\left(\tfrac{1}{2}, \tfrac{5}{2}\right)\Delta A$$

$$+ f\left(\tfrac{3}{2}, \tfrac{1}{2}\right)\Delta A + f\left(\tfrac{3}{2}, \tfrac{3}{2}\right)\Delta A + f\left(\tfrac{3}{2}, \tfrac{5}{2}\right)\Delta A$$

$$= \tfrac{9}{4}\left(1\right) + \tfrac{25}{4}\left(1\right) + \tfrac{41}{4}\left(1\right) + \tfrac{17}{4}\left(1\right) + \tfrac{33}{4}\left(1\right) + \tfrac{49}{4}\left(1\right) = \tfrac{87}{2} = 43.5$$

5. (a) Each subrectangle and its midpoint are shown in the figure. The area of
each subrectangle is $\Delta A = 2$, so we evaluate f at each midpoint and
estimate

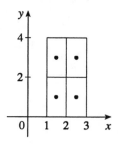

$$\iint_R f\left(x,y\right)\, dA \approx \sum_{i=1}^{2}\sum_{j=1}^{2} f\left(\overline{x}_i, \overline{y}_j\right)\Delta A$$

$$= f\left(1.5, 1\right)\Delta A + f\left(1.5, 3\right)\Delta A$$

$$+ f\left(2.5, 1\right)\Delta A + f\left(2.5, 3\right)\Delta A$$

$$= 1\left(2\right) + (-8)\left(2\right) + 5\left(2\right) + (-1)\left(2\right) = -6$$

(b) The subrectangles are shown in the figure. In each subrectangle, the sample point farthest from the origin is the upper right corner, and the area of each subrectangle is $\Delta A = \frac{1}{2}$. Thus we estimate

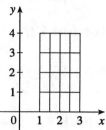

$$\iint_R f(x,y)\, dA \approx \sum_{i=1}^{4}\sum_{j=1}^{4} f(x_i, y_j)\,\Delta A$$

$$= f(1.5,1)\,\Delta A + f(1.5,2)\,\Delta A + f(1.5,3)\,\Delta A + f(1.5,4)\,\Delta A$$
$$+ f(2,1)\,\Delta A + f(2,2)\,\Delta A + f(2,3)\,\Delta A + f(2,4)\,\Delta A$$
$$+ f(2.5,1)\,\Delta A + f(2.5,2)\,\Delta A + f(2.5,3)\,\Delta A + f(2.5,4)\,\Delta A$$
$$+ f(3,1)\,\Delta A + f(3,2)\,\Delta A + f(3,3)\,\Delta A + f(3,4)\,\Delta A$$
$$= 1\left(\tfrac{1}{2}\right) + (-4)\left(\tfrac{1}{2}\right) + (-8)\left(\tfrac{1}{2}\right) + (-6)\left(\tfrac{1}{2}\right) + 3\left(\tfrac{1}{2}\right) + 0\left(\tfrac{1}{2}\right) + (-5)\left(\tfrac{1}{2}\right) + (-8)\left(\tfrac{1}{2}\right)$$
$$+ 5\left(\tfrac{1}{2}\right) + 3\left(\tfrac{1}{2}\right) + (-1)\left(\tfrac{1}{2}\right) + (-4)\left(\tfrac{1}{2}\right) + 8\left(\tfrac{1}{2}\right) + 6\left(\tfrac{1}{2}\right) + 3\left(\tfrac{1}{2}\right) + 0\left(\tfrac{1}{2}\right)$$
$$= -3.5$$

7. The values of $f(x,y) = \sqrt{52 - x^2 - y^2}$ get smaller as we move farther from the origin, so on any of the subrectangles in the problem, the function will have its largest value at the lower left corner of the subrectangle and its smallest value at the upper right corner, and any other value will lie between these two. So using these subrectangles we have $U < V < L$. (Note that this is true no matter how R is divided into subrectangles.)

9. With $m = n = 2$, we have $\Delta A = 4$. Using the contour map to estimate the value of f at the center of each subrectangle, we have

$$\iint_R f(x,y)\, dA \approx \sum_{i=1}^{2}\sum_{j=1}^{2} f(\overline{x}_i, \overline{y}_j)\,\Delta A = \Delta A\,[f(1,1) + f(1,3) + f(3,1) + f(3,3)]$$

$$\approx 4\,(27 + 4 + 14 + 17) = 248$$

11. $z = f(x,y) = 4 - 2y \geq 0$ for $0 \leq y \leq 1$. Thus the integral represents the volume of that part of the rectangular solid $[0,1] \times [0,1] \times [0,4]$ which lies below the plane $z = 4 - 2y$. So

$$\iint_R (4 - y)\, dA = (1)(1)(2) + \tfrac{1}{2}(1)(1)(2) = 3$$

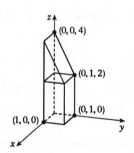

Section 12.1 Double Integrals over Rectangles

13. To calculate the estimates using a programmable calculator, we can use an algorithm similar to that of Exercise 5.1.5. In Maple, we can define the function $f(x,y) = e^{-x^2-y^2}$ (calling it f),load the `student` package, and then use thecommand

n	estimate
1	0.6065
4	0.5694
16	0.5606
64	0.5585
256	0.5579
1024	0.5578

```
middlesum(middlesum(f,x=0..1,m),
                        y=0..1,m);
```

to get the estimate with $n = m^2$ squares of equal size. Mathematica has no special Riemann sum command, but we can define f and then use nested Sum commands to calculate the estimates.

15. If we divide R into mn subrectangles, $\iint_R k\,dA \approx \sum\limits_{i=1}^{m} \sum\limits_{j=1}^{n} f\left(x_{ij}^*, y_{ij}^*\right) \Delta A$ for any choice of sample points $\left(x_{ij}^*, y_{ij}^*\right)$. But $f\left(x_{ij}^*, y_{ij}^*\right) = k$ always and $\sum\limits_{i=1}^{m} \sum\limits_{j=1}^{n} \Delta A = $ area of $R = (b-a)(d-c)$. Thus, no matter how we choose the sample points,

$$\sum_{i=1}^{m}\sum_{j=1}^{n} f\left(x_{ij}^*, y_{ij}^*\right) \Delta A = k \sum_{i=1}^{m}\sum_{j=1}^{n} \Delta A = k(b-a)(d-c) \text{ and so}$$

$$\iint_R k\,dA = \lim_{m,n\to\infty} \sum_{i=1}^{m}\sum_{j=1}^{n} f\left(x_{ij}^*, y_{ij}^*\right) \Delta A = \lim_{m,n\to\infty} k \sum_{i=1}^{m}\sum_{j=1}^{n} \Delta A$$

$$= \lim_{m,n\to\infty} k(b-a)(d-c) = k(b-a)(d-c)$$

Section 12.2 Iterated Integrals

1. $\int_0^2 \left(2xy - 3x^2\right) dy = \left[xy^2 - 3x^2 y\right]_{y=0}^{y=2} = 4x - 6x^2,$

$\int_0^1 \left(2xy - 3x^2\right) dx = \left[yx^2 - x^3\right]_{x=0}^{x=1} = y - 1$

3. $\int_0^4 \int_0^2 x\sqrt{y}\, dx\, dy = \int_0^4 \sqrt{y}\left[\frac{1}{2}x^2\right]_{x=0}^{x=2} dy = \int_0^4 2\sqrt{y}\, dy = \left[\frac{4}{3}y^{3/2}\right]_0^4 = \frac{32}{3}$

5. $\int_{-1}^1 \int_0^1 \left(x^3 y^2 + 3xy^2\right) dy\, dx = \int_{-1}^1 \left[\frac{1}{4}x^3 y^4 + xy^3\right]_{y=0}^{y=1} dx = \int_{-1}^1 \left[\frac{1}{4}x^3 + x\right] dx$

$$= \left[\frac{1}{16}x^4 + \frac{1}{2}x^2\right]_{-1}^1 = 0$$

Alternate Solution: Applying Fubini's Theorem, the integral equals

$\int_0^1 \int_{-1}^1 \left(x^3 y^2 + 3xy^2\right) dx\, dy = \int_0^1 \left[\frac{1}{4}y^2 x^4 + \frac{3}{2}y^2 x^2\right]_{x=-1}^{x=1} dy = \int_0^1 0\, dy = 0.$

7. $\int_0^{\ln 2} \int_0^{\ln 5} e^{2x-y}\, dx\, dy = \left(\int_0^{\ln 5} e^{2x}\, dx\right)\left(\int_0^{\ln 2} e^{-y}\, dy\right) = \left[\frac{1}{2}e^{2x}\right]_0^{\ln 5} \left[-e^{-y}\right]_0^{\ln 2}$

$$= \left(\frac{25}{2} - \frac{1}{2}\right)\left(-\frac{1}{2} + 1\right) = 6$$

9. $\int_1^2 \int_0^3 \left(2y^2 - 3xy^3\right) dy\, dx = \int_1^2 \left[\frac{2}{3}y^3 - \frac{3}{4}xy^4\right]_{y=0}^{y=3} dx = \int_1^2 \left(18 - \frac{243}{4}x\right) dx = \left[18x - \frac{243}{8}x^2\right]_1^2$

$$= -\frac{585}{8}$$

11. $\int_0^{\pi/6} \int_1^4 x \sin y\, dx\, dy = \left(\int_0^{\pi/6} \sin y\, dy\right)\left(\int_1^4 x\, dx\right) = \left[-\cos y\right]_0^{\pi/6} \left[\frac{1}{2}x^2\right]_1^4 = \left(1 - \frac{\sqrt{3}}{2}\right)\frac{15}{2}$

$$= \frac{15(2-\sqrt{3})}{4}$$

13. $\int_0^{\pi/6} \int_0^{\pi/3} x \sin(x+y)\, dy\, dx$

$$= \int_0^{\pi/6} \left[-x\cos(x+y)\right]_{y=0}^{y=\pi/3} dx = \int_0^{\pi/6}\left[x\cos x - x\cos\left(x + \frac{\pi}{3}\right)\right] dx$$

(by integrating by parts separately for each term)

$$= x\left[\sin x - \sin\left(x + \frac{\pi}{3}\right)\right]_0^{\pi/6} - \int_0^{\pi/6}\left[\sin x - \sin\left(x + \frac{\pi}{3}\right)\right] dx$$

$$= \frac{\pi}{6}\left[\frac{1}{2} - 1\right] - \left[-\cos x + \cos\left(x + \frac{\pi}{3}\right)\right]_0^{\pi/6} = -\frac{\pi}{12} - \left[-\frac{\sqrt{3}}{2} + 0 - \left(-1 + \frac{1}{2}\right)\right]$$

$$= \frac{\sqrt{3}-1}{2} - \frac{\pi}{12}$$

15. $z = f(x, y) = 4 - x - 2y \geq 0$ for $0 \leq x \leq 1$ and $0 \leq y \leq 1$. So the solid is the region in the first octant which lies below the plane $z = 4 - x - 2y$ and above $[0, 1] \times [0, 1]$.

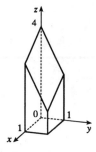

17. $V = \int_1^4 \int_{-1}^0 (2x + 5y + 1)\, dx\, dy = \int_1^4 \left[x^2 + 5xy + x\right]_{x=-1}^{x=0} dy = \int_1^4 5y\, dy = \frac{5}{2}y^2\big]_1^4 = \frac{75}{2}$

19. $V = \int_{-2}^2 \int_{-1}^1 \left(1 - \frac{1}{4}x^2 - \frac{1}{9}y^2\right) dx\, dy = 4\int_0^2 \int_0^1 \left(1 - \frac{1}{4}x^2 - \frac{1}{9}y^2\right) dx\, dy$

$= 4\int_0^2 \left[x - \frac{1}{12}x^3 - \frac{1}{9}y^2 x\right]_{x=0}^{x=1} dy = 4\int_0^2 \left(\frac{11}{12} - \frac{1}{9}y^2\right) dy = 4\left[\frac{11}{12}y - \frac{1}{27}y^3\right]_0^2$

$= 4 \cdot \frac{83}{54} = \frac{166}{27}$

21. Here we need the volume of the solid lying under the surface $z = x\sqrt{x^2 + y}$ and above the square $R = [0, 1] \times [0, 1]$ in the xy-plane.

$V = \int_0^1 \int_0^1 x\sqrt{x^2 + y}\, dx\, dy = \int_0^1 \frac{1}{3}\left[(x^2 + y)^{3/2}\right]_{x=0}^{x=1} dy = \frac{1}{3}\int_0^1 \left[(1 + y)^{3/2} - y^{3/2}\right] dy$

$= \frac{1}{3} \cdot \frac{2}{5}\left[(1 + y)^{5/2} - y^{5/2}\right]_0^1 = \frac{4}{15}\left(2\sqrt{2} - 1\right)$

23. In the first octant, $z \geq 0$ $\Rightarrow$ $y \leq 3$, so

$V = \int_0^3 \int_0^2 (9 - y^2)\, dx\, dy = \int_0^3 \left[9x - y^2 x\right]_{x=0}^{x=2} dy = \int_0^3 (18 - 2y^2)\, dy = \left[18y - \frac{2}{3}y^3\right]_0^3 = 36$

25. In Maple, we can calculate the integral by defining the integrand
as f and then using the command
`int(int(f,x=0..1),y=0..1);`.
In Mathematica, we can use the command
`Integrate[Integrate[f,{x,0,1}],{y,0,1}]`. We
find that $\iint_R x^5 y^3 e^{xy}\, dA = 21e - 57 \approx 0.0839$. We can use
`plot3d` (in Maple) or `Plot3d` (in Mathematica) to graph the
function.

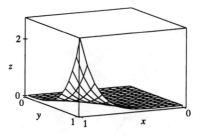

27. $A(R) = 2 \cdot 5 = 10$, so

$f_{\text{ave}} = \dfrac{1}{A(R)} \iint_R f(x, y)\, dA = \frac{1}{10}\int_0^5 \int_{-1}^1 x^2 y\, dx\, dy = \frac{1}{10}\int_0^5 \left[\frac{1}{3}x^3 y\right]_{x=-1}^{x=1} dy = \frac{1}{10}\int_0^5 \frac{2}{3}y\, dy$

$= \frac{1}{10}\left[\frac{1}{3}y^2\right]_0^5 = \frac{5}{6}$

29. Let $f(x, y) = \dfrac{x - y}{(x + y)^3}$. Then a CAS gives $\int_0^1 \int_0^1 f(x, y)\, dy\, dx = \frac{1}{2}$ and $\int_0^1 \int_0^1 f(x, y)\, dx\, dy = -\frac{1}{2}$.

To explain the seeming violation of Fubini's Theorem, note that f has an infinite discontinuity at $(0, 0)$
and thus does not satisfy the conditions of Fubini's Theorem. In fact, both iterated integrals involve
improper integrals which diverge at their lower limits of integration.

Section 12.3   Double Integrals over General Regions

1. $\int_0^1 \int_0^y x\,dx\,dy = \int_0^1 \left[\frac{1}{2}x^2\right]_{x=0}^{x=y} dy = \int_0^1 \frac{1}{2}y^2\,dy = \frac{1}{2}\left[\frac{1}{3}y^3\right]_0^1 = \frac{1}{6}$

3. $\int_0^1 \int_0^x \sin\left(x^2\right)\,dy\,dx = \int_0^1 \sin\left(x^2\right)\left[y\right]_{y=0}^{y=x}\,dx = \int_0^1 x\sin\left(x^2\right)\,dx = \frac{1}{2}\left[-\cos\left(x^2\right)\right]_0^1 = \frac{1}{2}\left(1 - \cos 1\right)$

5. $\int_0^1 \int_{x^2}^{\sqrt{x}} xy\,dy\,dx = \int_0^1 \left[\frac{1}{2}xy^2\right]_{y=x^2}^{y=\sqrt{x}}\,dx = \frac{1}{2}\int_0^1 \left(x^2 - x^5\right)\,dx = \frac{1}{2}\left[\frac{1}{3}x^3 - \frac{1}{6}x^6\right]_0^1 = \frac{1}{12}$

7. $\int_1^2 \int_y^{y^3} e^{x/y}\,dx\,dy = \int_1^2 \left[ye^{x/y}\right]_{x=y}^{x=y^3}\,dy = \int_1^2 \left(ye^{y^2} - ey\right)\,dy = \left[\frac{1}{2}e^{y^2} - \frac{1}{2}ey^2\right]_1^2 = \frac{1}{2}\left(e^4 - 4e\right)$

9. $\int_0^1 \int_0^{x^2} x\cos y\,dy\,dx = \int_0^1 \left[x\sin y\right]_{y=0}^{y=x^2}\,dx = \int_0^1 x\sin x^2\,dx = -\frac{1}{2}\cos x^2\Big]_0^1 = \frac{1}{2}\left(1 - \cos 1\right)$

11.

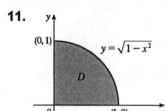

$\int_0^1 \int_0^{\sqrt{1-x^2}} xy\,dy\,dx = \int_0^1 \left[\frac{1}{2}xy^2\right]_{y=0}^{y=\sqrt{1-x^2}}\,dx$

$\qquad\qquad = \int_0^1 \frac{x - x^3}{2}\,dx$

$\qquad\qquad = \frac{1}{2}\left[\frac{x^2}{2} - \frac{x^4}{4}\right]_0^1 = \frac{1}{8}$

13.

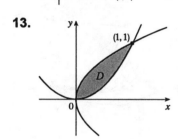

$V = \int_0^1 \int_{x^2}^{\sqrt{x}} \left(x^2 + y^2\right)\,dy\,dx = \int_0^1 \left[\left(x^2 y + \frac{y^3}{3}\right)\right]_{y=x^2}^{y=\sqrt{x}}\,dx$

$\qquad = \int_0^1 \left(x^{5/2} - x^4 + \frac{1}{3}x^{3/2} - \frac{1}{3}x^6\right)\,dx$

$\qquad = \left[\frac{2}{7}x^{7/2} - \frac{1}{5}x^5 + \frac{2}{15}x^{5/2} - \frac{1}{21}x^7\right]_0^1$

$\qquad = \frac{18}{105} = \frac{6}{35}$

15.

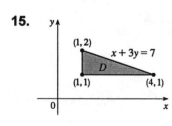

$V = \int_1^2 \int_1^{7-3y} xy\,dx\,dy$

$\qquad = \int_1^2 \left[\frac{yx^2}{2}\right]_{x=1}^{x=7-3y}\,dy$

$\qquad = \frac{1}{2}\int_1^2 \left(48y - 42y^2 + 9y^3\right)\,dy$

$\qquad = \frac{1}{2}\left[24y^2 - 14y^3 + \frac{9}{4}y^4\right]_1^2 = \frac{31}{8}$

17.

$V = \int_0^1 \int_0^{1-x} (1 - x - y)\,dy\,dx$

$\qquad = \int_0^1 \left[y - xy - \frac{y^2}{2}\right]_{y=0}^{y=1-x}\,dx$

$\qquad = \int_0^1 \left[(1-x)^2 - \frac{1}{2}(1-x)^2\right]\,dx$

$\qquad = \int_0^1 \frac{1}{2}(1-x)^2\,dx = \left[-\frac{1}{6}(1-x)^3\right]_0^1 = \frac{1}{6}$

19.

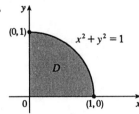

$$V = \int_0^1 \int_0^{\sqrt{1-x^2}} y \, dy \, dx$$

$$= \int_0^1 \left[\frac{y^2}{2}\right]_{y=0}^{y=\sqrt{1-x^2}} dx$$

$$= \int_0^1 \frac{1-x^2}{2} \, dx$$

$$= \tfrac{1}{2}\left[x - \tfrac{1}{3}x^3\right]_0^1 = \tfrac{1}{3}$$

21.

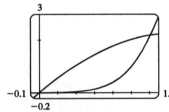

From the graph, it appears that the two curves intersect at $x = 0$ and at $x \approx 1.213$. Thus the desired integral is

$$\iint_D x \, dA \approx \int_0^{1.213} \int_{x^4}^{3x-x^2} x \, dy \, dx$$

$$= \int_0^{1.213} [xy]_{y=x^4}^{y=3x-x^2} \, dx$$

$$= \left[(x^3 - \tfrac{1}{4}x^4) - \tfrac{1}{6}x^6\right]_0^{1.213} \approx 0.713$$

23. The two bounding curves $y = x^3 - x$ and $y = x^2 + x$ intersect at the origin and at $x = 2$, with $x^2 + x > x^3 - x$ on $(0, 2)$. Using a CAS, we find that the volume is

$$V = \int_0^2 \int_{x^3-x}^{x^2+x} z \, dy \, dx = \int_0^2 \int_{x^3-x}^{x^2+x} \left(x^3 y^4 + xy^2\right) dy \, dx = \frac{13{,}984{,}735{,}616}{14{,}549{,}535}.$$

25.

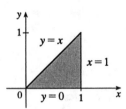

Because the region of integration is
$$D = \{(x,y) \mid 0 \le y \le x, 0 \le x \le 1\}$$
$$= \{(x,y) \mid y \le x \le 1, 0 \le y \le 1\}$$
we have
$$\int_0^1 \int_0^x f(x,y) \, dy \, dx = \iint_D f(x,y) \, dA$$
$$= \int_0^1 \int_y^1 f(x,y) \, dx \, dy$$

27.

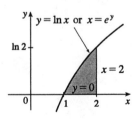

Because the region of integration is
$$D = \{(x,y) \mid 0 \le y \le \ln x, 1 \le x \le 2\}$$
$$= \{(x,y) \mid e^y \le x \le 2, 0 \le y \le \ln 2\}$$
we have
$$\int_1^2 \int_0^{\ln x} f(x,y) \, dy \, dx = \iint_D f(x,y) \, dA$$
$$= \int_0^{\ln 2} \int_{e^y}^2 f(x,y) \, dx \, dy$$

29.

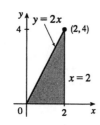

Because the region of integration is
$$D = \{(x,y) \mid y/2 \le x \le 2, 0 \le y \le 4\}$$
$$= \{(x,y) \mid 0 \le y \le 2x, 0 \le x \le 2\}$$
we have
$$\int_0^4 \int_{y/2}^2 f(x,y)\, dx\, dy = \iint_D f(x,y)\, dA$$
$$= \int_0^2 \int_0^{2x} f(x,y)\, dy\, dx$$

31.

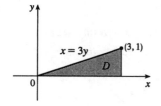

$$\int_0^1 \int_{3y}^3 e^{x^2}\, dx\, dy = \int_0^3 \int_0^{x/3} e^{x^2}\, dy\, dx$$
$$= \int_0^3 \left[e^{x^2} y \right]_{y=0}^{y=x/3} dx = \int_0^3 \left(\frac{x}{3}\right) e^{x^2}\, dx$$
$$= \frac{1}{6} e^{x^2} \Big]_0^3 = \frac{e^9 - 1}{6}$$

33.

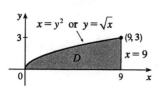

$$\int_0^3 \int_{y^2}^9 y \cos x^2\, dx\, dy = \int_0^9 \int_0^{\sqrt{x}} y \cos x^2\, dy\, dx$$
$$= \int_0^9 \cos x^2 \left[\frac{y^2}{2}\right]_{y=0}^{y=\sqrt{x}} dx$$
$$= \int_0^9 \frac{1}{2} x \cos x^2\, dx = \frac{1}{4} \sin x^2 \Big]_0^9$$
$$= \frac{1}{4} \sin 81$$

35.

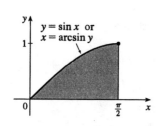

$$\int_0^1 \int_{\arcsin y}^{\pi/2} \cos x \sqrt{1 + \cos^2 x}\, dx\, dy$$
$$= \int_0^{\pi/2} \int_0^{\sin x} \cos x \sqrt{1 + \cos^2 x}\, dy\, dx$$
$$= \int_0^{\pi/2} \cos x \sqrt{1 + \cos^2 x}\, [y]_{y=0}^{y=\sin x}\, dx$$
$$= \int_0^{\pi/2} \cos x \sqrt{1 + \cos^2 x}\, \sin x\, dx$$
$$[\text{Let } u = \cos x,\ du = -\sin x\, dx,\ dx = du/(-\sin x)]$$
$$= \int_1^0 -u\sqrt{1 + u^2}\, du = -\frac{1}{3}\left(1 + u^2\right)^{3/2}\Big]_1^0$$
$$= \frac{1}{3}\left(\sqrt{8} - 1\right) = \frac{1}{3}\left(2\sqrt{2} - 1\right)$$

37. $D = \{(x,y) \mid 0 \le x \le 1, -x + 1 \le y \le 1\} \cup \{(x,y) \mid -1 \le x \le 0, x + 1 \le y \le 1\}$
$$\cup \{(x,y) \mid 0 \le x \le 1, -1 \le y \le x - 1\} \cup \{(x,y) \mid -1 \le x \le 0, -1 \le y \le -x - 1\},$$
all type I.
$$\iint_D x^2\, dA = \int_0^1 \int_{1-x}^1 x^2\, dy\, dx + \int_{-1}^0 \int_{x+1}^1 x^2\, dy\, dx + \int_0^1 \int_{-1}^{x-1} x^2\, dy\, dx + \int_{-1}^0 \int_{-1}^{-x-1} x^2\, dy\, dx$$
$$= 4 \int_0^1 \int_{1-x}^1 x^2\, dy\, dx \quad [\text{by symmetry of the regions and because } f(x,y) = x^2 \ge 0]$$
$$= 4 \int_0^1 x^3\, dx = 1$$

Section 12.3 Double Integrals over General Regions

39. For $D = [0,1] \times [0,1]$, $0 \le \sqrt{x^3 + y^3} \le \sqrt{2}$ and $A(D) = 1$, so $0 \le \iint_D \sqrt{x^3 + y^3}\, dA \le \sqrt{2}$.

41. Since $m \le f(x,y) \le M$, $\iint_D m\, dA \le \iint_D f(x,y)\, dA \le \iint_D M\, dA$
by (8) $\Rightarrow$ $m \iint_D 1\, dA \le \iint_D f(x,y)\, dA \le M \iint_D 1\, dA$ by (7) $\Rightarrow$
$mA(D) \le \iint_D f(x,y)\, dA \le MA(D)$ by (10).

43. $\iint_D \left(x^2 \tan x + y^3 + 4 \right)\, dA = \iint_D x^2 \tan x\, dA + \iint_D y^3\, dA + \iint_D 4\, dA$. But $x^2 \tan x$ is an odd
function of x and D is symmetric with respect to the y-axis, so $\iint_D x^2 \tan x\, dA = 0$. Similarly, y^3 is an
odd function of y and D is symmetric with respect to the x-axis, so $\iint_D y^3\, dA = 0$. Thus

$$\iint_D \left(x^2 \tan x + y^3 + 4 \right)\, dA = 4 \iint_D dA = 4\,(\text{area of } D) = 4 \cdot \pi \left(\sqrt{2} \right)^2 = 8\pi$$

45. Since $\sqrt{1 - x^2 - y^2} \ge 0$, we can interpret $\iint_D \sqrt{1 - x^2 - y^2}\, dA$ as the volume of the solid that lies
below the graph of $z = \sqrt{1 - x^2 - y^2}$ and above the region D in the xy-plane. $z = \sqrt{1 - x^2 - y^2}$ is
equivalent to $x^2 + y^2 + z^2 = 1$, $z \ge 0$ which meets the xy-plane in the circle $x^2 + y^2 = 1$, the boundary
of D. Thus, the solid is an upper hemisphere of radius 1 which has volume $\frac{1}{2} \left[\frac{4}{3}\pi \, (1)^3 \right] = \frac{2}{3}\pi$.

Section 12.4 Double Integrals in Polar Coordinates

1. The region R is more easily described by polar coordinates: $R = \{(r, \theta) \mid 0 \le r \le 2, 0 \le \theta \le 2\pi\}$.
Thus $\iint_R f(x, y)\, dA = \int_0^{2\pi} \int_0^2 f(r \cos\theta, r \sin\theta)\, r\, dr\, d\theta$.

3. The region R is more easily described by rectangular coordinates:
$R = \{(x, y) \mid -2 \le x \le 2, x \le y \le 2\}$. Thus $\iint_R f(x, y)\, dA = \int_{-2}^{2} \int_x^2 f(x, y)\, dy\, dx$.

5. The region R is more easily described by polar coordinates: $R = \{(r, \theta) \mid 2 \le r \le 5, 0 \le \theta \le 2\pi\}$.
Thus $\iint_R f(x, y)\, dA = \int_0^{2\pi} \int_2^5 f(r \cos\theta, r \sin\theta)\, r\, dr\, d\theta$.

7. $\iint_R x\, dA = \int_0^{2\pi} \int_0^5 r^2 \cos\theta\, dr\, d\theta = \left(\int_0^{2\pi} \cos\theta\, d\theta \right) \left(\int_0^5 r^2\, dr \right) = 0$

9. $\iint_R xy\, dA = \int_0^{\pi/2} \int_2^5 r^3 \cos\theta \sin\theta\, dr\, d\theta = \left(\int_0^{\pi/2} \frac{\sin 2\theta}{2}\, d\theta \right) \left(\int_2^5 r^3\, dr \right) = \frac{1}{2} \cdot \frac{5^4 - 2^4}{4} = \frac{609}{8}$

11. The circle $r = 1$ intersects the cardioid $r = 1 + \sin\theta$ when $1 = 1 + \sin\theta \;\Rightarrow\; \theta = 0$ or $\theta = \pi$, so

$$\iint_D \frac{1}{\sqrt{x^2 + y^2}}\, dA = \int_0^\pi \int_1^{1 + \sin\theta} \left(\frac{1}{r} \right) r\, dr\, d\theta$$

$$= \int_0^\pi [r]_{r=1}^{r=1+\sin\theta}\, d\theta = \int_0^\pi \sin\theta\, d\theta = [-\cos\theta]_0^\pi = 2$$

13. $V = \iint_{x^2 + y^2 \le 9} (x^2 + y^2)\, dA = \int_0^{2\pi} \int_0^3 r^3\, dr\, d\theta = 2\pi \left(\frac{81}{4} \right) = \frac{81\pi}{2}$

15. $V = 2 \iint_{x^2 + y^2 \le a^2} \sqrt{a^2 - x^2 - y^2}\, dA = 2 \int_0^{2\pi} \int_0^a \sqrt{a^2 - r^2}\, r\, dr\, d\theta = \frac{4\pi}{3} \left[-(a^2 - r^2)^{3/2} \right]_0^a$
$= \frac{4\pi}{3} a^3$

17. The cone $z = \sqrt{x^2 + y^2}$ intersects the sphere $x^2 + y^2 + z^2 = 1$ when $x^2 + y^2 + \left(\sqrt{x^2 + y^2} \right)^2 = 1$ or
$x^2 + y^2 = \frac{1}{2}$. So

$$V = \iint_{x^2 + y^2 \le 1/2} \left(\sqrt{1 - x^2 - y^2} - \sqrt{x^2 + y^2} \right) dA = \int_0^{2\pi} \int_0^{1/\sqrt{2}} \left(\sqrt{1 - r^2} - r \right) r\, dr\, d\theta$$

$$= \frac{2\pi}{3} \left[-(1 - r^2)^{3/2} - r^3 \right]_0^{1/\sqrt{2}} = \frac{2\pi}{3} \left(-\frac{1}{\sqrt{2}} + 1 \right) = \frac{\pi}{3} \left(2 - \sqrt{2} \right)$$

19. The given solid is the region inside the cylinder $x^2 + y^2 = 4$ between the surfaces
$z = \sqrt{64 - 4x^2 - 4y^2}$ and $z = -\sqrt{64 - 4x^2 - 4y^2}$. So

$$V = \iint_{x^2 + y^2 \le 4} \left[\sqrt{64 - 4x^2 - 4y^2} - \left(-\sqrt{64 - 4x^2 - 4y^2} \right) \right] dA$$

$$= \iint_{x^2 + y^2 \le 4} 2\sqrt{64 - 4x^2 - 4y^2}\, dA = 4 \int_0^{2\pi} \int_0^2 \sqrt{16 - r^2}\, r\, dr\, d\theta$$

$$= 8\pi \left[-\frac{1}{3} (16 - r^2)^{3/2} \right]_0^2 = \frac{8\pi}{3} \left(64 - 12^{3/2} \right) = \frac{8\pi}{3} \left(64 - 24\sqrt{3} \right)$$

21. From the sketch of the curve $r = \cos 2\theta$, we see that a loop is given by
the region $D = \left\{ (r, \theta) \mid -\frac{\pi}{4} \le \theta \le \frac{\pi}{4}, 0 \le r \le \cos 2\theta \right\}$. So the area is

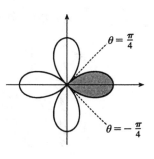

$$A(D) = \iint_D dA = \int_{-\pi/4}^{\pi/4} \int_0^{\cos 2\theta} r \, dr \, d\theta$$
$$= \int_{-\pi/4}^{\pi/4} \left[\tfrac{1}{2} r^2 \right]_{r=0}^{r=\cos 2\theta} d\theta = \tfrac{1}{2} \int_{-\pi/4}^{\pi/4} \cos^2 2\theta \, d\theta$$
$$= \tfrac{1}{4} \int_{-\pi/4}^{\pi/4} (1 + \cos 4\theta) \, d\theta$$
$$= \tfrac{1}{4} \left[\theta + \tfrac{1}{4} \sin 4\theta \right]_{-\pi/4}^{\pi/4} = \tfrac{\pi}{8}$$

23.

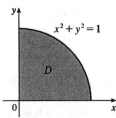

$$\int_0^{\pi/2} \int_0^1 r e^{r^2} \, dr \, d\theta = \tfrac{\pi}{2} \left[\tfrac{1}{2} e^{r^2} \right]_0^1 = \tfrac{1}{4} \pi (e - 1)$$

25.

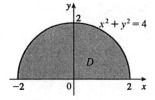

$$\int_0^\pi \int_0^2 \left(r^4 \cos^2 \theta \sin^2 \theta \right) r \, dr \, d\theta$$
$$= \int_0^\pi \int_0^2 \left(\tfrac{1}{4} r^5 \sin^2 2\theta \right) dr \, d\theta$$
$$= \tfrac{8}{3} \int_0^\pi \sin^2 2\theta \, d\theta$$
$$= \tfrac{8}{12} \left[2\theta - \sin 2\theta \cos 2\theta \right]_0^\pi = \tfrac{4\pi}{3}$$

27. The surface of the water in the pool is a circular disk D with radius 20 ft. If we place D on coordinate
axes with the origin at the center of D and define $f(x, y)$ to be the depth of the water at (x, y), then the
volume of water in the pool is the volume of the solid that lies above $D = \left\{ (x, y) \mid x^2 + y^2 \le 400 \right\}$ and
below the graph of $f(x, y)$. We can associate north with the positive y-direction, so we are given that the
depth is constant in the x-direction and the depth increases linearly in the y-direction from
$f(0, -20) = 2$ to $f(0, 20) = 7$. The trace in the yz-plane is a line segment from $(0, -20, 2)$ to
$(0, 20, 7)$. The slope of this line is $\frac{7-2}{20-(-20)} = \frac{1}{8}$, so an equation of the line is $z - 7 = \frac{1}{8}(y - 20)$ ⇒
$z = \frac{1}{8} y + \frac{9}{2}$. Since $f(x, y)$ is independent of x, $f(x, y) = \frac{1}{8} y + \frac{9}{2}$. Thus the volume is given by
$\iint_D f(x, y) \, dA$, which is most conveniently evaluated using polar coordinates. Then
$D = \left\{ (r, \theta) \mid 0 \le r \le 20, 0 \le \theta \le 2\pi \right\}$ and substituting $x = r \cos \theta$, $y = r \sin \theta$ the integral becomes

$$\int_0^{2\pi} \int_0^{20} \left(\tfrac{1}{8} r \sin \theta + \tfrac{9}{2} \right) r \, dr \, d\theta = \int_0^{2\pi} \left[\tfrac{1}{24} r^3 \sin \theta + \tfrac{9}{4} r^2 \right]_{r=0}^{r=20} d\theta$$
$$= \int_0^{2\pi} \left(\tfrac{1000}{3} \sin \theta + 900 \right) d\theta = \left[-\tfrac{1000}{3} \cos \theta + 900\theta \right]_0^{2\pi}$$
$$= 1800\pi$$

Thus the pool contains $1800\pi \approx 5655$ ft^3 of water.

29. $\int_{1/\sqrt{2}}^{1} \int_{\sqrt{1-x^2}}^{x} xy\, dy\, dx + \int_{1}^{\sqrt{2}} \int_{0}^{x} xy\, dy\, dx + \int_{\sqrt{2}}^{2} \int_{0}^{\sqrt{4-x^2}} xy\, dy\, dx = \int_{0}^{\pi/4} \int_{1}^{2} r^3 \cos\theta \sin\theta\, dr\, d\theta$

$$= \int_{0}^{\pi/4} \left[\frac{r^4}{4} \cos\theta \sin\theta \right]_{r=1}^{r=2} d\theta$$

$$= \frac{15}{4} \int_{0}^{\pi/4} \sin\theta \cos\theta\, d\theta$$

$$= \frac{15}{4} \left[\frac{\sin^2\theta}{2} \right]_{0}^{\pi/4} = \frac{15}{16}$$

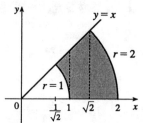

31. (a) We integrate by parts with $u = x$ and $dv = xe^{-x^2}\, dx$. Then $du = dx$ and $v = -\frac{1}{2}e^{-x^2}$, so

$$\int_{0}^{\infty} x^2 e^{-x^2}\, dx = \lim_{t\to\infty} \int_{0}^{t} x^2 e^{-x^2}\, dx = \lim_{t\to\infty} \left(-\frac{1}{2}xe^{-x^2} \Big]_{0}^{t} + \int_{0}^{t} \frac{1}{2}e^{-x^2}\, dx \right)$$

$$= \lim_{t\to\infty} \left(-\frac{1}{2}te^{-t^2} \right) + \frac{1}{2} \int_{0}^{\infty} e^{-x^2}\, dx = 0 + \frac{1}{2} \int_{0}^{\infty} e^{-x^2}\, dx \text{ (by l'Hospital's Rule)}$$

$$= \frac{1}{4} \int_{-\infty}^{\infty} e^{-x^2}\, dx \text{ (since } e^{-x^2} \text{ is an even function)} = \frac{1}{4}\sqrt{\pi} \text{ [by Exercise 30(c)]}$$

(b) Let $u = \sqrt{x}$. Then $u^2 = x \Rightarrow dx = 2u\, du \Rightarrow$

$$\int_{0}^{\infty} \sqrt{x}e^{-x}\, dx = \lim_{t\to\infty} \int_{0}^{t} \sqrt{x}e^{-x}\, dx = \lim_{t\to\infty} \int_{0}^{\sqrt{t}} ue^{-u^2} 2u\, du = 2 \int_{0}^{\infty} u^2 e^{-u^2}\, du$$

$$= 2 \left(\frac{1}{4}\sqrt{\pi} \right) \text{ [by part(a)]} = \frac{1}{2}\sqrt{\pi}$$

Section 12.5 Applications of Double Integrals

1. $Q = \iint_D \left(x^2 + 3y^2\right) dA = \int_0^2 \int_1^2 \left(x^2 + 3y^2\right) dy \, dx = \int_0^2 \left(x^2 + 7\right) dx = 14 + \frac{8}{3} = \frac{50}{3}$ C

3. $m = \int_0^1 \int_{x^2}^1 xy \, dy \, dx = \int_0^1 \left(\frac{1}{2}x - \frac{1}{2}x^5\right) dx = \frac{1}{4} - \frac{1}{12} = \frac{1}{6}$,

$\quad M_y = \int_0^1 \int_{x^2}^1 x^2 y \, dy \, dx = \int_0^1 \left(\frac{1}{2}x^2 - \frac{1}{2}x^6\right) dx = \frac{1}{6} - \frac{1}{14} = \frac{2}{21}$ and

$\quad M_x = \int_0^1 \int_{x^2}^1 xy^2 \, dy \, dx = \int_0^1 \left(\frac{1}{3}x - \frac{1}{3}x^7\right) dx = \frac{1}{6} - \frac{1}{24} = \frac{1}{8}$. Hence $m = \frac{1}{6}$, $(\overline{x}, \overline{y}) = \left(\frac{4}{7}, \frac{3}{4}\right)$.

5. $m = \int_0^2 \int_{x/2}^{3-x} (x+y) \, dy \, dx = \int_0^2 \left[x\left(3 - \frac{3}{2}x\right) + \frac{1}{2}(3-x)^2 - \frac{1}{8}x^2\right] dx = \int_0^2 \left[-\frac{9}{8}x^2 + \frac{9}{2}\right] dx = 6$,

$\quad M_y = \int_0^2 \int_{x/2}^{3-x} \left(x^2 + xy\right) dy \, dx = \int_0^2 \left[x^2 y + \frac{1}{2}xy^2\right]_{y=x/2}^{y=3-x} dx = \int_0^2 \left(\frac{9}{2}x - \frac{9}{8}x^3\right) dx = \frac{9}{2}$, and

$\quad M_x = \int_0^2 \int_{x/2}^{3-y} \left(xy + y^2\right) dy \, dx = \int_0^2 \left(9 - \frac{9}{2}x\right) dx = 9$. Hence $m = 6$, $(\overline{x}, \overline{y}) = \left(\frac{3}{4}, \frac{3}{2}\right)$.

7.

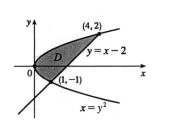

$m = \int_{-1}^2 \int_{y^2}^{y+2} 3 \, dx \, dy = \int_{-1}^2 \left(3y + 6 - 3y^2\right) dy = \frac{27}{2}$,

$\quad M_y = \int_{-1}^2 \int_{y^2}^{y+2} 3x \, dx \, dy = \int_{-1}^2 \frac{3}{2}\left[(y+2)^2 - y^4\right] dy$

$\quad\quad = \left[\frac{1}{2}(y+2)^3 - \frac{3}{10}y^5\right]_{-1}^2 = \frac{108}{5}$

and

$\quad M_x = \int_{-1}^2 \int_{y^2}^{y+2} 3y \, dx \, dy = \int_{-1}^2 \left(3y^2 + 6y - 3y^3\right) dy$

$\quad\quad = \left[y^3 + 3y^2 - \frac{3}{4}y^4\right]_{-1}^2 = \frac{27}{4}$

Hence $m = \frac{27}{2}$, $(\overline{x}, \overline{y}) = \left(\frac{8}{5}, \frac{1}{2}\right)$.

9. $\rho(x,y) = ky = kr\sin\theta$, $m = \int_0^{\pi/2} \int_0^1 kr^2 \sin\theta \, dr \, d\theta = \frac{1}{3}k \int_0^{\pi/2} \sin\theta \, d\theta = \frac{1}{3}k\left[-\cos\theta\right]_0^{\pi/2} = \frac{1}{3}k$,

$\quad M_y = \int_0^{\pi/2} \int_0^1 kr^3 \sin\theta \cos\theta \, dr \, d\theta = \frac{1}{4}k \int_0^{\pi/2} \sin\theta \cos\theta \, d\theta = \frac{1}{8}k\left[-\cos 2\theta\right]_0^{\pi/2} = \frac{1}{8}k$,

$\quad M_x = \int_0^{\pi/2} \int_0^1 kr^3 \sin^2\theta \, dr \, d\theta = \frac{1}{4}k \int_0^{\pi/2} \sin^2\theta \, d\theta = \frac{1}{8}k\left[\theta + \sin 2\theta\right]_0^{\pi/2} = \frac{\pi}{16}k$. Hence

$(\overline{x}, \overline{y}) = \left(\frac{3}{8}, \frac{3\pi}{16}\right)$.

11. Placing the vertex opposite the hypotenuse at $(0,0)$, $\rho(x,y) = k\left(x^2 + y^2\right)$. Then

$$m = \int_0^a \int_0^{a-x} k\left(x^2 + y^2\right) dy \, dx = k \int_0^a \left[ax^2 - x^3 + \frac{1}{3}(a-x)^3\right] dx$$

$$= k\left[\frac{1}{3}ax^3 - \frac{1}{4}x^4 - \frac{1}{12}(a-x)^4\right]_0^a = \frac{1}{6}ka^4$$

By symmetry,

$$M_y = M_x = \int_0^a \int_0^{a-x} ky\left(x^2 + y^2\right) dy \, dx = k \int_0^a \left[\frac{1}{2}(a-x)^2 x^2 + \frac{1}{4}(a-x)^4\right] dx$$

$$= k\left[\frac{1}{6}a^2 x^3 - \frac{1}{4}ax^4 + \frac{1}{10}x^5 - \frac{1}{20}(a-x)^5\right]_0^a = \frac{1}{15}ka^5$$

Hence $(\overline{x}, \overline{y}) = \left(\frac{2}{5}a, \frac{2}{5}a\right)$.

13. $I_x = \int_0^1 \int_{x^2}^1 y^2 (xy) \, dy \, dx = \int_0^1 \frac{1}{4} (x - x^9) \, dx = \frac{1}{8} - \frac{1}{40} = \frac{1}{10}$,

$I_y = \int_0^1 \int_{x^2}^1 x^3 y \, dy \, dx = \int_0^1 \frac{1}{2} (x^3 - x^7) \, dx = \frac{1}{8} - \frac{1}{16} = \frac{1}{16}$, $I_0 = I_x + I_y = \frac{13}{80}$.

15. $I_x = \int_{-1}^2 \int_{y^2}^{y+2} 3y^2 \, dx \, dy = \int_{-1}^2 (3y^3 + 6y^2 - 3y^4) \, dy = \left[\frac{3}{4} y^4 + 2y^3 - \frac{3}{5} y^5 \right]_{-1}^2 = \frac{189}{20}$,

$I_y = \int_{-1}^2 \int_{y^2}^{y+2} 3x^2 \, dx \, dy = \int_{-1}^2 \left[(y+2)^3 - y^6 \right] dy = \left[\frac{1}{4} (y+2)^4 - \frac{1}{7} y^7 \right]_{-1}^2 = \frac{1269}{28}$, and

$I_0 = I_x + I_y = \frac{1917}{35}$.

17. (a) $f(x,y)$ is a joint density function, so we know $\iint_{\mathbb{R}^2} f(x,y) \, dA = 1$. Since $f(x,y) = 0$ outside the rectangle $[0,1] \times [0,2]$, we can say

$$\iint_{\mathbb{R}^2} f(x,y) \, dA = \int_{-\infty}^{\infty} \int_{-\infty}^{\infty} f(x,y) \, dy \, dx = \int_0^1 \int_0^2 Cx(1+y) \, dy \, dx$$

$$= C \int_0^1 x \left[y + \frac{1}{2} y^2 \right]_{y=0}^{y=2} dx = C \int_0^1 4x \, dx = C \left[2x^2 \right]_0^1 = 2C$$

Then $2C = 1 \ \Rightarrow \ C = \frac{1}{2}$.

(b) $P(X \leq 1, Y \leq 1) = \int_{-\infty}^1 \int_{-\infty}^1 f(x,y) \, dy \, dx = \int_0^1 \int_0^1 \frac{1}{2} x (1+y) \, dy \, dx$

$$= \int_0^1 \frac{1}{2} x \left[y + \frac{1}{2} y^2 \right]_{y=0}^{y=1} dx = \int_0^1 \frac{1}{2} x \left(\frac{3}{2} \right) dx = \frac{3}{4} \left[\frac{1}{2} x^2 \right]_0^1 = \frac{3}{8} \text{ or } 0.375$$

(c) $P(X + Y \leq 1) = P((X,Y) \in D)$ where D is the triangular region shown in the figure. Thus

$P(X + Y \leq 1) = \iint_D f(x,y) \, dA = \int_0^1 \int_0^{1-x} \frac{1}{2} x (1+y) \, dy \, dx$

$$= \int_0^1 \frac{1}{2} x \left[y + \frac{1}{2} y^2 \right]_{y=0}^{y=1-x} dx = \int_0^1 \frac{1}{2} x \left(\frac{1}{2} x^2 - 2x + \frac{3}{2} \right) dx$$

$$= \frac{1}{4} \int_0^1 (x^3 - 4x^2 + 3x) \, dx = \frac{1}{4} \left[\frac{x^4}{4} - 4\frac{x^3}{3} + 3\frac{x^2}{2} \right]_0^1$$

$$= \frac{5}{48} \approx 0.1042$$

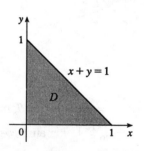

19. (a) $f(x,y) \geq 0$, so f is a joint density function if $\iint_{\mathbb{R}^2} f(x,y) \, dA = 1$. Here, $f(x,y) = 0$ outside the first quadrant, so

$$\iint_{\mathbb{R}^2} f(x,y) \, dA = \int_0^{\infty} \int_0^{\infty} 0.1 e^{-(0.5x + 0.2y)} \, dy \, dx = 0.1 \int_0^{\infty} \int_0^{\infty} e^{-0.5x} e^{-0.2y} \, dy \, dx$$

$$= 0.1 \int_0^{\infty} e^{-0.5x} \, dx \int_0^{\infty} e^{-0.2y} \, dy = 0.1 \lim_{t \to \infty} \int_0^t e^{-0.5x} \, dx \lim_{t \to \infty} \int_0^t e^{-0.2y} \, dy$$

$$= 0.1 \lim_{t \to \infty} \left[-2 e^{-0.5x} \right]_0^t \lim_{t \to \infty} \left[-5 e^{-0.2y} \right]_0^t$$

$$= 0.1 \lim_{t \to \infty} \left[-2 \left(e^{-0.5t} - 1 \right) \right] \lim_{t \to \infty} \left[-5 \left(e^{-0.2t} - 1 \right) \right]$$

$$= (0.1) \cdot (-2) (0 - 1) \cdot (-5) (0 - 1) = 1$$

Thus $f(x,y)$ is a joint density function.

(b) (i) No restriction is placed on X, so

$$P(Y \geq 1) = \int_{-\infty}^{\infty} \int_{1}^{\infty} f(x,y) \, dy \, dx = \int_{0}^{\infty} \int_{1}^{\infty} 0.1 e^{-(0.5x + 0.2y)} \, dy \, dx$$

$$= 0.1 \int_{0}^{\infty} e^{-0.5x} \, dx \int_{1}^{\infty} e^{-0.2y} \, dy = 0.1 \lim_{t \to \infty} \int_{0}^{t} e^{-0.5x} \, dx \lim_{t \to \infty} \int_{1}^{t} e^{-0.2y} \, dy$$

$$= 0.1 \lim_{t \to \infty} \left[-2e^{-0.5x}\right]_{0}^{t} \lim_{t \to \infty} \left[-5e^{-0.2y}\right]_{1}^{t}$$

$$= 0.1 \lim_{t \to \infty} \left[-2\left(e^{-0.5t} - 1\right)\right] \lim_{t \to \infty} \left[-5\left(e^{-0.2t} - e^{-0.2}\right)\right]$$

$$= (0.1) \cdot (-2)(0-1) \cdot (-5)\left(0 - e^{-0.2}\right) = e^{-0.2} \approx 0.8187$$

(ii) $P(X \leq 2, Y \leq 4) = \int_{-\infty}^{2} \int_{-\infty}^{4} f(x,y) \, dy \, dx = \int_{0}^{2} \int_{0}^{4} 0.1 e^{-(0.5x + 0.2y)} \, dy \, dx$

$$= 0.1 \int_{0}^{2} e^{-0.5x} \, dx \int_{0}^{4} e^{-0.2y} \, dy = 0.1 \left[-2e^{-0.5x}\right]_{0}^{2} \left[-5e^{-0.2y}\right]_{0}^{4}$$

$$= (0.1) \cdot (-2)\left(e^{-1} - 1\right) \cdot (-5)\left(e^{-0.8} - 1\right)$$

$$= \left(e^{-1} - 1\right)\left(e^{-0.8} - 1\right) = 1 + e^{-1.8} - e^{-0.8} - e^{-1} \approx 0.3481$$

(c) The expected value of X is given by

$$\mu_1 = \iint_{\mathbb{R}^2} x f(x,y) \, dA = \int_{0}^{\infty} \int_{0}^{\infty} x \left[0.1 e^{-(0.5x + 0.2y)}\right] \, dy \, dx$$

$$= 0.1 \int_{0}^{\infty} x e^{-0.5x} \, dx \int_{0}^{\infty} e^{-0.2y} \, dy = 0.1 \lim_{t \to \infty} \int_{0}^{t} x e^{-0.5x} \, dx \lim_{t \to \infty} \int_{0}^{t} e^{-0.2y} \, dy$$

To evaluate the first integral, we integrate by parts with $u = x$ and
$dv = e^{-0.5x} \, dx$ (or we can use Formula 96 in the Table of Integrals):
$\int x e^{-0.5x} \, dx = -2x e^{-0.5x} - \int -2 e^{-0.5x} \, dx = -2x e^{-0.5x} - 4e^{-0.5x} = -2(x+2) e^{-0.5x}$. Thus

$$\mu_1 = 0.1 \lim_{t \to \infty} \left[-2(x+2) e^{-0.5x}\right]_{0}^{t} \lim_{t \to \infty} \left[-5e^{-0.2y}\right]_{0}^{t}$$

$$= 0.1 \lim_{t \to \infty} (-2)\left[(t+2) e^{-0.5t} - 2\right] \lim_{t \to \infty} (-5)\left[e^{-0.2t} - 1\right]$$

$$= 0.1 (-2)\left(\lim_{t \to \infty} \frac{t+2}{e^{0.5t}} - 2\right)(-5)(-1) = 2 \text{ (by l'Hospital's Rule)}$$

The expected value of Y is given by

$$\mu_2 = \iint_{\mathbb{R}^2} y f(x,y) \, dA = \int_{0}^{\infty} \int_{0}^{\infty} y \left[0.1 e^{-(0.5x + 0.2y)}\right] \, dy \, dx$$

$$= 0.1 \int_{0}^{\infty} e^{-0.5x} \, dx \int_{0}^{\infty} y e^{-0.2y} \, dy = 0.1 \lim_{t \to \infty} \int_{0}^{t} e^{-0.5x} \, dx \lim_{t \to \infty} \int_{0}^{t} y e^{-0.2y} \, dy$$

To evaluate the second integral, we integrate by parts with $u = y$ and $dv = e^{-0.2y} \, dy$ (or
again we can use Formula 96 in the Table of Integrals) which gives
$\int y e^{-0.2y} \, dy = -5y e^{-0.2y} + \int 5 e^{-0.2y} \, dy = -5(y+5) e^{-0.2y}$. Then

$$\mu_2 = 0.1 \lim_{t \to \infty} \left[-2e^{-0.5x}\right]_{0}^{t} \lim_{t \to \infty} \left[-5(y+5) e^{-0.2y}\right]_{0}^{t}$$

$$= 0.1 \lim_{t \to \infty} \left[-2\left(e^{-0.5t} - 1\right)\right] \lim_{t \to \infty} \left(-5\left[(t+5) e^{-0.2t} - 5\right]\right)$$

$$= 0.1 (-2)(-1) \cdot (-5)\left(\lim_{t \to \infty} \frac{t+5}{e^{0.2t}} - 5\right) = 5 \text{ (by l'Hospital's Rule)}$$

21. The random variables X and Y are normally distributed with $\mu_1 = 45$, $\mu_2 = 20$, $\sigma_1 = 0.5$, and $\sigma_2 = 0.1$. The individual density functions for X and Y, then, are $f_1(x) = \frac{1}{0.5\sqrt{2\pi}} e^{-(x-45)^2/0.5}$ and $f_2(y) = \frac{1}{0.1\sqrt{2\pi}} e^{-(y-20)^2/0.02}$. Since X and Y are independent, the joint density function is the product

$$f(x,y) = f_1(x) f_2(y)$$

$$= \frac{1}{0.5\sqrt{2\pi}} e^{-(x-45)^2/0.5} \frac{1}{0.1\sqrt{2\pi}} e^{-(y-20)^2/0.02}$$

$$= \frac{10}{\pi} e^{-2(x-45)^2 - 50(y-20)^2}$$

Then

$$P(40 \leq X \leq 50, 20 \leq Y \leq 25) = \int_{40}^{50} \int_{20}^{25} f(x,y)\, dy\, dx$$

$$= \frac{10}{\pi} \int_{40}^{50} \int_{20}^{25} e^{-2(x-45)^2 - 50(y-20)^2}\, dy\, dx$$

Using a CAS or calculator to evaluate the integral, we get $P(40 \leq X \leq 50, 20 \leq Y \leq 25) \approx 0.500$.

23. (a) If $f(P, A)$ is the probability that an individual at A will be infected by an individual at P, and $k\, dA$ is the number of infected individuals in an element of area dA, then $f(P, A)\, k\, dA$ is the number of infections that should result from exposure of the individual at A to infected people in the element of area dA. Integration over D gives the number of infections of the person at A due to all the infected people in D. In rectangular coordinates (with the origin at the city's center), the exposure of a person at A is

$$E = \iint_D k f(P, A)\, dA$$

$$= k \iint_D \frac{20 - d(P, A)}{20}\, dA$$

$$= k \iint_D \left[1 - \frac{\sqrt{(x - x_0)^2 + (y - y_0)^2}}{20} \right] dx\, dy$$

(b) If $A = (0,0)$, then

$$E = k \iint_D \left[1 - \frac{1}{20}\sqrt{x^2 + y^2} \right] dx\,dy = k \int_0^{2\pi} \int_0^{10} \left(1 - \frac{r}{20} \right) r\,dr\,d\theta$$

$$= 2\pi k \left[\frac{r^2}{2} - \frac{r^3}{60} \right]_0^{10} = 2\pi k \left(50 - \tfrac{50}{3} \right) = \tfrac{200}{3}\pi k \approx 209k$$

For A at the edge of the city, it is convenient to use a polar coordinate system centered at A. Then the polar equation for the circular boundary of the city becomes $r = 20\cos\theta$ instead of $r = 10$, and the distance from A to a point P in the city is again r (see the figure.) So

$$E = k \int_{-\pi/2}^{\pi/2} \int_0^{20\cos\theta} \left(1 - \frac{r}{20} \right) r\,dr\,d\theta$$

$$= k \int_{-\pi/2}^{\pi/2} \left[\frac{r^2}{2} - \frac{r^3}{60} \right]_{r=0}^{r=20\cos\theta} d\theta$$

$$= k \int_{-\pi/2}^{\pi/2} \left(200\cos^2\theta - \tfrac{400}{3}\cos^3\theta \right) d\theta$$

$$= 200k \int_{-\pi/2}^{\pi/2} \left[\tfrac{1}{2} + \tfrac{1}{2}\cos 2\theta - \tfrac{2}{3}\left(1 - \sin^2\theta \right)\cos\theta \right] d\theta$$

$$= 200k \left[\tfrac{1}{2}\theta + \tfrac{1}{4}\sin 2\theta - \tfrac{2}{3}\sin\theta + \tfrac{2}{3}\cdot\tfrac{1}{3}\sin^3\theta \right]_{-\pi/2}^{\pi/2}$$

$$= 200k \left[\tfrac{\pi}{4} + 0 - \tfrac{2}{3} + \tfrac{2}{9} + \tfrac{\pi}{4} + 0 - \tfrac{2}{3} + \tfrac{2}{9} \right]$$

$$= 200k \left(\tfrac{\pi}{2} - \tfrac{8}{9} \right) \approx 136k$$

$r = 20\cos\theta$

Therefore the risk of infection is much lower at the edge of the city than in the middle, so it is better to live at the edge.

Section 12.6 Surface Area

1. Here $z = f(x, y) = 4 - x - 2y$ with $0 \le x^2 + y^2 \le 4$. Thus, by (6),

$$A(S) = \iint_D \sqrt{1 + (-1)^2 + (-2)^2}\, dA = \sqrt{6} \iint_{x^2 + y^2 \le 4} dA = 4\sqrt{6}\pi.$$

3. $z = f(x, y) = y^2 - x^2$ with $1 \le x^2 + y^2 \le 4$. Then

$$
\begin{aligned}
A(S) &= \iint_D \sqrt{1 + 4x^2 + 4y^2}\, dA = \int_0^{2\pi} \int_1^2 \sqrt{1 + 4r^2}\, r\, dr\, d\theta \\
&= 4\pi \left(\tfrac{1}{24}\right) \left(1 + 4r^2\right)^{3/2}\Big]_1^2 \\
&= \tfrac{\pi}{6} \left(17\sqrt{17} - 5\sqrt{5}\right)
\end{aligned}
$$

5. A parametric representation of the surface is $x = x$, $y = 4x + z^2$, $z = z$ with $0 \le x \le 1, 0 \le z \le 1$. Hence $\mathbf{r}_x \times \mathbf{r}_z = 4\mathbf{i} - \mathbf{j} + 2z\mathbf{k}$.

Note: In general, if $y = f(x, z)$ then $\mathbf{r}_z \times \mathbf{r}_x = -\dfrac{\partial f}{\partial x}\mathbf{i} + \mathbf{j} - \dfrac{\partial f}{\partial x}\mathbf{k}$ and

$$A(S) = \iint_D \sqrt{1 + \left(\dfrac{\partial f}{\partial x}\right)^2 + \left(\dfrac{\partial f}{\partial z}\right)^2}\, dA.\ \text{Then}$$

$$
\begin{aligned}
A(S) &= \int_0^1 \int_0^1 \sqrt{17 + 4z^2}\, dx\, dz = \int_0^1 \sqrt{17 + 4z^2}\, dz \\
&= \tfrac{1}{2}\left(z\sqrt{17 + 4z^2} + \tfrac{17}{2}\ln\left|2z + \sqrt{4z^2 + 17}\right|\right)\Big]_0^1 \\
&= \tfrac{\sqrt{21}}{2} + \tfrac{17}{4}\left[\ln\left(2 + \sqrt{21}\right) - \ln\sqrt{17}\right]
\end{aligned}
$$

7. Let $A(S_1)$ be the surface area of that portion of the surface which lies above the plane $z = 0$. Then $A(S) = 2A(S_1)$. Following Example 1, a parametric representation of S_1 is $x = a\sin\phi\cos\theta$, $y = a\sin\phi\sin\theta$, $z = a\cos\phi$ and $|\mathbf{r}_\phi \times \mathbf{r}_\theta| = a^2\sin\phi$. For D, $0 \le \phi \le \tfrac{\pi}{2}$ and for each fixed ϕ, $\left(x - \tfrac{1}{2}a\right)^2 + y^2 \le \left(\tfrac{1}{2}a\right)^2$ or $\left[a\sin\phi\cos\theta - \tfrac{1}{2}a\right]^2 + a^2\sin^2\phi\sin^2\theta \le (a/2)^2$ implies $a^2\sin^2\phi - a^2\sin\phi\cos\theta \le 0$ or $\sin\phi\left(\sin\phi - \cos\theta\right) \le 0$. But $0 \le \phi \le \tfrac{\pi}{2}$, so $\cos\theta \ge \sin\phi$ or $\sin\left(\tfrac{\pi}{2} + \theta\right) \ge \sin\phi$ or $\phi - \tfrac{\pi}{2} \le \theta \le \tfrac{\pi}{2} - \phi$. Hence $D = \left\{(\phi, \theta) \mid 0 \le \phi \le \tfrac{\pi}{2}, \phi - \tfrac{\pi}{2} \le \theta \le \tfrac{\pi}{2} - \phi\right\}$. Then

$$
\begin{aligned}
A(S_1) &= \int_0^{\pi/2} \int_{\phi - (\pi/2)}^{(\pi/2) - \phi} a^2\sin\phi\, d\theta\, d\phi = a^2 \int_0^{\pi/2}\left(\pi - 2\phi\right)\sin\phi\, d\phi \\
&= a^2\left[(-\pi\cos\phi) - 2\left(-\phi\cos\phi + \sin\phi\right)\right]_0^{\pi/2} \\
&= a^2\left(\pi - 2\right)
\end{aligned}
$$

Thus $A(S) = 2a^2(\pi - 2)$.

Alternate Solution: Working on S_1 we could parametrize the portion of the sphere by $x = x$, $y = y$,

$z = \sqrt{a^2 - x^2 - y^2}$. Then $|\mathbf{r}_x \times \mathbf{r}_y| = \sqrt{1 + \dfrac{x^2}{a^2 - x^2 - y^2} + \dfrac{y^2}{a^2 - x^2 - y^2}} = \dfrac{a}{\sqrt{a^2 - x^2 - y^2}}$ and

$$A(S_1) = \iint_{0 \le (x-(a/2))^2 + y^2 \le (a/2)^2} \dfrac{a}{\sqrt{a^2 - x^2 - y^2}}\, dA = \int_{-\pi/2}^{\pi/2} \int_0^{a\cos\theta} \dfrac{a}{\sqrt{a^2 - r^2}}\, r\, dr\, d\theta$$

$$= \int_{-\pi/2}^{\pi/2} -a\left(a^2 - r^2\right)^{1/2}\Big]_{r=0}^{r=a\cos\theta} d\theta = \int_{-\pi/2}^{\pi/2} a^2\left[1 - \left(1 - \cos^2\theta\right)^{1/2}\right] d\theta$$

$$= \int_{-\pi/2}^{\pi/2} a^2\left(1 - |\sin\theta|\right) d\theta = 2a^2 \int_0^{\pi/2} (1 - \sin\theta)\, d\theta = 2a^2 \left(\tfrac{\pi}{2} - 1\right)$$

Thus $A(S) = 4a^2\left(\tfrac{\pi}{2} - 1\right) = 2a^2(\pi - 2)$.

Notes:

(1) Perhaps working in spherical coordinates is the most obvious approach here. However, you must be careful in setting up D.

(2) In the alternate solution, you can avoid having to use $|\sin\theta|$ byworking in the first octant and then multiplying by 8. However,if you set up S_1 as above and arrived at $A(S_1) = a^2\pi$, you now see your error.

9. $\mathbf{r}_u = \langle v, 1, 1\rangle$, $\mathbf{r}_v = \langle u, 1, -1\rangle$ and $\mathbf{r}_u \times \mathbf{r}_v = \langle -2, u+v, v-u\rangle$. Then

$$A(S) = \iint_{u^2 + v^2 \le 1} \sqrt{4 + 2u^2 + 2v^2}\, dA = \int_0^{2\pi} \int_0^1 r\sqrt{4 + 2r^2}\, dr\, d\theta$$

$$= 2\pi\left(\tfrac{1}{6}\right)\left(4 + 2r^2\right)^{3/2}\Big]_0^1 = \tfrac{\pi}{3}\left(6\sqrt{6} - 8\right) = \pi\left(2\sqrt{6} - \tfrac{8}{3}\right)$$

11. (a) The midpoints of the four squares are $\left(\tfrac{1}{4}, \tfrac{1}{4}\right)$, $\left(\tfrac{1}{4}, \tfrac{3}{4}\right)$, $\left(\tfrac{3}{4}, \tfrac{1}{4}\right)$, and $\left(\tfrac{3}{4}, \tfrac{3}{4}\right)$; the derivatives of the function $f(x,y) = x^2 + y^2$ are $f_x(x,y) = 2x$ and $f_y(x,y) = 2y$, so the Midpoint Rule gives

$$A(S) = \int_0^1 \int_0^1 \sqrt{[f_x(x,y)]^2 + [f_y(x,y)]^2 + 1}\, dy\, dx$$

$$\approx \tfrac{1}{4}\left(\sqrt{\left[2\left(\tfrac{1}{4}\right)\right]^2 + \left[2\left(\tfrac{1}{4}\right)\right]^2 + 1} + \sqrt{\left[2\left(\tfrac{1}{4}\right)\right]^2 + \left[2\left(\tfrac{3}{4}\right)\right]^2 + 1}\right.$$

$$\left. + \sqrt{\left[2\left(\tfrac{3}{4}\right)\right]^2 + \left[2\left(\tfrac{1}{4}\right)\right]^2 + 1} + \sqrt{\left[2\left(\tfrac{3}{4}\right)\right]^2 + \left[2\left(\tfrac{3}{4}\right)\right]^2 + 1}\right)$$

$$= \tfrac{1}{4}\left(\sqrt{\tfrac{3}{2}} + 2\sqrt{\tfrac{7}{2}} + \sqrt{\tfrac{11}{2}}\right) \approx 1.8279$$

(b) A CAS estimates the integral to be

$A(S) = \int_0^1 \int_0^1 \sqrt{f_x^2 + f_y^2 + 1}\, dy\, dx = \int_0^1 \int_0^1 \sqrt{4x^2 + 4y^2 + 1}\, dy\, dx \approx 1.8616$. This agrees with the Midpoint estimate only in the first decimal place.

13. (a) $x = a \sin u \cos v$, $y = b \sin u \sin v$, $z = c \cos u$ $\Rightarrow$ **(b)**

$$\frac{x^2}{a^2} + \frac{y^2}{b^2} + \frac{z^2}{c^2} = (\sin u \cos v)^2 + (\sin u \sin v)^2 + (\cos u)^2$$

$$= \sin^2 u + \cos^2 u = 1$$

and since the ranges of u and v are sufficient to generate the
entire graph, the parametric equations represent an ellipsoid.

(c) From the parametric equations (with $a = 1$, $b = 2$, and $c = 3$), we
calculate $\mathbf{r}_u = \cos u \cos v \, \mathbf{i} + 2 \cos u \sin v \, \mathbf{j} - 3 \sin u \, \mathbf{k}$ and

$\mathbf{r}_v = -\sin u \sin v \, \mathbf{i} + 2 \sin u \cos v \, \mathbf{j}$. So $\mathbf{r}_u \times \mathbf{r}_v = 6 \sin^2 u \cos v \, \mathbf{i} + 3 \sin^2 u \sin v \, \mathbf{j} + 2 \sin u \cos u \, \mathbf{k}$,
and the surface area is given by

$$A\,(S) = \int_0^{2\pi} \int_0^\pi |\mathbf{r}_u \times \mathbf{r}_v| \, du \, dv = \int_0^{2\pi} \int_0^\pi \sqrt{36 \sin^4 u \cos^2 v + 9 \sin^4 u \sin^2 v + 4 \cos^2 u \sin^2 u} \, du \, dv$$

15. If we revolve the curve $y = f\,(x)$, $a \le x \le b$ about the x-axis, where $f\,(x) \ge 0$, then from Equations 3
in Section 10.5 we know we can parametrize the surface using $x = x$, $y = f\,(x) \cos \theta$, and
$z = f\,(x) \sin \theta$, where $a \le x \le b$ and $0 \le \theta \le 2\pi$. Thus we can say the surface is represented by
$\mathbf{r}\,(x, \theta) = x \, \mathbf{i} + f\,(x) \cos \theta \, \mathbf{j} + f\,(x) \sin \theta \, \mathbf{k}$, with $a \le x \le b$ and $0 \le \theta \le 2\pi$. Then by (4), the surface
area is given by $A\,(S) = \iint_D |\mathbf{r}_x \times \mathbf{r}_\theta| \, dA$ where D is the rectangular parameter region $[a, b] \times [0, 2\pi]$.
Here, $\mathbf{r}_x\,(x, \theta) = \mathbf{i} + f'\,(x) \cos \theta \, \mathbf{j} + f'\,(x) \sin \theta \, \mathbf{k}$ and $\mathbf{r}_\theta\,(x, \theta) = -f\,(x) \sin \theta \, \mathbf{j} + f\,(x) \cos \theta \, \mathbf{k}$. So

$$\mathbf{r}_x \times \mathbf{r}_\theta = \begin{vmatrix} \mathbf{i} & \mathbf{j} & \mathbf{k} \\ 1 & f'\,(x) \cos \theta & f'\,(x) \sin \theta \\ 0 & -f\,(x) \sin \theta & f\,(x) \cos \theta \end{vmatrix}$$

$$= \left[f\,(x) f'\,(x) \cos^2 \theta + f\,(x) f'\,(x) \sin^2 \theta \right] \mathbf{i} - f\,(x) \cos \theta \, \mathbf{j} - f\,(x) \sin \theta \, \mathbf{k}$$

$$= f\,(x) f'\,(x) \, \mathbf{i} - f\,(x) \cos \theta \, \mathbf{j} - f\,(x) \sin \theta \, \mathbf{k} \text{ and}$$

$$|\mathbf{r}_x \times \mathbf{r}_\theta| = \sqrt{[f\,(x) f'\,(x)]^2 + [f\,(x)]^2 \cos^2 \theta + [f\,(x)]^2 \sin^2 \theta}$$

$$= \sqrt{[f\,(x)]^2 \left([f'\,(x)]^2 + 1 \right)} = f\,(x) \sqrt{1 + [f'\,(x)]^2} \text{ [since } f\,(x) \ge 0]. \text{ Thus}$$

$$A\,(S) = \iint_D |\mathbf{r}_x \times \mathbf{r}_\theta| \, dA = \int_a^b \int_0^{2\pi} f\,(x) \sqrt{1 + [f'\,(x)]^2} \, d\theta \, dx$$

$$= \int_a^b f\,(x) \sqrt{1 + [f'\,(x)]^2} \, [\theta]_0^{2\pi} \, dx = 2\pi \int_a^b f\,(x) \sqrt{1 + [f'\,(x)]^2} \, dx$$

17. $y = \sqrt{x}$ $\Rightarrow$ $1 + \left(\dfrac{dy}{dx}\right)^2 = 1 + \left(\dfrac{1}{2\sqrt{x}}\right)^2 = 1 + \dfrac{1}{4x}$. So

$$S = \int_4^9 2\pi y \sqrt{1 + \left(\frac{dy}{dx}\right)^2} \, dx = \int_4^9 2\pi \sqrt{x} \sqrt{1 + \frac{1}{4x}} \, dx = 2\pi \int_4^9 \left(x + \tfrac{1}{4}\right) dx$$

$$= 2\pi \left[\tfrac{2}{3} \left(x + \tfrac{1}{4}\right)^{3/2} \right]_4^9 = \tfrac{4\pi}{3} \left[\tfrac{1}{8} \left(4x + 1\right)^{3/2} \right]_4^9 = \tfrac{\pi}{6} \left(37\sqrt{37} - 17\sqrt{17} \right)$$

Section 12.7 Triple Integrals

1. $\int_0^1 \int_{-1}^2 \int_0^3 xyz^2 \, dz \, dx \, dy = \int_0^1 \int_{-1}^2 xy\,(9) \, dx \, dy = 9 \int_0^1 x \left(2 - \frac{1}{2}\right) dx = 9 \left(\frac{3}{4} \cdot 1^2\right) = \frac{27}{4}$

3. $\int_0^1 \int_0^z \int_0^y xyz \, dx \, dy \, dz = \int_0^1 \int_0^z \left(\frac{1}{2} y^3 z\right) dy \, dz = \int_0^1 \frac{1}{8} z^5 \, dz = \frac{1}{48} z^6 \Big]_0^1 = \frac{1}{48}$

5. $\int_0^\pi \int_0^2 \int_0^{\sqrt{4-z^2}} z \sin y \, dx \, dz \, dy = \int_0^\pi \int_0^2 z\sqrt{4-z^2} \sin y \, dz \, dy = \int_0^\pi \left[-\frac{1}{3}\left(4-z^2\right)^{3/2}\right]_{z=0}^{z=2} \sin y \, dy$

$\qquad = \int_0^\pi \frac{8}{3} \sin y \, dy = -\frac{8}{3} \cos y \big]_0^\pi = \frac{16}{3}$

7. $\int_0^1 \int_0^{2z} \int_0^{z+2} yz \, dx \, dy \, dz = \int_0^1 \int_0^{2z} yz\,(z+2) \, dy \, dz = \int_0^1 \left(2z^4 + 4z^3\right) dz = \frac{7}{5}$

9. $\int_0^1 \int_0^{x^2} \int_0^{x+2y} y \, dz \, dy \, dx = \int_0^1 \int_0^{x^2} \left(yx + 2y^2\right) dy \, dx = \int_0^1 \left[\frac{1}{2} xy^2 + \frac{2}{3} y^3\right]_{y=0}^{y=x^2} dx$

$\qquad = \int_0^1 \left(\frac{1}{2} x^5 + \frac{2}{3} x^6\right) dx = \left[\frac{1}{12} x^6 + \frac{2}{21} x^7\right]_0^1 = \frac{5}{28}$

11.

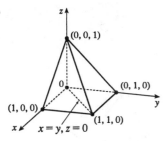

By symmetry $\iiint_E z \, dV = 2 \iiint_{E'} z \, dV$ where E' is the part of E to the left [as viewed from $(10, 10, 0)$] of the plane $x = y$. So

$\iiint_E z \, dV = \int_0^1 \int_y^1 \int_0^{1-x} 2z \, dz \, dx \, dy = \int_0^1 \int_y^1 (1 - x)^2 \, dx \, dy$

$\qquad = \int_0^1 \left[-\frac{1}{3} (1 - x)^3\right]_{x=y}^{x=1} dy = \int_0^1 \frac{1}{3} (1 - y)^3 \, dy$

$\qquad = \frac{1}{12} (1 - y)^4 \Big]_0^1 = \frac{1}{12}$

13.

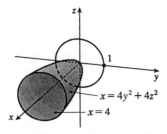

The projection E on the yz-plane is the disk $y^2 + z^2 \le 1$. Using polar coordinates $y = r \cos \theta$ and $z = r \sin \theta$, we get

$\iiint_E x \, dV = \iint_D \left[\int_{4y^2 + 4z^2}^4 x \, dx\right] dA$

$\qquad = \frac{1}{2} \iint_D \left[4^2 - \left(4y^2 + 4z^2\right)^2\right] dA = 8 \int_0^{2\pi} \int_0^1 \left(1 - r^4\right) r \, dr \, d\theta$

$\qquad = 8 \int_0^{2\pi} d\theta \int_0^1 \left(r - r^5\right) dr = 8\,(2\pi) \left[\frac{1}{2} r^2 - \frac{1}{6} r^6\right]_0^1 = \frac{16\pi}{3}$

15. The plane $2x + 3y + 6z = 12$ intersects the xy-plane when $2x + 3y + 6\,(0) = 12 \ \Rightarrow \ y = 4 - \frac{2}{3} x$.

So $E = \left\{(x, y, z) \mid 0 \le x \le 6,\ 0 \le y \le 4 - \frac{2}{3} x,\ 0 \le z \le \frac{1}{6} (12 - 2x - 3y)\right\}$ and

$V = \int_0^6 \int_0^{4 - 2x/3} \int_0^{(12 - 2x - 3y)/6} dz \, dy \, dx = \frac{1}{6} \int_0^6 \int_0^{4 - 2x/3} (12 - 2x - 3y) \, dy \, dx$

$\qquad = \frac{1}{6} \int_0^6 \left[\frac{(12 - 2x)^2}{3} - \frac{3}{2} \frac{12 - 2x}{9}\right] dx = \frac{1}{36} \int_0^6 (12 - 2x)^2 \, dx = \left[\frac{1}{36} \left(-\frac{1}{6}\right) (12 - 2x)^3\right]_0^6 = 8$

17.

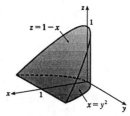

$z = 1 - x$

$x = y^2$

$$V = \int_0^1 \int_{-\sqrt{x}}^{\sqrt{x}} \int_0^{1-x} dz\,dy\,dx$$
$$= \int_0^1 \int_{-\sqrt{x}}^{\sqrt{x}} (1-x)\,dy\,dx$$
$$= \int_0^1 2\sqrt{x}\,(1-x)\,dx$$
$$= \int_0^1 2\left(\sqrt{x} - x^{3/2}\right)dx$$
$$= 2\left(\tfrac{2}{3} - \tfrac{2}{5}\right) = \tfrac{8}{15}$$

19. (a) The wedge can be described as the region

$$D = \{(x,y,z)\mid y^2 + z^2 \le 1, 0 \le x \le 1, 0 \le y \le x\}$$
$$= \left\{(x,y,z)\mid 0 \le x \le 1, 0 \le y \le x, 0 \le z \le \sqrt{1-y^2}\right\}$$

So the integral expressing the volume of the wedge is

$$\iiint_D dV = \int_0^1 \int_0^x \int_0^{\sqrt{1-y^2}} dz\,dy\,dx.$$

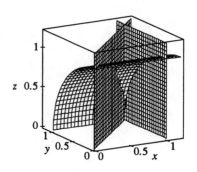

(b) A CAS gives $\int_0^1 \int_0^x \int_0^{\sqrt{1-y^2}} dz\,dy\,dx = \tfrac{\pi}{4} - \tfrac{1}{3}$. (Or use
Formulas 30 and 87 from the Table of Integrals.)

21. $E = \{(x,y,z)\mid 0 \le x \le 1, 0 \le z \le 1 - x, 0 \le y \le 2 - 2z\}$

23.

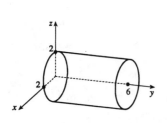

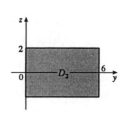

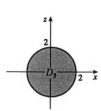

If D_1, D_2, D_3 are the projections of E on the xy-, yz-, and xz-planes, then
$D_1 = \{(x,y)\mid -2 \le x \le 2, 0 \le y \le 6\}$, $D_2 = \{(y,z)\mid -2 \le z \le 2, 0 \le y \le 6\}$,
$D_3 = \{(x,z)\mid x^2 + z^2 \le 4\}$. Therefore

$$E = \left\{(x,y,z)\mid -\sqrt{4-x^2} \le z \le \sqrt{4-x^2}, -2 \le x \le 2, 0 \le y \le 6\right\}$$
$$= \left\{(x,y,z)\mid -\sqrt{4-z^2} \le x \le \sqrt{4-z^2}, -2 \le z \le 2, 0 \le y \le 6\right\}$$

$$\iiint_E f(x,y,z)\,dV = \int_{-2}^2 \int_0^6 \int_{-\sqrt{4-x^2}}^{\sqrt{4-x^2}} f(x,y,z)\,dz\,dy\,dx = \int_0^6 \int_{-2}^2 \int_{-\sqrt{4-x^2}}^{\sqrt{4-x^2}} f(x,y,z)\,dz\,dx\,dy$$
$$= \int_0^6 \int_{-2}^2 \int_{-\sqrt{4-z^2}}^{\sqrt{4-z^2}} f(x,y,z)\,dx\,dz\,dy = \int_{-2}^2 \int_0^6 \int_{-\sqrt{4-z^2}}^{\sqrt{4-z^2}} f(x,y,z)\,dx\,dy\,dz$$
$$= \int_{-2}^2 \int_{-\sqrt{4-x^2}}^{\sqrt{4-x^2}} \int_0^6 f(x,y,z)\,dy\,dz\,dx = \int_{-2}^2 \int_{-\sqrt{4-z^2}}^{\sqrt{4-z^2}} \int_0^6 f(x,y,z)\,dy\,dx\,dz$$

25.

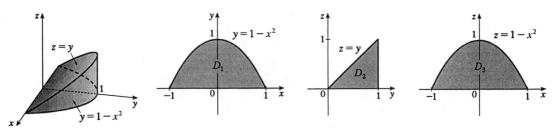

If D_1, D_2, and D_3 are the projections of E on the xy-, yz-, and xz-planes, then

$$D_1 = \{(x,y) \mid -1 \leq x \leq 1, 0 \leq y \leq 1 - x^2\}$$
$$= \left\{(x,y) \mid 0 \leq y \leq 1, -\sqrt{1-y} \leq x \leq \sqrt{1-y}\right\},$$
$$D_2 = \{(y,z) \mid 0 \leq y \leq 1, 0 \leq z \leq y\}$$
$$= \{(y,z) \mid 0 \leq z \leq 1, z \leq y \leq 1\}, \text{ and}$$
$$D_3 = \{(x,z) \mid -1 \leq x \leq 1, 0 \leq z \leq 1 - x^2\}$$
$$= \left\{(x,z) \mid 0 \leq z \leq 1, -\sqrt{1-z} \leq x \leq \sqrt{1-z}\right\}.$$

Therefore

$$E = \{(x,y,z) \mid -1 \leq x \leq 1, 0 \leq y \leq 1 - x^2, 0 \leq z \leq y\}$$
$$= \left\{(x,y,z) \mid 0 \leq y \leq 1, -\sqrt{1-y} \leq x \leq \sqrt{1-y}, 0 \leq z \leq y\right\}$$
$$= \left\{(x,y,z) \mid 0 \leq y \leq 1, 0 \leq z \leq y, -\sqrt{1-y} \leq x \leq \sqrt{1-y}\right\}$$
$$= \left\{(x,y,z) \mid 0 \leq z \leq 1, z \leq y \leq 1, -\sqrt{1-y} \leq x \leq \sqrt{1-y}\right\}$$
$$= \{(x,y,z) \mid -1 \leq x \leq 1, 0 \leq z \leq 1 - x^2, z \leq y \leq 1 - x^2\}$$
$$= \left\{(x,y,z) \mid 0 \leq z \leq 1, -\sqrt{1-z} \leq x \leq \sqrt{1-z}, z \leq y \leq 1 - x^2\right\}$$

Then

$$\iiint_E f(x,y,z)\,dV = \int_{-1}^{1} \int_0^{1-x^2} \int_0^y f(x,y,z)\,dz\,dy\,dx$$
$$= \int_0^1 \int_{-\sqrt{1-y}}^{\sqrt{1-y}} \int_0^y f(x,y,z)\,dz\,dx\,dy$$
$$= \int_0^1 \int_0^y \int_{-\sqrt{1-y}}^{\sqrt{1-y}} f(x,y,z)\,dx\,dz\,dy$$
$$= \int_0^1 \int_z^1 \int_{-\sqrt{1-y}}^{\sqrt{1-y}} f(x,y,z)\,dx\,dy\,dz$$
$$= \int_{-1}^1 \int_0^{1-x^2} \int_z^{1-x^2} f(x,y,z)\,dy\,dz\,dx$$
$$= \int_0^1 \int_{-\sqrt{1-z}}^{\sqrt{1-z}} \int_z^{1-x^2} f(x,y,z)\,dy\,dx\,dz$$

27.

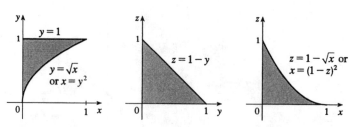

The diagrams show the projections of E on the xy-, yz-, and xz-planes. Therefore

$$\int_0^1 \int_{\sqrt{x}}^1 \int_0^{1-y} f(x,y,z)\,dz\,dy\,dx = \int_0^1 \int_0^{y^2} \int_0^{1-y} f(x,y,z)\,dz\,dx\,dy$$

$$= \int_0^1 \int_0^{1-z} \int_0^{y^2} f(x,y,z)\,dx\,dy\,dz$$

$$= \int_0^1 \int_0^{1-y} \int_0^{y^2} f(x,y,z)\,dx\,dz\,dy$$

$$= \int_0^1 \int_0^{1-\sqrt{x}} \int_{\sqrt{x}}^{1-z} f(x,y,z)\,dy\,dz\,dx$$

$$= \int_0^1 \int_0^{(1-z)^2} \int_{\sqrt{x}}^{1-z} f(x,y,z)\,dy\,dx\,dz$$

29.

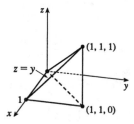 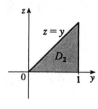 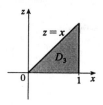

$\int_0^1 \int_y^1 \int_0^y f(x,y,z)\,dz\,dx\,dy = \iiint_E f(x,y,z)\,dV$ where

$E = \{(x,y,z) \mid 0 \le z \le y, y \le x \le 1, 0 \le y \le 1\}$. If D_1, D_2, and D_3 are the projections of E on the xy-, yz- and xz-planes then $D_1 = \{(x,y) \mid 0 \le y \le 1, y \le x \le 1\} = \{(x,y) \mid 0 \le x \le 1, 0 \le y \le x\}$, $D_2 = \{(y,z) \mid 0 \le y \le 1, 0 \le z \le y\} = \{(y,z) \mid 0 \le z \le 1, z \le y \le 1\}$, and $D_3 = \{(x,z) \mid 0 \le x \le 1, 0 \le z \le x\} = \{(x,z) \mid 0 \le z \le 1, z \le x \le 1\}$. Thus we also have

$E = \{(x,y,z) \mid 0 \le x \le 1, 0 \le y \le x, 0 \le z \le y\} = \{(x,y,z) \mid 0 \le y \le 1, 0 \le z \le y, y \le x \le 1\}$

$= \{(x,y,z) \mid 0 \le z \le 1, z \le y \le 1, y \le x \le 1\} = \{(x,y,z) \mid 0 \le x \le 1, 0 \le z \le x, z \le y \le x\}$

$= \{(x,y,z) \mid 0 \le z \le 1, z \le x \le 1, z \le y \le x\}$. Then

$$\int_0^1 \int_y^1 \int_0^y f(x,y,z)\,dz\,dx\,dy = \int_0^1 \int_0^x \int_0^y f(x,y,z)\,dz\,dy\,dx = \int_0^1 \int_0^y \int_y^1 f(x,y,z)\,dx\,dz\,dy$$

$$= \int_0^1 \int_z^1 \int_y^1 f(x,y,z)\,dx\,dy\,dz = \int_0^1 \int_0^x \int_z^x f(x,y,z)\,dy\,dz\,dx$$

$$= \int_0^1 \int_z^1 \int_z^x f(x,y,z)\,dy\,dx\,dz$$

31. $m = \int_0^1 \int_0^{x^2} \int_0^{x+2y} 2 \, dz \, dy \, dx = 2 \int_0^1 \int_0^{x^2} (x+2y) \, dy \, dx = 2 \int_0^1 (x^3 + x^4) \, dx = \frac{9}{10}$,

$M_{yz} = \int_0^1 \int_0^{x^2} \int_0^{x+2y} 2x \, dz \, dy \, dx = \int_0^1 \int_0^{x^2} 2 (x^2 + 2xy) \, dy \, dx = \int_0^1 2 (x^4 + x^5) \, dx = \frac{11}{15}$, $M_{xz} = \frac{1}{1}$

and

$$M_{xy} = \int_0^1 \int_0^{x^2} \int_0^{x+2y} 2z \, dz \, dy \, dx = \int_0^1 \int_0^{x^2} 2 (x+2y)^2 /2 \, dy \, dx$$

$$= \int_0^1 2 \left[\tfrac{1}{2}x^2 y + xy^2 + \tfrac{2}{3}y^3\right]_{y=0}^{y=x^2} dx = \int_0^1 2 \left(\tfrac{1}{2}x^4 + x^5 + \tfrac{2}{3}x^6\right) dx = \frac{76}{105}$$

Hence $(\bar{x}, \bar{y}, \bar{z}) = \left(\frac{22}{27}, \frac{25}{63}, \frac{152}{189}\right)$.

33. $m = \int_0^a \int_0^a \int_0^a (x^2 + y^2 + z^2) \, dx \, dy \, dz = \int_0^a \int_0^a \left[\tfrac{1}{3}a^3 + a (y^2 + z^2)\right] dy \, dz$

$= \int_0^a \left[\tfrac{2}{3}a^4 + a^2 z^2\right] dz = \tfrac{2}{3}a^5 + \tfrac{1}{3}a^5 = a^5$

$M_{yz} = \int_0^a \int_0^a \int_0^a \left[x^3 + x (y^2 + z^2)\right] dx \, dy \, dz = \int_0^a \int_0^a \left[\tfrac{1}{4}a^4 + \tfrac{1}{2}a^2 (y^2 + z^2)\right] dy \, dz$

$= \int_0^a \left[\tfrac{1}{4}a^5 + \tfrac{1}{6}a^5 + \tfrac{1}{2}a^3 z^2\right] dz = \tfrac{1}{4}a^6 + \tfrac{1}{3}a^6 = \tfrac{7}{12}a^6$

$= M_{xz} = M_{xy}$ by symmetry of E and $\rho(x, y, z)$

Hence $(\bar{x}, \bar{y}, \bar{z}) = \left(\tfrac{7}{12}a, \tfrac{7}{12}a, \tfrac{7}{12}a\right)$.

35. (a) $m = \int_{-1}^1 \int_{-\sqrt{1-y^2}}^{\sqrt{1-y^2}} \int_{4y^2 + 4z^2}^4 (x^2 + y^2 + z^2) \, dx \, dz \, dy$

(b) $(\bar{x}, \bar{y}, \bar{z})$ where $\bar{x} = m^{-1} \int_{-1}^1 \int_{-\sqrt{1-y^2}}^{\sqrt{1-y^2}} \int_{4y^2 + 4z^2}^4 x (x^2 + y^2 + z^2) \, dx \, dz \, dy$,

$\bar{y} = m^{-1} \int_{-1}^1 \int_{-\sqrt{1-y^2}}^{\sqrt{1-y^2}} \int_{4y^2 + 4z^2}^4 y (x^2 + y^2 + z^2) \, dx \, dz \, dy$, and

$\bar{z} = m^{-1} \int_{-1}^1 \int_{-\sqrt{1-y^2}}^{\sqrt{1-y^2}} \int_{4y^2 + 4z^2}^4 z (x^2 + y^2 + z^2) \, dx \, dz \, dy$

(c) $I_z = \int_{-1}^1 \int_{-\sqrt{1-y^2}}^{\sqrt{1-y^2}} \int_{4y^2 + 4z^2}^4 (x^2 + y^2) (x^2 + y^2 + z^2) \, dx \, dz \, dy$

37. (a) $m = \int_0^1 \int_0^{\sqrt{1-x^2}} \int_0^y (1 + x + y + z) \, dz \, dy \, dx = \frac{3\pi}{32} + \frac{11}{24}$

(b) $(\bar{x}, \bar{y}, \bar{z}) = \left(\left(m^{-1} \int_0^1 \int_0^{\sqrt{1-x^2}} \int_0^y x (1 + x + y + z) \, dz \, dy \, dx, \right. \right.$

$m^{-1} \int_0^1 \int_0^{\sqrt{1-x^2}} \int_0^y y (1 + x + y + z) \, dz \, dy \, dx,$

$\left. \left. m^{-1} \int_0^1 \int_0^{\sqrt{1-x^2}} \int_0^y z (1 + x + y + z) \, dz \, dy \, dx \right) \right.$

$= \left(\frac{28}{9\pi + 44}, \frac{30\pi + 128}{45\pi + 220}, \frac{45\pi + 208}{135\pi + 660} \right)$

(c) $I_z = \int_0^1 \int_0^{\sqrt{1-x^2}} \int_0^y (x^2 + y^2) (1 + x + y + z) \, dz \, dy \, dx = \frac{68 + 15\pi}{240}$

39. $I_x = \int_0^L \int_0^L \int_0^L k (y^2 + z^2) \, dz \, dy \, dx = kL \int_0^L \left[y^2 L + \tfrac{1}{3}L^3\right] dy = kL \left[\tfrac{1}{3}L^4 + \tfrac{1}{3}L^4\right] = \tfrac{2}{3}kL^5$.

By symmetry, $I_x = I_y = I_z = \tfrac{2}{3}kL^5$.

41. (a) $f(x, y, z)$ is a joint density function, so we know $\iiint_{\mathbb{R}^3} f(x, y, z)\, dV = 1$. Here we have

$$\iiint_{\mathbb{R}^3} f(x, y, z)\, dV = \int_{-\infty}^{\infty} \int_{-\infty}^{\infty} \int_{-\infty}^{\infty} f(x, y, z)\, dz\, dy\, dx = \int_0^2 \int_0^2 \int_0^2 Cxyz\, dz\, dy\, dx$$

$$= C \int_0^2 x\, dx \int_0^2 y\, dy \int_0^2 z\, dz = C \left[\frac{x^2}{2}\right]_0^2 \left[\frac{y^2}{2}\right]_0^2 \left[\frac{z^2}{2}\right]_0^2$$

$$= 8C$$

Then we must have $8C = 1 \Rightarrow C = \frac{1}{8}$.

(b) $P(X \le 1, Y \le 1, Z \le 1) = \int_{-\infty}^1 \int_{-\infty}^1 \int_{-\infty}^1 f(x, y, z)\, dz\, dy\, dx$

$$= \int_0^1 \int_0^1 \int_0^1 \tfrac{1}{8} xyz\, dz\, dy\, dx = \tfrac{1}{8} \int_0^1 x\, dx \int_0^1 y\, dy \int_0^1 z\, dz$$

$$= \frac{1}{8} \left[\frac{x^2}{2}\right]_0^1 \left[\frac{y^2}{2}\right]_0^1 \left[\frac{z^2}{2}\right]_0^1 = \frac{1}{8} \left(\frac{1}{2}\right)^3 = \frac{1}{64}$$

(c) $P(X + Y + Z \le 1) = P((X, Y, Z) \in E)$ where E is the solid region in the first octant bounded by the coordinate planes and the plane $x + y + z = 1$. The plane $x + y + z = 1$ meets the xy-plane in the line $x + y = 1$, so we have

$$P(X + Y + Z \le 1) = \iiint_E f(x, y, z)\, dV = \int_0^1 \int_0^{1-x} \int_0^{1-x-y} \tfrac{1}{8} xyz\, dz\, dy\, dx$$

$$= \tfrac{1}{8} \int_0^1 \int_0^{1-x} xy \left[\tfrac{1}{2} z^2\right]_{z=0}^{z=1-x-y} dy\, dx$$

$$= \tfrac{1}{16} \int_0^1 \int_0^{1-x} xy (1 - x - y)^2\, dy\, dx$$

$$= \tfrac{1}{16} \int_0^1 \int_0^{1-x} \left[(x^3 - 2x^2 + x) y + (2x^2 - 2x) y^2 + xy^3\right] dy\, dx$$

$$= \tfrac{1}{16} \int_0^1 \left[(x^3 - 2x^2 + x) \tfrac{1}{2} y^2 + (2x^2 - 2x) \tfrac{1}{3} y^3 + x \left(\tfrac{1}{4} y^4\right)\right]_{y=0}^{y=1-x} dx$$

$$= \tfrac{1}{192} \int_0^1 (x - 4x^2 + 6x^3 - 4x^4 + x^5)\, dx = \tfrac{1}{192} \left(\tfrac{1}{30}\right) = \tfrac{1}{5760}$$

43. $V(E) = L^3$,

$$f_{\text{ave}} = \frac{1}{L^3} \int_0^L \int_0^L \int_0^L xyz\, dx\, dy\, dz = \frac{1}{L^3} \int_0^L x^2\, dx \int_0^L y^2\, dy \int_0^L z^2\, dz = \frac{1}{L^3} \frac{L^2}{2} \frac{L^2}{2} \frac{L^2}{2} = \frac{L^3}{8}$$

45. The triple integral will attain its maximum when the integrand $1 - x^2 - 2y^2 - 3z^2$ is positive in the region E and negative everywhere else. For if E contains some region F where the integrand is negative, the integral could be increased by excluding F from E, and if E fails to contain some part G of the region where the integrand is positive, the integral could be increased by including G in E. So we require that $x^2 + 2y^2 + 3z^2 \le 1$. This describes the region bounded by the ellipsoid $x^2 + 2y^2 + 3z^2 = 1$.

Section 12.8 Triple Integrals in Cylindrical and Spherical Coordinates

1.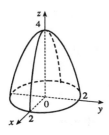
$$\int_0^{2\pi}\int_0^2\int_0^{4-r^2} r\,dz\,dr\,d\theta = \int_0^{2\pi}\int_0^2 (4r - r^3)\,dr\,d\theta$$
$$= \int_0^{2\pi}\left[2r^2 - \tfrac14 r^4\right]_{r=0}^{r=2} d\theta$$
$$= \int_0^{2\pi} (8-4)\,d\theta = 8\pi$$

3.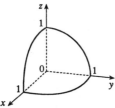
$$\int_0^{\pi/2}\int_0^{\pi/2}\int_0^1 \rho^2 \sin\phi\,d\rho\,d\theta\,d\phi = \int_0^{\pi/2}\int_0^{\pi/2} \tfrac13 \sin\phi\,d\theta\,d\phi$$
$$= \tfrac13 \int_0^{\pi/2} \tfrac{\pi}{2}\sin\phi\,d\phi$$
$$= \tfrac{\pi}{6}\left[-\cos\phi\right]_0^{\pi/2} = \tfrac{\pi}{6}$$

5. The solid E is most conveniently described if we use cylindrical coordinates:
$E = \{(r,\theta,z)\mid 0\le\theta\le\tfrac{\pi}{2}, 0\le r\le 3, 0\le z\le 2\}$. Then
$$\iiint_E f(x,y,z)\,dV = \int_0^{\pi/2}\int_0^3\int_0^2 f(r\cos\theta, r\sin\theta, z)\,r\,dz\,dr\,d\theta.$$

7. $\iiint_E (x^2+y^2)\,dV = \int_{-1}^2\int_0^{2\pi}\int_0^2 (r^2)\,r\,dr\,d\theta\,dz = (3)(2\pi)\left[\tfrac14 r^4\right]_0^2 = 24\pi$

9. $\iiint_E y\,dV = \int_0^{2\pi}\int_1^2\int_0^{2+r\cos\theta} r^2\sin\theta\,dz\,dr\,d\theta = \int_0^{2\pi}\int_1^2 \left[2r^2\sin\theta + r^3\cos\theta\right] dr\,d\theta$
$= \int_0^{2\pi}\left[\tfrac13(2r^3\sin\theta) + \tfrac14(r^4\cos\theta)\right]_{r=1}^{r=2} d\theta = \int_0^{2\pi}\left[\tfrac{14}{3}\sin\theta + \tfrac{15}{4}\cos\theta\right] d\theta = 0$

11. $\iiint_E x^2\,dV = \int_0^{2\pi}\int_0^1\int_0^{2r} r^2\cos^2\theta\,r\,dz\,dr\,d\theta = \int_0^{2\pi}\int_0^1 \left[r^3\cos^2\theta z\right]_{z=0}^{z=2r} dr\,d\theta$
$= \int_0^{2\pi}\int_0^1 2r^4\cos^2\theta\,dr\,d\theta = \int_0^{2\pi}\left[\tfrac25 r^5\cos^2\theta\right]_{r=0}^{r=1} d\theta = \tfrac25\int_0^{2\pi}\cos^2\theta\,d\theta$
$= \dfrac25\displaystyle\int_0^{2\pi}\dfrac{1+\cos2\theta}{2}\,d\theta = \tfrac15\left[\theta + \tfrac12\sin2\theta\right]_0^{2\pi} = \dfrac{2\pi}{5}$

13. The paraboloid $z = 4x^2 + 4y^2$ intersects the plane $z = a$ when $a = 4x^2 + 4y^2$ or $x^2 + y^2 = \tfrac14 a$. So, in
cylindrical coordinates, $E = \{(r,\theta,z)\mid 0\le r\le\tfrac12\sqrt{a}, 0\le\theta\le 2\pi, 4r^2\le z\le a\}$. Thus
$$m = \int_0^{2\pi}\int_0^{\sqrt{a}/2}\int_{4r^2}^a Kr\,dz\,dr\,d\theta = 2\pi K\int_0^{\sqrt{a}/2}(ar - 4r^3)\,dr = 2\pi K\left[\tfrac12 ar^2 - r^4\right]_0^{\sqrt{a}/2} = \tfrac18 a^2\pi K$$
Since the region is homogeneous and symmetric, $M_{yz} = M_{xz} = 0$ and
$$M_{xy} = \int_0^{2\pi}\int_0^{\sqrt{a}/2}\int_{4r^2}^a Krz\,dz\,dr\,d\theta = 2\pi K\int_0^{\sqrt{a}/2}\left(\tfrac12 a^2 r - 8r^5\right) dr = 2\pi K\left[\tfrac14 a^2 r^2 - \tfrac43 r^6\right]_0^{\sqrt{a}/2}$$
$$= \tfrac{1}{12} a^3\pi K$$
Hence $(\overline{x},\overline{y},\overline{z}) = (0,0,\tfrac23 a)$.

15. $\iiint_B (x^2+y^2+z^2)\,dV = \int_0^\pi\int_0^{2\pi}\int_0^1 \rho^4\sin\phi\,d\rho\,d\theta\,d\phi = 2\pi\int_0^\pi \tfrac15\sin\phi\,d\phi = \tfrac{2\pi}{5}\left[-\cos\phi\right]_0^\pi = \tfrac{4\pi}{5}$

17. $\iiint_E y^2\,dV = \int_0^{\pi/2}\int_0^{\pi/2}\int_0^1 (\rho^2\sin^2\phi\sin^2\theta)\,(\rho^2\sin\phi)\,d\rho\,d\phi\,d\theta$

$\qquad = \int_0^{\pi/2}\int_0^{\pi/2}\int_0^1 \rho^4\sin^3\phi\sin^2\theta\,d\rho\,d\phi\,d\theta = \int_0^{\pi/2}\int_0^{\pi/2}\frac{1}{5}\sin^3\phi\sin^2\theta\,d\phi\,d\theta$

$\qquad = \int_0^{\pi/2}\frac{2}{15}\sin^2\theta\,d\theta = \frac{2}{15}\left[\frac{1}{2}\theta - \frac{1}{4}\sin 2\theta\right]_0^{\pi/2} = \frac{\pi}{30}$

19. $\iiint_E \sqrt{x^2+y^2+z^2}\,dV = \int_0^{2\pi}\int_0^{\pi/6}\int_0^2 \rho^3\sin\phi\,d\rho\,d\phi\,d\theta$

$\qquad = 8\pi\int_0^{\pi/6}\sin\phi\,d\phi = 8\pi\left[-\cos\phi\right]_0^{\pi/6} = 8\pi\left(1-\frac{\sqrt3}{2}\right) = 4\pi\left(2-\sqrt3\right)$

21. (a) Since $\rho = 4\cos\phi$ implies $\rho^2 = 4\rho\cos\phi$, the equation is that of a sphere of radius 2 with center at $(0,0,2)$. Thus

$$V = \int_0^{2\pi}\int_0^{\pi/3}\int_0^{4\cos\phi}\rho^2\sin\phi\,d\rho\,d\phi\,d\theta = 2\pi\int_0^{\pi/3}\sin\phi\left(\frac{64}{3}\cos^3\phi\right)d\phi$$

$$= \frac{32}{3}\pi\left[-\cos^4\phi\right]_0^{\pi/3} = 10\pi$$

(b) By the symmetry of the problem $M_{yz} = M_{xz} = 0$. Then

$$M_{xy} = \int_0^{2\pi}\int_0^{\pi/3}\int_0^{4\cos\phi}\rho^3\cos\phi\sin\phi\,d\rho\,d\phi\,d\theta = 2\pi\int_0^{\pi/3}\cos\phi\sin\phi\left(64\cos^4\phi\right)d\phi$$

$$= 128\pi\left[-\frac{1}{6}\cos^6\phi\right]_0^{\pi/3} = 21\pi$$

Hence $(\overline{x},\overline{y},\overline{z}) = (0,0,2.1)$.

23. (a) The density function is $\rho(x,y,z) = K$, a constant, and by the symmetry of the problem $M_{xz} = M_{yz} = 0$. Then

$M_{xy} = \int_0^{2\pi}\int_0^{\pi/2}\int_0^a K\rho^3\sin\phi\cos\phi\,d\rho\,d\phi\,d\theta = \frac{1}{2}\pi Ka^4\int_0^{\pi/2}\sin\phi\cos\phi\,d\phi = \frac{1}{8}\pi Ka^4$. But the mass is K (volume of the hemisphere) $= \frac{2}{3}\pi Ka^3$, so the centroid is $\left(0,0,\frac{3}{8}a\right)$.

(b) Place the center of the base at $(0,0,0)$; the density function is $\rho(x,y,z) = K$. By symmetry, the moments of inertia about any two such diameters will be equal, so we just need to find I_x:

$$I_x = \int_0^{2\pi}\int_0^{\pi/2}\int_0^a (K\rho^2\sin\phi)\,\rho^2\left(\sin^2\phi\sin^2\theta + \cos^2\phi\right)d\rho\,d\phi\,d\theta$$

$$= K\int_0^{2\pi}\int_0^{\pi/2}\left(\sin^3\phi\sin^2\theta + \sin\phi\cos^2\phi\right)\left(\frac{1}{5}a^5\right)d\phi\,d\theta$$

$$= \frac{1}{5}Ka^5\int_0^{2\pi}\left[\sin^2\theta\left(-\cos\phi + \frac{1}{3}\cos^3\phi\right) + \left(-\frac{1}{3}\cos^3\phi\right)\right]_{\phi=0}^{\phi=\pi/2}d\theta$$

$$= \frac{1}{5}Ka^5\int_0^{2\pi}\left[\frac{2}{3}\sin^2\theta + \frac{1}{3}\right]d\theta = \frac{1}{5}Ka^5\left[\frac{2}{3}\left(\frac{1}{2}\theta - \frac{1}{4}\sin 2\theta\right) + \frac{1}{3}\theta\right]_0^{2\pi}$$

$$= \frac{1}{5}Ka^5\left[\frac{2}{3}(\pi-0) + \frac{1}{3}(2\pi-0)\right] = \frac{4}{15}Ka^5\pi$$

25. In spherical coordinates $z = \sqrt{x^2+y^2}$ becomes $\cos\phi = \sin\phi$ or $\phi = \frac{\pi}{4}$. Then

$V = \int_0^{2\pi}\int_0^{\pi/4}\int_0^1 \rho^2\sin\phi\,d\rho\,d\phi\,d\theta = 2\pi\int_0^{\pi/4}\sin\phi\,d\phi\left(\int_0^1\rho^2\,d\rho\right) = \frac{1}{3}\pi\left(2-\sqrt2\right)$,

$M_{xy} = \int_0^{2\pi}\int_0^{\pi/4}\int_0^1 \rho^3\sin\phi\cos\phi\,d\rho\,d\phi\,d\theta = 2\pi\left[-\frac{1}{4}\cos 2\phi\right]_0^{\pi/4}\left(\frac{1}{4}\right) = \frac{\pi}{8}$ and by symmetry

$M_{yz} = M_{xz} = 0$. Hence $(\overline{x},\overline{y},\overline{z}) = \left(0,0,\dfrac{3}{8(2-\sqrt2)}\right)$.

27. In cylindrical coordinates the paraboloid is given by $z = r^2$ and the plane by $z = 2r \sin\theta$ and they intersect in the circle $r = 2\sin\theta$. Then $\iiint_E z\, dV = \int_0^\pi \int_0^{2\sin\theta} \int_{r^2}^{2r\sin\theta} rz\, dz\, dr\, d\theta = \frac{5\pi}{6}$ (using a CAS).

29. $\int_{-1}^1 \int_{-\sqrt{1-x^2}}^{\sqrt{1-x^2}} \int_{x^2+y^2}^{2-x^2-y^2} \left(x^2+y^2\right)^{3/2} dz\, dy\, dx = \int_0^{2\pi} \int_0^1 \int_{r^2}^{2-r^2} \left(r^2\right)^{3/2} r\, dz\, dr\, d\theta$

$= 2\pi \int_0^1 \left[r^4 z\right]_{z=r^2}^{z=2-r^2} dr = 2\pi \int_0^1 \left(2r^4 - r^6 - r^6\right) dr = 4\pi \left[\frac{1}{5}r^5 - \frac{1}{7}r^7\right]_0^1 = \frac{8\pi}{35}$

31. If E is the solid enclosed by the surface $\rho = 1 + \frac{1}{5}\sin 6\theta \sin 5\phi$, it can be described in spherical coordinates as $E = \left\{(\rho, \theta, \phi) \mid 0 \le \rho \le 1 + \frac{1}{5}\sin 6\theta \sin 5\phi,\ 0 \le \theta \le 2\pi,\ 0 \le \phi \le \pi\right\}$. Its volume is given by $V(E) = \iiint_E dV = \int_0^\pi \int_0^{2\pi} \int_0^{1+(\sin 6\theta \sin 5\phi)/5} \rho^2 \sin\phi\, d\rho\, d\theta\, d\phi = \frac{136\pi}{99}$ (using a CAS).

33. (a) The mountain comprises a solid conical region C. The work done in lifting a small volume of material ΔV with density $g\,(P)$ to a height $h\,(P)$ above sea level is $h\,(P)\,g\,(P)\,\Delta V$. Summing over the whole mountain we get $W = \iiint_C h\,(P)\,g\,(P)\,dV$.

(b) Here C is a solid right circular cone with radius $R = 62{,}000$ ft, height $H = 12{,}400$ ft, and density $g\,(P) = 200$ lb/ft^3 at all points P in C. We use cylindrical coordinates:

$W = \int_0^{2\pi} \int_0^H \int_0^{R(1-z/H)} z \cdot 200 r\, dr\, dz\, d\theta$

$= 2\pi \int_0^H 200z \left[\frac{1}{2}r^2\right]_{r=0}^{r=R(1-z/H)} dz$

$= 400\pi \int_0^H z\frac{R^2}{2}\left(1 - \frac{z}{H}\right)^2 dz$

$= 200\pi R^2 \int_0^H \left(z - \frac{2z^2}{H} + \frac{z^3}{H^2}\right) dz$

$= 200\pi R^2 \left[\frac{z^2}{2} - \frac{2z^3}{3H} + \frac{z^4}{4H^2}\right]_0^H$

$= 200\pi R^2 \left(\frac{H^2}{2} - \frac{2H^2}{3} + \frac{H^2}{4}\right) = \frac{50}{3}\pi R^2 H^2$

$= \frac{50}{3}\pi (62{,}000)^2 (12{,}400)^2 \approx 3.1 \times 10^{19}$ ft-lb

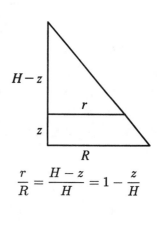

$\frac{r}{R} = \frac{H-z}{H} = 1 - \frac{z}{H}$

Section 12.9   Change of Variables in Multiple Integrals

1. $x = u + 4v$, $y = 3u - 2v$.

The Jacobian is $\dfrac{\partial (x,y)}{\partial (u,v)} = \begin{vmatrix} \partial x/\partial u & \partial x/\partial v \\ \partial y/\partial u & \partial y/\partial v \end{vmatrix} = \begin{vmatrix} 1 & 4 \\ 3 & -2 \end{vmatrix} = 1(-2) - 4(3) = -14.$

3. $\dfrac{\partial (x,y)}{\partial (u,v)} = \begin{vmatrix} \partial x/\partial u & \partial x/\partial v \\ \partial y/\partial u & \partial y/\partial v \end{vmatrix} = \begin{vmatrix} 2e^{2u} \cos v & -e^{2u} \sin v \\ 2e^{2u} \sin v & e^{2u} \cos v \end{vmatrix} = 2e^{4u} (\cos^2 v + \sin^2 v) = 2e^{4u}$

5. $\dfrac{\partial (x,y,z)}{\partial (u,v,w)} = \begin{vmatrix} 1 & 1 & 1 \\ 1 & 1 & -1 \\ 1 & -1 & 1 \end{vmatrix} = 1(1-1) - 1(1+1) + 1(-1-1) = -4$

7. S_1: $v = 0$, $0 \le u \le 2$, so $x = u$, $y = 2u$ and $y = 2x$. S_2: $u = 2$, $0 \le v \le 1$, so $x = 2 - 2v$, $y = 4 - v$ and $x = 2y - 6$. S_3: $v = 1$, $0 \le u \le 2$, so $x = u - 2$, $y = 2u - 1$ and $y = 2x + 3$. S_4: $u = 0$, $0 \le v \le 1$, so $x = -2v$, $y = -v$ and $2y = x$.

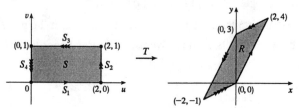

9. S_1: $0 \le u \le 1$, $v = 0$, $x = 4u$, $y = 0$ so $0 \le x \le 4$, $y = 0$ is the image of the first side.
S_2: $u + v = 1$, $x = 4(1 - v) + 3v = 4 - v$ $\Leftrightarrow$ $v = 4 - x$, $y = 4v = 4(4 - x) = 16 - 4x$,
$3 \le x \le 4$, so $y = 16 - 4x$, $3 \le x \le 4$ is the image of the second side.
S_3: $u = 0$, $0 \le v \le 1$, $x = 3v$, $y = 4v = \frac{4}{3}x$, $0 \le x \le 3$, so $y = \frac{4}{3}x$, $0 \le x \le 3$ is the image of the third side.

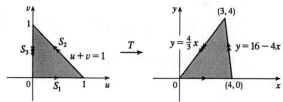

11. $\dfrac{\partial (x,y)}{\partial (u,v)} = \begin{vmatrix} 1/3 & 1/3 \\ -2/3 & 1/3 \end{vmatrix} = \frac{1}{3}$ and $3x + 4y = (u + v) + \frac{4}{3}(v - 2u) = \frac{1}{3}(7v - 5u)$. Then S is the
region bounded by the lines $u = 0$, $\frac{1}{3}(v - 2) = \frac{1}{3}(u + v) - 2$ or $u = 2$,
$\frac{1}{3}(v - 2u) = -\frac{2}{3}(u + v)$ or $v = 0$, and $\frac{1}{3}(v - 2u) = 3 - \frac{2}{3}(u + v)$ or $v = 3$. Thus
$\iint_R (3x + 4y) \, dA = \int_0^3 \int_0^2 \frac{1}{3}(7v - 5u) \left(\frac{1}{3} \, du \, dv \right) = \frac{1}{9} \int_0^3 (14v - 10) \, dv = \frac{1}{9}(33) = \frac{11}{3}.$

13. $\dfrac{\partial (x,y)}{\partial (u,v)} = \begin{vmatrix} 2 & 0 \\ 0 & 3 \end{vmatrix} = 6$, $x^2 = 4u^2$ and the planar ellipse $9x^2 + 4y^2 \le 36$ is the image of the disk
$u^2 + v^2 \le 1$. Thus $\iint_R x^2 \, dA = \iint_{u^2 + v^2 \le 1} (4u^2)(6) \, du \, dv = \int_0^{2\pi} \int_0^1 (24r^2 \cos^2 \theta) \, r \, dr \, d\theta = 6\pi.$

15. $\dfrac{\partial (x, y)}{\partial (u, v)} = \begin{vmatrix} 1/v & -u/v^2 \\ 0 & 1 \end{vmatrix} = \dfrac{1}{v}$, $xy = u$, $y = x$ is the image of the parabola $v^2 = u$, $y = 3x$ is the

image of the parabola $v^2 = 3u$, and the hyperbolas $xy = 1$, $xy = 3$ are the images of the lines $u = 1$

and $u = 3$ respectively. Thus

$$\iint_R xy \, dA = \int_1^3 \int_{\sqrt{u}}^{\sqrt{3u}} u \, (1/v) \, dv \, du = \int_1^3 u \left(\ln \sqrt{3u} - \ln \sqrt{u} \right) du$$

$$= \int_1^3 u \ln \sqrt{3} \, du = 4 \ln \sqrt{3} = 2 \ln 3$$

17. (a) $\dfrac{\partial (x, y, z)}{\partial (u, v, w)} = \begin{vmatrix} a & 0 & 0 \\ 0 & b & 0 \\ 0 & 0 & c \end{vmatrix} = abc$ and the solid enclosed by the

ellipsoid is the image of the ball $u^2 + v^2 + w^2 \leq 1$. So

$$\iiint_E dV = \iiint_{u^2 + v^2 + w^2 \leq 1} abc \, du \, dv \, dw = (abc) \, (\text{volume of the ball}) = \tfrac{4}{3} \pi abc.$$

(b) If we approximate the surface of Earth by the ellipsoid $\dfrac{x^2}{6378^2} + \dfrac{y^2}{6378^2} + \dfrac{z^2}{6356^2} = 1$, then we can

estimate the volume of Earth by finding the volume of the solid E enclosed by the ellipsoid. From

part (a), this is $\iiint_E dV = \tfrac{4}{3} \pi (6378) (6378) (6356) \approx 1.083 \times 10^{12} \text{ km}^3$.

19. Letting $u = 2x - y$ and $v = 3x + y$, we have $x = \tfrac{1}{5} (u + v)$, $y = \tfrac{1}{5} (2v - 3u)$. Then

$\dfrac{\partial (x, y)}{\partial (u, v)} = \begin{vmatrix} 1/5 & 1/5 \\ -3/5 & 2/5 \end{vmatrix} = \dfrac{1}{5}$ and

$$\iint_R xy \, dA = \int_{-2}^1 \int_{-3}^1 \dfrac{(u + v) (2v - 3u)}{25} \left(\dfrac{1}{5} \right) du \, dv = \tfrac{1}{125} \int_{-2}^1 \int_{-3}^1 \left(2v^2 - uv - 3u^2 \right) du \, dv$$

$$= \tfrac{1}{125} \int_{-2}^1 \left(8v^2 + 4v - 28 \right) dv = -\tfrac{66}{125}$$

21. Letting $u = y - x$, $v = y + x$, we have $y = \tfrac{1}{2} (u + v)$, $x = \tfrac{1}{2} (v - u)$. Then

$\dfrac{\partial (x, y)}{\partial (u, v)} = \begin{vmatrix} -1/2 & 1/2 \\ 1/2 & 1/2 \end{vmatrix} = -\dfrac{1}{2}$ and R is the image of the trapezoidal region with vertices $(-1, 1)$,

$(-2, 2)$, $(2, 2)$, and $(1, 1)$. Thus

$$\iint_R \cos \dfrac{y - x}{y + x} \, dA = \int_1^2 \int_{-v}^v \left| -\dfrac{1}{2} \right| \cos \dfrac{u}{v} \, du \, dv = \dfrac{1}{2} \int_1^2 \left[v \sin \dfrac{u}{v} \right]_{u=-v}^{u=v} dv$$

$$= \tfrac{1}{2} \int_1^2 2v \sin (1) \, dv = \tfrac{3}{2} \sin 1$$

23. Let $u = x + y$ and $v = -x + y$. Then $u + v = 2y \Rightarrow y = \tfrac{1}{2} (u + v)$ and

$u - v = 2x \Rightarrow x = \tfrac{1}{2} (u - v)$. $\dfrac{\partial (x, y)}{\partial (u, v)} = \begin{vmatrix} 1/2 & -1/2 \\ 1/2 & 1/2 \end{vmatrix} = \dfrac{1}{2}$. Now

$|u| = |x + y| \leq |x| + |y| \leq 1 \Rightarrow -1 \leq u \leq 1$, and

$|v| = |-x + y| \leq |x| + |y| \leq 1 \Rightarrow -1 \leq v \leq 1$.

R is the image of the square region with vertices $(1, 1)$, $(1, -1)$, $(-1, -1)$, and

$(-1, 1)$. So $\iint_R e^{x+y} \, dA = \tfrac{1}{2} \int_{-1}^1 \int_{-1}^1 e^u \, du \, dv = \tfrac{1}{2} \cdot 2 \cdot e^u \Big]_{-1}^1 = e - e^{-1}$.

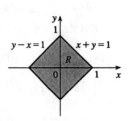

Chapter 12 Review

Concept Check

1. (a) A double Riemann sum of f is $\sum_{i=1}^{m} \sum_{j=1}^{n} f\left(x_{ij}^{*}, y_{ij}^{*}\right) \Delta A$, where ΔA is the area of each subrectangle and $f\left(x_{ij}^{*}, y_{ij}^{*}\right)$ is a sample point in each subrectangle. If $f\left(x, y\right) \geq 0$, this sum represents an approximation to the volume of the solid that lies above the rectangle R and below the graph of f.

(b) $\iint_{R} f\left(x, y\right) dA = \lim_{m,n \to \infty} \sum_{i=1}^{m} \sum_{j=1}^{n} f\left(x_{ij}^{*}, y_{ij}^{*}\right) \Delta A$

(c) If $f\left(x, y\right) \geq 0$, $\iint_{R} f\left(x, y\right) dA$ represents the volume of the solid that lies above the rectangle R and below the surface $z = f\left(x, y\right)$. If f takes on both positive and negative values, $\iint_{R} f\left(x, y\right) dA$ is the difference of the volume above R but below the surface $z = f\left(x, y\right)$ and the volume below R but above the surface $z = f\left(x, y\right)$.

(d) We usually evaluate $\iint_{R} f\left(x, y\right) dA$ as an iterated integral according to Fubini's Theorem (see page 851).

(e) The Midpoint Rule for Double Integrals says that we approximate the double integral $\iint_{R} f\left(x, y\right) dA$ by the double Riemann sum $\sum_{i=1}^{m} \sum_{j=1}^{n} f\left(\overline{x}_{i}, \overline{y}_{j}\right) \Delta A$ where the sample points $\left(\overline{x}_{i}, \overline{y}_{j}\right)$ are the centers of the subrectangles.

(f) $f_{\text{ave}} = \dfrac{1}{A\left(R\right)} \iint_{R} f\left(x, y\right) dA$ where $A\left(R\right)$ is the area of R.

2. (a) See (1) and (2) and the accompanying discussion in Section 12.3.

(b) See (3) and the preceding discussion in Section 12.3.

(c) See (5) and the preceding discussion in Section 12.3.

(d) See (6)–(11) in Section 12.3.

3. We may want to change from rectangular to polar coordinates in a double integral if the region R of integration is more easily described in polar coordinates. To accomplish this, we use
$\iint_{R} f\left(x, y\right) dA = \int_{\alpha}^{\beta} \int_{a}^{b} f\left(r \cos \theta, r \sin \theta\right) r \, dr \, d\theta$ where R is given by $0 \leq a \leq r \leq b$, $\alpha \leq \theta \leq \beta$.

4. (a) $m = \iint_{D} \rho\left(x, y\right) dA$

(b) $M_{x} = \iint_{D} y\rho\left(x, y\right) dA$, $M_{y} = \iint_{D} x\rho\left(x, y\right) dA$

(c) The center of mass is $\left(\overline{x}, \overline{y}\right)$ where $\overline{x} = \dfrac{M_{y}}{m}$ and $\overline{y} = \dfrac{M_{x}}{m}$.

(d) $I_{x} = \iint_{D} y^{2}\rho\left(x, y\right) dA$, $I_{y} = \iint_{D} x^{2}\rho\left(x, y\right) dA$, $I_{O} = \iint_{D} \left(x^{2} + y^{2}\right) \rho\left(x, y\right) dA$

5. (a) $P\left(a \le X \le b, c \le Y \le d\right) = \int_a^b \int_c^d f\left(x, y\right) dy\, dx$

(b) $f\left(x, y\right) \ge 0$ and $\iint_{\mathbb{R}^2} f\left(x, y\right) dA = 1$.

(c) The expected value of X is $\mu_1 = \iint_{\mathbb{R}^2} x f\left(x, y\right) dA$; the expected value of Y is $\mu_2 = \iint_{\mathbb{R}^2} y f\left(x, y\right) dA$.

6. (a) $A\left(S\right) = \iint_D |\mathbf{r}_u \times \mathbf{r}_v|\, dA$

(b) $A\left(S\right) = \iint_D \sqrt{1 + \left(\dfrac{\partial z}{\partial x}\right)^2 + \left(\dfrac{\partial z}{\partial y}\right)^2}\, dA$

(c) $A\left(S\right) = 2\pi \int_a^b f\left(x\right) \sqrt{1 + \left[f'\left(x\right)\right]^2}\, dx$

7. (a) $\iiint_B f\left(x, y, z\right) dV = \displaystyle\lim_{l,m,n \to \infty} \sum_{i=1}^{l} \sum_{j=1}^{m} \sum_{k=1}^{n} f\left(x_{ijk}^*, y_{ijk}^*, z_{ijk}^*\right) \Delta V$

(b) We usually evaluate $\iiint_B f\left(x, y, z\right) dV$ as an iterated integral according to Fubini's Theorem for Triple Integrals (see page 885).

(c) See the equation preceding (5) on page 886 and the associated discussion.

(d) See (5) and (6) and the accompanying discussion in Section 12.7.

(e) See (10) and the accompanying discussion in Section 12.7.

(f) See (11) and the preceding discussion in Section 12.7.

8. (a) $m = \iiint_E \rho\left(x, y, z\right) dV$

(b) $M_{yz} = \iiint_E x\rho\left(x, y, z\right) dV$, $M_{xz} = \iiint_E y\rho\left(x, y, z\right) dV$, $M_{xy} = \iiint_E z\rho\left(x, y, z\right) dV$.

(c) The center of mass is $\left(\bar{x}, \bar{y}, \bar{z}\right)$ where $\bar{x} = \dfrac{M_{yz}}{m}$, $\bar{y} = \dfrac{M_{xz}}{m}$, and $\bar{z} = \dfrac{M_{xy}}{m}$.

(d) $I_x = \iiint_E \left(y^2 + z^2\right) \rho\left(x, y, z\right) dV$, $I_y = \iiint_E \left(x^2 + z^2\right) \rho\left(x, y, z\right) dV$, $I_z = \iiint_E \left(x^2 + y^2\right) \rho\left(x, y, z\right) dV$.

9. (a) See (2) and the accompanying discussion in Section 12.8.

(b) See (4) and the accompanying discussion in Section 12.8.

(c) We may want to change from rectangular to cylindrical or spherical coordinates in a triple integral if the region E of integration is more easily described in cylindrical or spherical coordinates or if the triple integral is easier to evaluate using cylindrical or spherical coordinates.

10. (a) $\dfrac{\partial\left(x, y\right)}{\partial\left(u, v\right)} = \begin{vmatrix} \partial x/\partial u & \partial x/\partial v \\ \partial y/\partial u & \partial y/\partial v \end{vmatrix} = \dfrac{\partial x}{\partial u}\dfrac{\partial y}{\partial v} - \dfrac{\partial x}{\partial v}\dfrac{\partial y}{\partial u}$

(b) See (9) and the accompanying discussion in Section 12.9.

(c) See (13) and the accompanying discussion in Section 12.9.

True-False Quiz

1. This is true by Fubini's Theorem.

3. $\iint_D \sqrt{4 - x^2 - y^2}\, dA =$ the volume under the surface $x^2 + y^2 + z^2 = 4$ and above the xy-plane
$= \frac{1}{2}$ (the volume of the sphere $x^2 + y^2 + z^2 = 4$) $= \frac{1}{2} \cdot \frac{4}{3}\pi (2)^3 = \frac{16}{3}\pi$

5. The volume enclosed by the cone $z = \sqrt{x^2 + y^2}$ and the plane $z = 2$ is, using cylindrical coordinates,
$V = \int_0^{2\pi} \int_0^2 \int_r^2 r\, dz\, dr\, d\theta \neq \int_0^{2\pi} \int_0^2 \int_r^2 dz\, dr\, d\theta$, so the assertion is false.

Exercises

1. As shown in the contour map, we divide R into 9 equally sized subsquares, each with area $\Delta A = 1$.
Then we approximate $\iint_R f(x, y)\, dA$ by a Riemann sum with $m = n = 3$ and the sample points the
upper right corners of each square, so

$$\iint_R f(x, y)\, dA \approx \sum_{i=1}^{3} \sum_{j=1}^{3} f(x_i, y_j)\, \Delta A$$

$$= \Delta A \left[f(1, 1) + f(1, 2) + f(1, 3) + f(2, 1) + f(2, 2) \right.$$
$$\left. + f(2, 3) + f(3, 1) + f(3, 2) + f(3, 3) \right]$$

Using the contour lines to estimate the function values, we have

$$\iint_R f(x, y)\, dA \approx 1 \left[2.7 + 4.7 + 8.0 + 4.7 + 6.7 + 10.0 + 6.7 + 8.6 + 11.9 \right] \approx 64.0$$

3. $\int_{-2}^{2} \int_0^4 (4x^3 + 3xy^2)\, dx\, dy = \int_{-2}^{2} (256 + 24y^2)\, dy = 2 \left[256y + 8y^3 \right]_0^2 = 1152$

5. $\int_1^2 \int_0^{x^2} \frac{1}{x+y}\, dy\, dx = \int_1^2 \left[\ln|x + y| \right]_{y=0}^{y=x^2}\, dx = \int_1^2 \left[\ln(1 + x) \right]\, dx = \left[(1 + x) \ln(1 + x) - (1 + x) \right]_1^2$
$= \ln\left(\frac{27}{4}\right) - 1$

7. $\int_0^1 \int_0^{x^2} \int_0^y y^2 z\, dz\, dy\, dx = \int_0^1 \int_0^{x^2} \frac{1}{2} y^4\, dy\, dx = \int_0^1 \frac{1}{10} x^{10}\, dx = \frac{1}{110}$

9. The region R is more easily described by polar coordinates: $R = \{(r, \theta) \mid 2 \le r \le 4, 0 \le \theta \le \pi\}$. Thus
$\iint_R f(x, y)\, dA = \int_0^{\pi} \int_2^4 f(r \cos\theta, r \sin\theta)\, r\, dr\, d\theta$.

11.

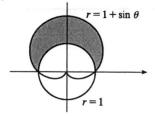

$r = 1 + \sin\theta$

$r = 1$

The region whose area is given by $\int_0^{\pi} \int_1^{1 + \sin\theta} r\, dr\, d\theta$ is

$\{(r, \theta) \mid 0 \le \theta \le \pi, 1 \le r \le 1 + \sin\theta\}$, which is the region

outside the circle $r = 1$ and inside the cardioid $r = 1 + \sin\theta$.

13.

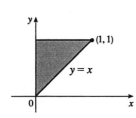

$$\int_0^1 \int_x^1 e^{x/y}\,dy\,dx = \int_0^1 \int_0^y e^{x/y}\,dx\,dy$$
$$= \int_0^1 (ey - y)\,dy = \tfrac{1}{2}(e-1)$$

15. $\displaystyle \int_2^4 \int_0^1 \frac{1}{(x-y)^2}\,dx\,dy = \int_2^4 \left(-\frac{1}{y} - \frac{1}{1-y}\right)dy = [-\ln y + \ln|1-y|]_2^4$
$$= -\ln 4 + \ln 3 + \ln 2 = \ln \tfrac{3}{2}$$

17. The curves $y^2 = x^3$ and $y = x$ intersect when $x^3 = x$, that is when $x = 0$ and $x = 1$ (note that $x \neq -1$ since $x^3 = y^2 \;\Rightarrow\; x \geq 0$.) So $\int_0^1 \int_{x^{3/2}}^x xy\,dy\,dx = \int_0^1 \left[\frac{1}{2}x^3 - \frac{1}{2}x^4\right]dx = \left[\frac{1}{8}x^4 - \frac{1}{10}x^5\right]_0^1 = \frac{1}{40}.$

19. $\int_0^1 \int_0^{1-y^2} (xy + 2x + 3y)\,dx\,dy = \int_0^1 \left[(y+2)\frac{1}{2}\left(1-y^2\right)^2 + 3y\left(1-y^2\right)\right]dy$
$$= \frac{1}{2}y^5 + y^4 - 4y^3 - 2y^2 + \frac{7}{2}y + 1 = \frac{1}{5} + \frac{1}{12} - \frac{2}{3} + \frac{7}{4} = \frac{41}{30}$$

21.

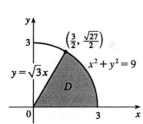

$$\iint_D \left(x^2 + y^2\right)^{3/2}\,dA = \int_0^{\pi/3}\int_0^3 \left(r^2\right)^{3/2} r\,dr\,d\theta$$
$$= \frac{\pi}{3}\frac{3^5}{5} = \frac{81\pi}{5}$$

23. $\iiint_E x^2 z\,dV = \int_0^2 \int_0^{2x} \int_0^x x^2 z\,dz\,dy\,dx = \int_0^2 \int_0^{2x} \frac{1}{2}x^4\,dy\,dx = \int_0^2 x^5\,dx = \frac{1}{6}\cdot 2^6 = \frac{32}{3}$

25. $\iiint_E y^2 z^2\,dV = \int_{-1}^1 \int_{-\sqrt{1-y^2}}^{\sqrt{1-y^2}} \int_0^{1-y^2-z^2} y^2 z^2\,dx\,dz\,dy = \int_{-1}^1 \int_{-\sqrt{1-y^2}}^{\sqrt{1-y^2}} y^2 z^2\left(1-y^2-z^2\right)dz\,dy$
$$= \int_0^{2\pi}\int_0^1 \left(r^2 \cos^2\theta\right)\left(r^2 \sin^2\theta\right)\left(1-r^2\right)r\,dr\,d\theta = \int_0^{2\pi}\int_0^1 \frac{1}{4}\sin^2 2\theta\left(r^5 - r^7\right)dr\,d\theta$$
$$= \int_0^{2\pi} \frac{1}{8}\left(1 - \cos 4\theta\right)\left[\frac{1}{6}r^6 - \frac{1}{8}r^8\right]_{r=0}^{r=1} d\theta = \frac{1}{192}\left[\theta - \frac{1}{4}\sin 4\theta\right]_0^{2\pi} = \frac{2\pi}{192} = \frac{\pi}{96}$$

27. $\iiint_E yz\,dV = \int_{-2}^2 \int_0^{\sqrt{4-x^2}} \int_0^y yz\,dz\,dy\,dx = \int_{-2}^2 \int_0^{\sqrt{4-x^2}} \frac{1}{2}y^3\,dy\,dx = \int_0^\pi \int_0^2 \frac{1}{2}r^3 \sin^3\theta r\,dr\,d\theta$
$$= \frac{16}{5}\int_0^\pi \sin^3\theta\,d\theta = \frac{16}{5}\left[-\cos\theta + \frac{1}{3}\cos^3\theta\right]_0^\pi = \frac{64}{15}$$

29. $V = \int_0^2 \int_1^4 \left(x^2 + 4y^2\right)dy\,dx = \int_0^2 \left(3x^2 + 84\right)dx = 176$

31.

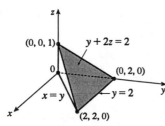

$$V = \int_0^2 \int_0^y \int_0^{(2-y)/2} dz\, dx\, dy = \int_0^2 \int_0^y \left(1 - \tfrac{1}{2}y\right) dx\, dy$$
$$= \int_0^2 \left(y - \tfrac{1}{2}y^2\right) dy = \tfrac{2}{3}$$

33. Using the wedge above the plane $z = 0$ and below the plane $z = mx$ and noting that we have the same volume for $m < 0$ as for $m > 0$ (so use $m > 0$), we have

$$V = 2\int_0^{a/3}\int_0^{\sqrt{a^2-9y^2}} mx\, dx\, dy = 2\int_0^{a/3} \tfrac{1}{2}m\left(a^2 - 9y^2\right) dy = m\left[a^2 y - 3y^3\right]_0^{a/3}$$
$$= m\left(\tfrac{1}{3}a^3 - \tfrac{1}{9}a^3\right) = \tfrac{2}{9}ma^3$$

35. (a) $m = \int_0^1 \int_0^{1-y^2} y\, dx\, dy = \int_0^1 \left(y - y^3\right) dy = \tfrac{1}{2} - \tfrac{1}{4} = \tfrac{1}{4}$

(b) $M_y = \int_0^1 \int_0^{1-y^2} xy\, dx\, dy = \int_0^1 \tfrac{1}{2}y\left(1 - y^2\right)^2 dy = -\tfrac{1}{12}\left(1 - y^2\right)^3\Big]_0^1 = \tfrac{1}{12}$,

$M_x = \int_0^1 \int_0^{1-y^2} y^2\, dx\, dy = \int_0^1 \left(y^2 - y^4\right) dy = \tfrac{2}{15}$. Hence $(\bar{x}, \bar{y}) = \left(\tfrac{1}{3}, \tfrac{8}{15}\right)$.

(c) $I_x = \int_0^1 \int_0^{1-y^2} y^3\, dx\, dy = \int_0^1 \left(y^3 - y^5\right) dy = \tfrac{1}{12}$,

$I_y = \int_0^1 \int_0^{1-y^2} yx^2\, dx\, dy = \int_0^1 \tfrac{1}{3}y\left(1 - y^2\right)^3 dy = -\tfrac{1}{24}\left(1 - y^2\right)^4\Big]_0^1 = \tfrac{1}{24}$

37. (a) The equation of the cone with the suggested orientation is $(h - z) = \tfrac{h}{a}\sqrt{x^2 + y^2}$, $0 \le z \le h$. Then $V = \tfrac{1}{3}\pi a^2 h$ is the volume of one frustum of a cone; by symmetry $M_{yz} = M_{zz} = 0$; and

$$M_{xy} = \iint_{x^2+y^2\le a^2}\int_0^{h-(h/a)\sqrt{x^2+y^2}} z\, dz\, dA = \int_0^{2\pi}\int_0^a\int_0^{(h/a)(a-r)} rz\, dz\, dr\, d\theta$$
$$= \pi\int_0^a r\frac{h^2}{a^2}(a-r)^2\, dr = \frac{\pi h^2}{a^2}\int_0^a \left(a^2 r - 2ar^2 + r^3\right) dr$$
$$= \frac{\pi h^2}{a^2}\left(\frac{a^4}{2} - \frac{2a^4}{3} + \frac{a^4}{4}\right) = \frac{\pi h^2 a^2}{12}$$

Hence the centroid is $(\bar{x}, \bar{y}, \bar{z}) = \left(0, 0, \tfrac{1}{4}h\right)$.

(b) $I_z = \int_0^{2\pi}\int_0^a\int_0^{(h/a)(a-r)} r^3\, dz\, dr\, d\theta = 2\pi\int_0^a \frac{h}{a}\left(ar^3 - r^4\right) dr = \frac{2\pi h}{a}\left(\frac{a^5}{4} - \frac{a^5}{5}\right) = \frac{\pi a^4 h}{10}$

39. Let D represent the given triangle; then D can be described as the area enclosed by the x- and y-axes and the line $y = 2 - 2x$, or equivalently $D = \{(x, y) \mid 0 \le x \le 1, 0 \le y \le 2 - 2x\}$. We want to find the surface area of the part of the graph of $z = x^2 + y$ that lies over D, so using Equation 12.6.6 we have

$$A\left(S\right) = \iint_D \sqrt{1 + \left(\frac{\partial z}{\partial x}\right)^2 + \left(\frac{\partial z}{\partial y}\right)^2}\, dA = \iint_D \sqrt{1 + (2x)^2 + (1)^2}\, dA$$

$$= \int_0^1 \int_0^{2-2x} \sqrt{2 + 4x^2}\, dy\, dx = \int_0^1 \sqrt{2 + 4x^2}\, [y]_0^{2-2x}\, dx = \int_0^1 (2 - 2x)\sqrt{2 + 4x^2}\, dx$$

$$= \int_0^1 2\sqrt{2 + 4x^2}\, dx - \int_0^1 2x\sqrt{2 + 4x^2}\, dx$$

Using Formula 21 in the Table of Integrals with $a = \sqrt{2}$, $u = 2x$, and $du = 2dx$,
$\int 2\sqrt{2 + 4x^2}\, dx = x\sqrt{2 + 4x^2} + \ln\left(2x + \sqrt{2 + 4x^2}\right)$. If we substitute $u = 2 + 4x^2$ in the second integral, then $du = 8x\, dx$ and $\int 2x\sqrt{2 + 4x^2}\, dx = \frac{1}{4}\int \sqrt{u}\, du = \frac{1}{4} \cdot \frac{2}{3}u^{3/2} = \frac{1}{6}\left(2 + 4x^2\right)^{3/2}$. Thus

$$A\left(S\right) = \left[x\sqrt{2 + 4x^2} + \ln\left(2x + \sqrt{2 + 4x^2}\right) - \tfrac{1}{6}\left(2 + 4x^2\right)^{3/2}\right]_0^1$$

$$= \sqrt{6} + \ln\left(2 + \sqrt{6}\right) - \tfrac{1}{6}(6)^{3/2} - \ln\sqrt{2} + \tfrac{\sqrt{2}}{3}$$

$$= \ln\tfrac{2 + \sqrt{6}}{\sqrt{2}} + \tfrac{\sqrt{2}}{3} = \ln\left(\sqrt{2} + \sqrt{3}\right) + \tfrac{\sqrt{2}}{3} \approx 1.6176$$

41.

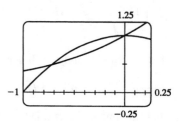

(0, 2)

$(\sqrt{2}, \sqrt{2})$

D

0   (2, 0) x

$$\int_0^{\sqrt{2}} \int_y^{\sqrt{4-y^2}} \frac{1}{1 + x^2 + y^2}\, dx\, dy = \int_0^{\pi/4} \int_0^2 \frac{1}{1 + r^2} r\, dr\, d\theta$$

$$= \tfrac{1}{2}\int_0^{\pi/4} \ln\left|1 + r^2\right|\big|_0^2\, d\theta$$

$$= \tfrac{1}{2}\int_0^{\pi/4} \ln 5\, d\theta = \tfrac{\pi}{8}\ln 5$$

43. From the graph, it appears that $1 - x^2 = e^x$ at $x \approx -0.71$ and at $x = 0$, with $1 - x^2 > e^x$ on $(-0.71, 0)$. So the desired integral is

$$\iint_D y^2\, dA \approx \int_{-0.71}^0 \int_{e^x}^{1-x^2} y^2\, dy\, dx$$

$$= \tfrac{1}{3}\int_{-0.71}^0 \left[(1 - x^2)^3 - e^{3x}\right]\, dx$$

$$= \tfrac{1}{3}\left[x - x^3 + \tfrac{3}{5}x^5 - \tfrac{1}{7}x^7 - \tfrac{1}{3}e^{3x}\right]_{-0.71}^0 \approx 0.0512$$

1.25

−1   0.25

−0.25

45. (a) $f(x,y)$ is a joint density function, so we know that $\iint_{\mathbb{R}^2} f(x,y)\, dA = 1$. Since $f(x,y) = 0$ outside the rectangle $[0,3] \times [0,2]$, we can say

$$\iint_{\mathbb{R}^2} f(x,y)\, dA = \int_{-\infty}^{\infty}\int_{-\infty}^{\infty} f(x,y)\, dy\, dx = \int_0^3 \int_0^2 C(x+y)\, dy\, dx$$

$$= C\int_0^3 \left[xy + \tfrac{1}{2}y^2\right]_{y=0}^{y=2} dx = C\int_0^3 (2x+2)\, dx = C\left[x^2 + 2x\right]_0^3 = 15C$$

Then $15C = 1 \;\Rightarrow\; C = \tfrac{1}{15}$.

(b) $P(X \le 2, Y \ge 1) = \int_{-\infty}^{2}\int_1^{\infty} f(x,y)\, dy\, dx = \int_0^2 \int_1^2 \tfrac{1}{15}(x+y)\, dy\, dx = \tfrac{1}{15}\int_0^2 \left[xy + \tfrac{1}{2}y^2\right]_{y=1}^{y=2} dx$

$= \tfrac{1}{15}\int_0^2 \left(x + \tfrac{3}{2}\right) dx = \tfrac{1}{15}\left[\tfrac{1}{2}x^2 + \tfrac{3}{2}x\right]_0^2 = \tfrac{1}{3}$

(c) $P(X+Y \le 1) = P((X,Y) \in D)$ where D is the triangular region shown in the figure. Thus

$$P(X+Y \le 1) = \iint_D f(x,y)\, dA = \int_0^1 \int_0^{1-x} \tfrac{1}{15}(x+y)\, dy\, dx$$

$$= \tfrac{1}{15}\int_0^1 \left[xy + \tfrac{1}{2}y^2\right]_{y=0}^{y=1-x} dx$$

$$= \tfrac{1}{15}\int_0^1 \left[x(1-x) + \tfrac{1}{2}(1-x)^2\right] dx$$

$$= \tfrac{1}{30}\int_0^1 (1-x^2)\, dx = \tfrac{1}{30}\left[x - \tfrac{1}{3}x^3\right]_0^1 = \tfrac{1}{45}$$

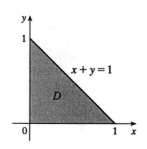

47.

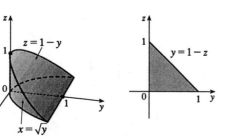

$\int_{-1}^{1} \int_{x^2}^{1} \int_0^{1-y} f(x,y,z)\, dz\, dy\, dx$

$= \int_0^1 \int_0^{1-z} \int_{-\sqrt{y}}^{\sqrt{y}} f(x,y,z)\, dx\, dy\, dz$

49. Since $u = x - y$, $v = x + y$, $x = \tfrac{1}{2}(u+v)$ and $y = \tfrac{1}{2}(v-u)$. Thus

$$\frac{\partial(x,y)}{\partial(u,v)} = \begin{vmatrix} 1/2 & 1/2 \\ -1/2 & 1/2 \end{vmatrix} = \frac{1}{2} \text{ and}$$

$$\iint_R \frac{x-y}{x+y}\, dA = \int_2^4 \int_{-2}^0 \frac{u}{v}\left(\tfrac{1}{2}\, du\, dv\right)$$

$$= -\int_2^4 \frac{dv}{v} = -\ln 2$$

51. Let $u = y - x$ and $v = y + x$ so $x = y - u = (v-x) - u \;\Rightarrow\; x = \tfrac{1}{2}(v-u)$ and

$y = v - \tfrac{1}{2}(v-u) = \tfrac{1}{2}(v+u)$. $\left|\dfrac{\partial(x,y)}{\partial(u,v)}\right| = \left|\dfrac{\partial x}{\partial u}\dfrac{\partial y}{\partial v} - \dfrac{\partial x}{\partial v}\dfrac{\partial y}{\partial u}\right| = \left|-\tfrac{1}{2}\left(\tfrac{1}{2}\right) - \tfrac{1}{2}\left(\tfrac{1}{2}\right)\right| = \left|-\tfrac{1}{2}\right| = \tfrac{1}{2}$.

R is the image under this transformation of the square with vertices $(u,v) = (0,0)$, $(-2,0)$, $(0,2)$, and $(-2,2)$. So

$$\iint_R xy\, dA = \int_0^2 \int_{-2}^0 \frac{v^2 - u^2}{4}\left(\frac{1}{2}\right) du\, dv = \tfrac{1}{8}\int_0^2 \left[v^2 u - \tfrac{1}{3}u^3\right]_{u=-2}^{u=0} dv = \tfrac{1}{8}\int_0^2 \left(2v^2 - \tfrac{8}{3}\right) dv$$

$$= \tfrac{1}{8}\left[\tfrac{2}{3}v^3 - \tfrac{8}{3}v\right]_0^2 = 0$$

This result could have been anticipated by symmetry, since the integrand is an odd function of y and R is symmetric about the x-axis.

Focus on Problem Solving

1.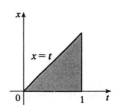

Let $R = \bigcup_{i=1}^{5} R_i$, where

$R_i = \{(x, y) \mid x + y \geq i + 2, x + y < i + 3, 1 \leq x \leq 3, 2 \leq y \leq 5\}$.

$\iint_R [\![x + y]\!] \, dA = \sum_{i=1}^{5} \iint_{R_i} [\![x + y]\!] \, dA = \sum_{i=1}^{5} [\![x + y]\!] \iint_{R_i} dA$, since

$[\![x + y]\!] = \text{constant} = i + 2$ for $(x, y) \in R_i$. Therefore

$\iint_R [\![x + y]\!] \, dA = \sum_{i=1}^{5} (i + 2) [A(R_i)]$

$= 3A(R_1) + 4A(R_2) + 5A(R_3) + 6A(R_4) + 7A(R_5)$

$= 3\left(\tfrac{1}{2}\right) + 4\left(\tfrac{3}{2}\right) + 5(2) + 6\left(\tfrac{3}{2}\right) + 7\left(\tfrac{1}{2}\right) = 30$

3. $f_{\text{ave}} = \dfrac{1}{b - a} \int_a^b f(x) \, dx = \dfrac{1}{1 - 0} \int_0^1 \left[\int_x^1 \cos(t^2) \, dt \right] dx$

$= \int_0^1 \int_x^1 \cos(t^2) \, dt \, dx$

$= \int_0^1 \int_0^t \cos(t^2) \, dx \, dt$ (changing the order of integration)

$= \int_0^1 t \cos(t^2) \, dt = \tfrac{1}{2} \sin(t^2)\big]_0^1 = \tfrac{1}{2} \sin 1$

5. Since $|xy| < 1$, except at $(1, 1)$, the formula for the sum of a geometric series gives

$\dfrac{1}{1 - xy} = \sum_{n=0}^{\infty} (xy)^n$, so

$\int_0^1 \int_0^1 \dfrac{1}{1 - xy} \, dx \, dy = \int_0^1 \int_0^1 \sum_{n=0}^{\infty} (xy)^n \, dx \, dy = \sum_{n=0}^{\infty} \int_0^1 \int_0^1 (xy)^n \, dx \, dy$

$= \sum_{n=0}^{\infty} \left[\int_0^1 x^n \, dx \right] \left[\int_0^1 y^n \, dy \right] = \sum_{n=0}^{\infty} \dfrac{1}{n+1} \cdot \dfrac{1}{n+1}$

$= \sum_{n=0}^{\infty} \dfrac{1}{(n+1)^2} = \dfrac{1}{1^2} + \dfrac{1}{2^2} + \dfrac{1}{3^2} + \cdots = \sum_{n=1}^{\infty} \dfrac{1}{n^2}$

7. (a) Since $|xyz| < 1$ except at $(1, 1, 1)$, the formula for the sum of a geometric series gives

$\dfrac{1}{1 - xyz} = \sum_{n=0}^{\infty} (xyz)^n$, so

$\int_0^1 \int_0^1 \int_0^1 \dfrac{1}{1 - xyz} \, dx \, dy \, dz = \int_0^1 \int_0^1 \int_0^1 \sum_{n=0}^{\infty} (xyz)^n \, dx \, dy \, dz = \sum_{n=0}^{\infty} \int_0^1 \int_0^1 \int_0^1 (xyz)^n \, dx \, dy \, dz$

$= \sum_{n=0}^{\infty} \left[\int_0^1 x^n \, dx \right] \left[\int_0^1 y^n \, dy \right] \left[\int_0^1 z^n \, dz \right] = \sum_{n=0}^{\infty} \dfrac{1}{n+1} \cdot \dfrac{1}{n+1} \cdot \dfrac{1}{n+1}$

$= \sum_{n=0}^{\infty} \dfrac{1}{(n+1)^3} = \dfrac{1}{1^3} + \dfrac{1}{2^3} + \dfrac{1}{3^3} + \cdots = \sum_{n=1}^{\infty} \dfrac{1}{n^3}$.

(b) Since $|-xyz| < 1$, except at $(1, 1, 1)$, the formula for the sum of a geometric series gives

$$\frac{1}{1+xyz} = \sum_{n=0}^{\infty} (-xyz)^n, \text{ so}$$

$$\int_0^1 \int_0^1 \int_0^1 \frac{1}{1+xyz} \, dx \, dy \, dz = \int_0^1 \int_0^1 \int_0^1 \sum_{n=0}^{\infty} (-xyz)^n \, dx \, dy \, dz = \sum_{n=0}^{\infty} \int_0^1 \int_0^1 \int_0^1 (-xyz)^n \, dx \, dy \, dz$$

$$= \sum_{n=0}^{\infty} (-1)^n \left[\int_0^1 x^n \, dx \right] \left[\int_0^1 y^n \, dy \right] \left[\int_0^1 z^n \, dz \right]$$

$$= \sum_{n=0}^{\infty} (-1)^n \frac{1}{n+1} \cdot \frac{1}{n+1} \cdot \frac{1}{n+1}$$

$$= \sum_{n=0}^{\infty} \frac{(-1)^n}{(n+1)^3} = \frac{1}{1^3} - \frac{1}{2^3} + \frac{1}{3^3} - \cdots = \sum_{n=1}^{\infty} \frac{(-1)^{n-1}}{n^3}$$

To evaluate this sum, we first write out a few terms: $s = 1 - \dfrac{1}{2^3} + \dfrac{1}{3^3} - \dfrac{1}{4^3} + \dfrac{1}{5^3} - \dfrac{1}{6^3} \approx 0.8998$.

Notice that $a_7 = \dfrac{1}{7^3} < 0.003$. By the Alternating Series Estimation Theorem from Section 8.4, we have $|s - s_6| \le a_7 < 0.003$. This error of 0.003 will not affect the second decimal place, so we have $s \approx 0.90$.

9. $\int_0^x \int_0^y \int_0^z f(t) \, dt \, dz \, dy = \iiint_E f(t) \, dV$, where

$E = \{(t, z, y) \mid 0 \le t \le z, 0 \le z \le y, 0 \le y \le x\}$. If we let D be the

projection of E on the yt-plane then $D = \{(y, t) \mid 0 \le t \le x, t \le y \le x\}$.

And we see from the diagram that

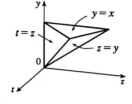

$E = \{(t, z, y) \mid t \le z \le y, t \le y \le x, 0 \le t \le x\}$. So

$$\int_0^x \int_0^y \int_0^z f(t) \, dt \, dz \, dy = \int_0^x \int_t^x \int_t^y f(t) \, dz \, dy \, dt = \int_0^x \left[\int_t^x (y - t) f(t) \, dy \right] dt$$

$$= \int_0^x \left[\left(\tfrac{1}{2} y^2 - ty \right) f(t) \right]_{y=t}^{y=x} \, dt = \int_0^x \left[\tfrac{1}{2} x^2 - tx - \tfrac{1}{2} t^2 + t^2 \right] f(t) \, dt$$

$$= \int_0^x \left[\tfrac{1}{2} x^2 - tx + \tfrac{1}{2} t^2 \right] f(t) \, dt = \int_0^x \tfrac{1}{2} \left(x^2 - 2tx + t^2 \right) f(t) \, dt$$

$$= \tfrac{1}{2} \int_0^x (x - t)^2 f(t) \, dt$$

Chapter 13   Vector Calculus

Section 13.1   Vector Fields

1. $\mathbf{F}(x, y) = x\,\mathbf{i} + y\,\mathbf{j}$

The length of the vector $x\,\mathbf{i} + y\,\mathbf{j}$ is the distance from $(0, 0)$ to (x, y). Flow lines are rays emanating from the origin.

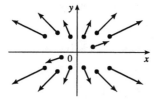

3. $\mathbf{F}(x, y) = y\,\mathbf{i} + \mathbf{j}$

The length of the vector $y\,\mathbf{i} + \mathbf{j}$ is $\sqrt{y^2 + 1}$. Flow lines are parabolas opening about the x-axis.

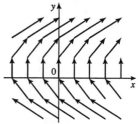

5. $\mathbf{F}(x, y) = \dfrac{y\,\mathbf{i} + x\,\mathbf{j}}{\sqrt{x^2 + y^2}}$

The length of the vector $\dfrac{y\,\mathbf{i} + x\,\mathbf{j}}{\sqrt{x^2 + y^2}}$ is 1.

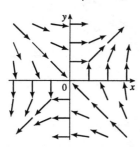

7. $\mathbf{F}(x, y, z) = \mathbf{j}$

All vectors in this field are parallel to the y-axis and have length 1.

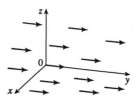

9. $\mathbf{F}(x, y, z) = y\,\mathbf{j}$

The length of $\mathbf{F}(x, y, z)$ is $|y|$. No vectors emanate from the xz-plane since $y = 0$ there. In each plane $y = b$, all the vectors are identical.

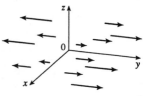

11. $\mathbf{F}(x, y) = \langle y, x \rangle$ corresponds to graph III, since in the first quadrant all the vectors have positive x- and y-components, in the second quadrant all vectors have positive x-components and negative y-components, in the third quadrant all vectors have negative x- and y-components, and in the fourth quadrant all vectors have negative x-components and positive y-components.

13. $\mathbf{F}(x, y) = \langle \sin x, \sin y \rangle$ corresponds to graph II, since the vector field is the same on each square of the form $[2n\pi, 2(n+1)\pi] \times [2m\pi, 2(m+1)\pi]$, m, n any integers.

15. $\mathbf{F}(x, y, z) = \mathbf{i} + 2\mathbf{j} + 3\mathbf{k}$ corresponds to graph IV, since all vectors have identical length and direction.

17. $\mathbf{F}(x, y, z) = x\mathbf{i} + y\mathbf{j} + 3\mathbf{k}$ corresponds to graph III; the projection of each vector onto the xy-plane is $x\mathbf{i} + y\mathbf{j}$, which points away from the origin, and the vectors point generally upward because their z-components are all 3.

19.

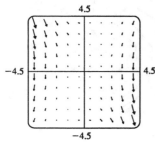

The vector field seems to have very short vectors near the line $y = 2x$. For $\mathbf{F}(x, y) = \langle 0, 0 \rangle$ we must have $y^2 - 2xy = 0$ and $3xy - 6x^2 = 0$. The first equation holds if $y = 0$ or $y = 2x$, and the second holds if $x = 0$ or $y = 2x$. So both equations hold [and thus $\mathbf{F}(x, y) = 0$] along the line $y = 2x$.

21. $\nabla f(x, y) = \langle f_x, f_y \rangle = \langle 3e^{3x} \cos 4y, -4e^{3x} \sin 4y \rangle$

23. $\nabla f(x, y, z) = \langle f_x, f_y, f_z \rangle = \langle y^2, 2xy - z^3, -3yz^2 \rangle$

25. $f(x, y) = x^2 - \frac{1}{2}y^2$, $\nabla f(x, y) = 2x\,\mathbf{i} - y\,\mathbf{j}$

The length of $\nabla f(x, y)$ is $\sqrt{4x^2 + y^2}$, and $\nabla f(x, y)$ terminates on the x-axis at the point $(3x, 0)$.

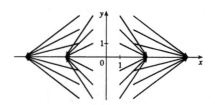

27. We graph ∇f along with a contour map of f. The graph shows that the gradient vectors are perpendicular to the level curves. Also, the gradient vectors point in the direction in which f is increasing and are longer where the level curves are closer together.

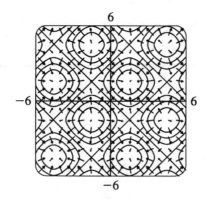

29. $f(x,y) = xy \Rightarrow \nabla f(x,y) = y\mathbf{i} + x\mathbf{j}$. In the first quadrant, both components of each vector are positive, while in the third quadrant both components are negative. However, in the second quadrant each vector's x-component is positive while its y-component is negative (and vice versa in the fourth quadrant). Thus, ∇f is graph IV.

31. $f(x,y) = x^2 + y^2 \Rightarrow \nabla f(x,y) = 2x\mathbf{i} + 2y\mathbf{j}$. Thus, each vector $\nabla f(x,y)$ has the same direction and twice the length of the position vector of the point (x,y), so the vectors all point directly away from the origin and their lengths increase as we move away from the origin. Hence, ∇f is graph II.

33. (a) We sketch the vector field $\mathbf{F}(x,y) = x\mathbf{i} - y\mathbf{j}$ along with several approximate flow lines.

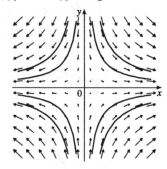

The flow lines appear to be hyperbolas with shape similar to the graph of $y = \pm 1/x$, so we might guess that the flow lines have equations $y = C/x$.

(b) If $x = x(t)$ and $y = y(t)$ are parametric equations of a flow line, then the velocity vector of the flow line at the point (x,y) is $x'(t)\mathbf{i} + y'(t)\mathbf{j}$. Since the velocity vectors coincide with the vectors in the vector field, we have $x'(t)\mathbf{i} + y'(t)\mathbf{j} = x\mathbf{i} - y\mathbf{j} \Rightarrow dx/dt = x$, $dy/dt = -y$. To solve these differential equations, we know $dx/dt = x \Rightarrow dx/x = dt \Rightarrow \ln|x| = t + C \Rightarrow x = \pm e^{t+C} = Ae^t$ for some constant A, and $dy/dt = -y \Rightarrow dy/y = -dt \Rightarrow \ln|y| = -t + K \Rightarrow y = \pm e^{-t+K} = Be^{-t}$ for some constant B. Therefore $xy = Ae^t Be^{-t} = AB = $ constant. If the flow line passes through $(1,1)$ then $(1)(1) = $ constant $= 1 \Rightarrow xy = 1 \Rightarrow y = 1/x$, $x > 0$.

Section 13.2　Line Integrals

1. $x = t^3$ and $y = t$, $0 \le t \le 1$, so

$$\int_C x\,ds = \int_0^1 (t^3) \sqrt{\left(\frac{dx}{dt}\right)^2 + \left(\frac{dy}{dt}\right)^2}\,dt = \int_0^1 (t^3)\sqrt{(3t^2)^2 + (1)^2}\,dt = \int_0^1 (t^3)\sqrt{9t^4 + 1}\,dt$$

$$= \tfrac{1}{54}\left(9t^4 + 1\right)^{3/2}\Big]_0^1 = \tfrac{1}{54}\left(10^{3/2} - 1\right)$$

3. $x = 4\cos t$, $y = 4\sin t$, $-\frac{\pi}{2} \le t \le \frac{\pi}{2}$.

$$\int_C xy^4\,ds = \int_{-\pi/2}^{\pi/2}(4\cos t)(4\sin t)^4\sqrt{(-4\sin t)^2 + (4\cos t)^2}\,dt$$

$$= \int_{-\pi/2}^{\pi/2} 4^5 \cos t \sin^4 t \sqrt{16\left(\sin^2 t + \cos^2 t\right)}\,dt$$

$$= 4^5\int_{-\pi/2}^{\pi/2}\left(\sin^4 t \cos t\right)(4)\,dt = (4)^6\left[\tfrac{1}{5}\sin^5 t\right]_{-\pi/2}^{\pi/2} = \tfrac{2\cdot 4^6}{5} = 1638.4$$

5.

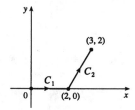

$C = C_1 + C_2$

On C_1: $x = x$, $y = 0$ $\Rightarrow$ $dy = 0\,dx$, $0 \le x \le 2$.

On C_2: $x = x$, $y = 2x - 4$ $\Rightarrow$ $dy = 2\,dx$, $2 \le x \le 3$.

Then

$$\int_C xy\,dx + (x - y)\,dy = \int_{C_1} xy\,dx + (x - y)\,dy + \int_{C_2} xy\,dx + (x - y)\,dy$$

$$= \int_0^2 (0 + 0)\,dx + \int_2^3\left[(2x^2 - 4x) + (-x + 4)(2)\right]dx$$

$$= \int_2^3 (2x^2 - 6x + 8)\,dx = \tfrac{17}{3}$$

7. $x = 2t$, $y = 3\sin t$, $z = 3\cos t$, $0 \le t \le \pi/2$.

$$\int_C xyz\,ds = \int_0^{\pi/2}(2t)(3\sin t)(3\cos t)\sqrt{\left(\frac{dx}{dt}\right)^2 + \left(\frac{dy}{dt}\right)^2 + \left(\frac{dz}{dt}\right)^2}\,dt$$

$$= \int_0^{\pi/2}(18t\sin t\cos t)\sqrt{(2)^2 + (3\cos t)^2 + (-3\sin t)^2}\,dt$$

$$= \int_0^{\pi/2}(18t\sin t\cos t)\sqrt{4 + 9\left(\cos^2 t + \sin^2 t\right)}\,dt$$

$$= 18\sqrt{13}\int_0^{\pi/2}(t\sin t\cos t)\,dt = 18\sqrt{13}\int_0^{\pi/2}\tfrac{1}{2}t\sin 2t\,dt$$

$$= 9\sqrt{13}\left[-\tfrac{1}{2}t\cos 2t + \tfrac{1}{4}\sin 2t\right]_0^{\pi/2} = \tfrac{9\sqrt{13}}{4}\pi$$

9. $x = -t + 1, y = 3t, z = 5t + 1, 0 \le t \le 1.$

$\int_C xy^2 z \, ds = \int_0^1 (1-t)(3t)^2(5t+1)\sqrt{(-1)^2 + (3)^2 + (5)^2} \, dt = \int_0^1 (1-t)(9t^2)(5t+1)\sqrt{35} \, dt$

$\qquad = 9\sqrt{35} \int_0^1 (t^2 + 4t^3 - 5t^4) \, dt = 3\sqrt{35}$

11.

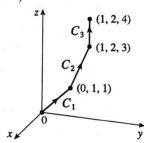

On C_1: $x = 0 \Rightarrow dx = 0 \, dt, y = t \Rightarrow dy = dt, z = t \Rightarrow dz = dt, 0 \le t \le 1.$
On C_2: $x = t \Rightarrow dx = dt, y = t+1 \Rightarrow dy = dt, z = 2t+1 \Rightarrow dz = 2 \, dt, 0 \le t \le 1.$
On C_3: $x = 1 \Rightarrow dx = 0 \, dt, y = 2 \Rightarrow dy = 0 \, dt, z = t+3 \Rightarrow dz = dt, 0 \le t \le 1.$
Then

$\int_C z^2 \, dx - z \, dy + 2y \, dz = \int_0^1 (0 - t + 2t) \, dt + \int_0^1 \left[(2t+1)^2 - (2t+1) + 2(t+1)(2) \right] dt$

$\qquad\qquad\qquad\qquad + \int_0^1 (0 + 0 + 4) \, dt$

$\qquad\qquad\qquad = \frac{1}{2} + \left[\frac{4}{3}t^3 + 3t^2 + 4t \right]_0^1 + 4 = \frac{77}{6}$

13. (a) Along the line $x = -3$, the vectors of $\mathbf{F}$ have positive y-components, so since the path goes upward, the integrand $\mathbf{F} \cdot \mathbf{T}$ is always positive. Therefore $\int_{C_1} \mathbf{F} \cdot d\mathbf{r} = \int_{C_1} \mathbf{F} \cdot \mathbf{T} \, ds$ is positive.

(b) All of the (nonzero) field vectors along the circle with radius 3 are pointed in the clockwise direction, that is, opposite the direction to the path. So $\mathbf{F} \cdot \mathbf{T}$ is negative, and therefore $\int_{C_2} \mathbf{F} \cdot d\mathbf{r} = \int_{C_2} \mathbf{F} \cdot \mathbf{T} \, ds$ is negative.

15. $\mathbf{F}(\mathbf{r}(t)) = (t^3)^2 (t^4) \mathbf{i} - (t^3)(t^4) \mathbf{j} = t^{10} \mathbf{i} - t^7 \mathbf{j}, \mathbf{r}'(t) = 3t^2 \mathbf{i} + 4t^3 \mathbf{j}.$

$\int_C \mathbf{F} \cdot d\mathbf{r} = \int_0^1 (t^{10} \mathbf{i} - t^7 \mathbf{j}) \cdot (3t^2 \mathbf{i} + 4t^3 \mathbf{j}) \, dt = \int_0^1 \left[(t^{10})(3t^2) + (-t^7)(4t^3) \right] dt$

$\qquad = \int_0^1 (3t^{12} - 4t^{10}) \, dt = \frac{3}{13} - \frac{4}{11} = -\frac{19}{143}$

17. $\int_C \mathbf{F} \cdot d\mathbf{r} = \int_0^1 \langle \sin t^3, \cos(-t^2), t^4 \rangle \cdot \langle 3t^2, -2t, 1 \rangle \, dt$

$\qquad = \int_0^1 (3t^2 \sin t^3 - 2t \cos t^2 + t^4) \, dt = \left[-\cos t^3 - \sin t^2 + \frac{1}{5}t^5 \right]_0^1$

$\qquad = \frac{6}{5} - \cos 1 - \sin 1$

19. We graph $\mathbf{F}(x, y) = (x - y)\mathbf{i} + xy\mathbf{j}$ and the curve C. We see that
most of the vectors starting on C point in roughly the same direction
as C, so for these portions of C the tangential component $\mathbf{F} \cdot \mathbf{T}$ is
positive. Although some vectors in the third quadrant which start on C
point in roughly the opposite direction, and hence give negative
tangential components, it seems reasonable that the effect of these
portions of C is outweighed by the positive tangential components.
Thus, we would expect $\int_C \mathbf{F} \cdot d\mathbf{r} = \int_C \mathbf{F} \cdot \mathbf{T}\, ds$ to be positive.

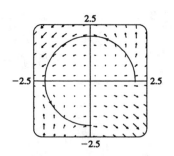

To verify, we evaluate $\int_C \mathbf{F} \cdot d\mathbf{r}$. The curve C can be represented by $\mathbf{r}(t) = 2\cos t\, \mathbf{i} + 2\sin t\, \mathbf{j}$,
$0 \le t \le \frac{3\pi}{2}$, so $\mathbf{F}(\mathbf{r}(t)) = (2\cos t - 2\sin t)\mathbf{i} + 4\cos t \sin t\, \mathbf{j}$ and $\mathbf{r}'(t) = -2\sin t\, \mathbf{i} + 2\cos t\, \mathbf{j}$. Then

$$\int_C \mathbf{F} \cdot d\mathbf{r} = \int_0^{3\pi/2} \mathbf{F}(\mathbf{r}(t)) \cdot \mathbf{r}'(t)\, dt$$

$$= \int_0^{3\pi/2} [-2\sin t\, (2\cos t - 2\sin t) + 2\cos t\, (4\cos t \sin t)]\, dt$$

$$= 4\int_0^{3\pi/2} (\sin^2 t - \sin t \cos t + 2\sin t \cos^2 t)\, dt = 3\pi + \tfrac{2}{3} \text{ (using a CAS)}$$

21. (a) $\int_C \mathbf{F} \cdot d\mathbf{r} = \int_0^1 \left\langle e^{t^2 - 1}, t^5 \right\rangle \cdot \left\langle 2t, 3t^2 \right\rangle\, dt = \int_0^1 \left(2te^{t^2 - 1} + 3t^7 \right)\, dt$

$= \left[e^{t^2 - 1} + \tfrac{3}{8} t^8 \right]_0^1 = \tfrac{11}{8} - 1/e$

(b) $\mathbf{r}(0) = \mathbf{0}$, $\mathbf{F}(\mathbf{r}(0)) = \left\langle e^{-1}, 0 \right\rangle$; $\mathbf{r}\left(\tfrac{1}{\sqrt{2}}\right) = \left\langle \tfrac{1}{2}, \tfrac{1}{2\sqrt{2}} \right\rangle$, $\mathbf{F}\left(\mathbf{r}\left(\tfrac{1}{\sqrt{2}}\right)\right) = \left\langle e^{-1/2}, \tfrac{1}{4\sqrt{2}} \right\rangle$;
$\mathbf{r}(1) = \langle 1, 1 \rangle$, $\mathbf{F}(\mathbf{r}(1)) = \langle 1, 1 \rangle$.

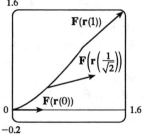

In order to generate the graph with Maple, we use the PLOT
command (not to be confused with the `plot` command) to define
each of the vectors. For example,
`v1:=PLOT(CURVES([[0,0],[evalf(1/exp(1)),0]]));`
generates the vector from the vector field at the point $(0, 0)$ (but
without an arrowhead) and gives it the name `v1`. To show
everything on the same screen, we use the `display` command.
In Mathematica, we use `ListPlot` (with the `PlotJoined -> True` option) to generate the
vectors, and then `Show` to show everything on the same screen.

23. The part of the astroid that lies in the quadrant is parametrized by $x = \cos^3 t$,
$y = \sin^3 t$, $0 \le t \le \frac{\pi}{2}$. Now $\dfrac{dx}{dt} = 3\cos^2 t\, (-\sin t)$ and $\dfrac{dy}{dt} = 3\sin^2 t \cos t$, so

$$\sqrt{\left(\frac{dx}{dt}\right)^2 + \left(\frac{dy}{dt}\right)^2} = \sqrt{9\cos^4 t \sin^2 t + 9\sin^4 t \cos^2 t} = 3\cos t \sin t \sqrt{\cos^2 t + \sin^2 t} = 3\cos t \sin t.$$

Therefore $\int_C x^3 y^5\, ds = \int_0^{\pi/2} \cos^9 t \sin^{15} t\, (3\cos t \sin t)\, dt = \frac{945}{16{,}777{,}216}\pi.$

25. We use the parametrization $x = 2\cos t$, $y = 2\sin t$, $-\frac{\pi}{2} \le t \le \frac{\pi}{2}$. Then

$$ds = \sqrt{\left(\frac{dx}{dt}\right)^2 + \left(\frac{dy}{dt}\right)^2}\, dt = \sqrt{(-2\sin t)^2 + (2\cos t)^2}\, dt = 2\, dt, \text{ so}$$

$m = \int_C k\, ds = 2k \int_{-\pi/2}^{\pi/2} dt = 2k\,(\pi)$,

$\bar{x} = \frac{1}{2\pi k}\int_C xk\, ds = \frac{1}{2\pi}\int_{-\pi/2}^{\pi/2}(2\cos t)\, 2\, dt = \frac{1}{2\pi}[4\sin t]_{-\pi/2}^{\pi/2} = \frac{4}{\pi}$,

$\bar{y} = \frac{1}{2\pi k}\int_C yk\, ds = \frac{1}{2\pi}\int_{-\pi/2}^{\pi/2}(2\sin t)\, 2\, dt = 0$. Hence $(\bar{x}, \bar{y}) = \left(\frac{4}{\pi}, 0\right)$.

27. (a) $\bar{x} = \dfrac{1}{m}\displaystyle\int_C x\rho\,(x,y,z)\, ds$, $\bar{y} = \dfrac{1}{m}\displaystyle\int_C y\rho\,(x,y,z)\, ds$, $\bar{z} = \dfrac{1}{m}\displaystyle\int_C z\rho\,(x,y,z)\, ds$ where

$m = \int_C \rho\,(x,y,z)\, ds$.

(b) $m = \int_C k\, ds = k\int_0^{2\pi}\sqrt{4\sin^2 t + 4\cos^2 t + 9}\, dt = k\sqrt{13}\int_0^{2\pi} dt = 2\pi k\sqrt{13}$,

$\bar{x} = \dfrac{1}{2\pi k\sqrt{13}}\displaystyle\int_0^{2\pi} k2\sqrt{13}\sin t\, dt = 0, \bar{y} = \dfrac{1}{2\pi k\sqrt{13}}\displaystyle\int_0^{2\pi} k2\sqrt{13}\cos t\, dt = 0$,

$\bar{z} = \dfrac{1}{2\pi k\sqrt{13}}\displaystyle\int_0^{2\pi} \left(k\sqrt{13}\right)(3t)\, dt = \dfrac{3}{2\pi}\left(2\pi^2\right) = 3\pi$. Hence $(\bar{x}, \bar{y}, \bar{z}) = (0, 0, 3\pi)$.

29. *Note:* The statement of the exercise in the first printing of the text should have referred to the wire in Example 3.

From Example 3, $\rho\,(x,y) = k\,(1-y)$, $x = \cos t$, $y = \sin t$, and $ds = dt$, $0 \le t \le \pi$ $\Rightarrow$

$\quad I_x = \int_C y^2 \rho\,(x,y)\, ds = \int_0^\pi \sin^2 t\,[k\,(1-\sin t)]\, dt = k\int_0^\pi \left(\sin^2 t - \sin^3 t\right) dt$

$\quad\quad = \frac{1}{2}k\int_0^\pi (1-\cos 2t)\, dt - k\int_0^\pi \left(1-\cos^2 t\right)\sin t\, dt \quad$ (Let $u = \cos t$, $du = -\sin t\, dt$
$\quad\quad\quad\quad\quad\quad$ in the second integral)

$\quad\quad = k\left[\frac{\pi}{2} + \int_1^{-1}\left(1-u^2\right) du\right] = k\left(\frac{\pi}{2} - \frac{4}{3}\right)$

$\quad I_y = \int_C x^2 \rho\,(x,y)\, ds = k\int_0^\pi \cos^2 t\,(1-\sin t)\, dt = \frac{k}{2}\int_0^\pi (1+\cos 2t)\, dt - k\int_0^\pi \cos^2 t\sin t\, dt$

$\quad\quad = k\left(\frac{\pi}{2} - \frac{2}{3}\right)$, using the same substitution as above.

31. $W = \int_C \mathbf{F}\cdot d\mathbf{r} = \int_0^{2\pi}\langle t-\sin t, 3-\cos t\rangle\cdot\langle 1-\cos t, \sin t\rangle\, dt$

$\quad = \int_0^{2\pi}(t - t\cos t - \sin t + \sin t\cos t + 3\sin t - \sin t\cos t)\, dt = 2\pi^2$

33. $W = \int_C \mathbf{F}\cdot d\mathbf{r} = \int_0^1 \langle t^6, -t^5, -t^7\rangle\cdot\langle 2t, -3t^2, 4t^3\rangle\, dt = \int_0^1 \left(5t^7 - 4t^{10}\right) dt = \frac{5}{8} - \frac{4}{11} = \frac{23}{88}$

35. Let $\mathbf{F} = 185\,\mathbf{k}$. To parametrize the staircase, let

$\quad x = 20\cos t, y = 20\sin t, z = \frac{90}{6\pi}t = \frac{15}{\pi}t, 0 \le t \le 6\pi \Rightarrow$

$\quad W = \int_C \mathbf{F}\cdot d\mathbf{r} = \int_0^{6\pi}\langle 0, 0, 185\rangle\cdot\langle -20\sin t, 20\cos t, \frac{15}{\pi}\rangle\, dt = (185)\frac{15}{\pi}\int_0^{6\pi} dt = (185)\,(90)$

$\quad\quad \approx 1.67 \times 10^4$ ft-lb

37. The work done in moving the object is $\int_C \mathbf{F} \cdot d\mathbf{r} = \int_C \mathbf{F} \cdot \mathbf{T}\, ds$. We can approximate this integral by dividing C into 7 segments of equal length $\Delta s = 2$ and approximating $\mathbf{F} \cdot \mathbf{T}$, that is, the tangential component of force, at a point (x_i^*, y_i^*) on each segment. Since C is composed of straight line segments, $\mathbf{F} \cdot \mathbf{T}$ is the scalar projection of each force vector onto C. If we choose (x_i^*, y_i^*) to be the point on the segment closest to the origin, then the work done is

$$\int_C \mathbf{F} \cdot \mathbf{T}\, ds \approx \sum_{i=1}^{7} [\mathbf{F}(x_i^*, y_i^*) \cdot \mathbf{T}(x_i^*, y_i^*)]\, \Delta s$$

$$= [2 + 2 + 2 + 2 + 1 + 1 + 1]\,(2) = 22$$

Thus, we estimate the work done to be approximately 22 J.

Section 13.3 The Fundamental Theorem for Line Integrals

1. C appears to be a smooth curve, and since ∇f is continuous, we know f is differentiable. Then Theorem 2 says that the value of $\int_C \nabla f \cdot d\mathbf{r}$ is simply the difference of the values of f at the terminal and initial points of C. From the graph, this is $50 - 10 = 40$.

3. $\partial (2x - 3y) / \partial y = -3 = \partial (2y - 3x) / \partial x$ and the domain of $\mathbf{F}$ is $\mathbb{R}^2$ which is open and simply-connected, so $\mathbf{F}$ is conservative. Thus there exists a function f such that $\nabla f = \mathbf{F}$, that is, $f_x (x, y) = 2x - 3y$ and $f_y (x, y) = 2y - 3x$. But $f_x (x, y) = 2x - 3y$ implies $f (x, y) = x^2 - 3yx + g (y)$ and differentiating both sides of this equation with respect to y gives $f_y (x, y) = -3x + g' (y)$. Thus $2y - 3x = -3x + g' (y)$ so $g' (y) = 2y$ and $g (y) = y^2 + K$ where K is a constant. Hence $f (x, y) = x^2 - 3xy + y^2 + K$ is a potential function for $\mathbf{F}$.

5. $\partial (x^2 + y) / \partial y = 1$, $\partial (x^2) / \partial x = 2x$ and these are not equal, so $\mathbf{F}$ is not conservative.

7. $\partial (1 + 4x^3 y^3) / \partial y = 12x^3 y^2 = \partial (3x^4 y^2) / \partial x$ and the domain of $\mathbf{F}$ is $\mathbb{R}^2$ which is open and simply-connected. Thus $\mathbf{F}$ is conservative so there exists a function f such that $\nabla f = \mathbf{F}$. Then $f_x (x, y) = 1 + 4x^3 y^3$ implies $f (x, y) = x + x^4 y^3 + g (y)$ and $f_y (x, y) = 3x^4 y^3 + g' (y)$. But $f_y (x, y) = 3x^4 y^2$ implies $g (y) = K$. Hence a potential function for $\mathbf{F}$ is $f (x, y) = x + x^4 y^3 + K$.

9. $\partial (ye^x + \sin y) / \partial y = e^x + \cos y = \partial (e^x + x \cos y) / \partial x$ and the domain of $\mathbf{F}$ is $\mathbb{R}^2$. Hence $\mathbf{F}$ is conservative so there exists a function f such that $\nabla f = \mathbf{F}$. Then $f_x (x, y) = ye^x + \sin y$ implies $f (x, y) = ye^x + x \sin y + g (y)$ and $f_y (x, y) = e^x + x \cos y + g' (y)$. But $f_y (x, y) = e^x + x \cos y$ so $g (y) = K$ and $f (x, y) = ye^x + x \sin y + K$ is a potential function for $\mathbf{F}$.

11. (a) $\mathbf{F}$ has continuous first-order partial derivatives and $\dfrac{\partial}{\partial y} (2xy) = 2x = \dfrac{\partial}{\partial x} (x^2)$ on $\mathbb{R}^2$, which is open and simply-connected. Thus, $\mathbf{F}$ is conservative by Theorem 6. Then we know that the line integral of $\mathbf{F}$ is independent of path; in particular, the value of $\int_C \mathbf{F} \cdot d\mathbf{r}$ depends only on the endpoints of C. Since all three curves have the same initial and terminal points, $\int_C \mathbf{F} \cdot d\mathbf{r}$ will have the same value for each curve.

(b) We first find a potential function f, so that $\nabla f = \mathbf{F}$. We know $f_x (x, y) = 2xy$ and $f_y (x, y) = x^2$. Integrating $f_x (x, y)$ with respect to x, we have $f (x, y) = x^2 y + g (y)$. Differentiating both sides with respect to y gives $f_y (x, y) = x^2 + g' (y)$, so we must have $x^2 + g' (y) = x^2 \Rightarrow g' (y) = 0 \Rightarrow g (y) = K$, a constant. Thus $f (x, y) = x^2 y + K$. All three curves start at $(1, 2)$ and end at $(3, 2)$, so by Theorem 2, $\int_C \mathbf{F} \cdot d\mathbf{r} = f (3, 2) - f (1, 2) = 18 - 2 = 16$ for each curve.

13. (a) $f_x (x, y) = 2xy^3$ implies $f (x, y) = x^2 y^3 + g (y)$ and $f_y (x, y) = 3x^2 y^2 + g' (y)$. But $f_y (x, y) = 3x^2 y^2$ so $f (x, y) = x^2 y^3$ (setting $K = 0$).

(b) Since $\mathbf{r} (0) = \langle 0, 1 \rangle$ and $\mathbf{r} \left(\frac{\pi}{2} \right) = \langle 1, \frac{1}{4} (\pi^2 + 4) \rangle$,
$\int_C \mathbf{F} \cdot d\mathbf{r} = f \left(1, \frac{1}{4} (\pi^2 + 4) \right) - f (0, 1) = \frac{1}{64} (\pi^2 + 4)^3$.

15. (a) $f_x(x,y,z) = y$ implies $f(x,y,z) = xy + g(y,z)$ and $f_y(x,y,z) = x + \partial g/\partial y$. But
$f_y(x,y,z) = x + z$ so $\partial g/\partial y = z$ and $g(y,z) = yz + h(z)$. Thus $f(x,y,z) = xy + yz + h(z)$
and $f_z(x,y,z) = y + h'(z)$. But $f_z(x,y,z) = y$ so $h'(z) = 0$ or $h(z) = K$. Hence
$f(x,y,z) = xy + yz$ (setting $K = 0$).

(b) $\int_C \mathbf{F} \cdot d\mathbf{r} = f(8,3,-1) - f(2,1,4) = 21 - 6 = 15$

17. (a) $f_x(x,y,z) = 2xz + \sin y$ implies $f(x,y,z) = x^2z + x\sin y + g(y,z)$ and
$f_y(x,y,z) = x\cos y + g_y(y,z)$. But $f_y(x,y,z) = x\cos y$ so $g_y(y,z) = 0$ and
$f(x,y,z) = x^2z + x\sin y + h(z)$. Thus $f_z(x,y,z) = x^2 + h'(z)$. But $f_z(x,y,z) = x^2$ so
$h'(z) = 0$ and $f(x,y,z) = x^2z + x\sin y$ (setting $K = 0$).

(b) $\mathbf{r}(0) = \langle 1,0,0 \rangle$, $\mathbf{r}(2\pi) = \langle 1,0,2\pi \rangle$. Thus $\int_C \mathbf{F} \cdot d\mathbf{r} = f(1,0,2\pi) - f(1,0,0) = 2\pi$.

19. Here $\mathbf{F}(x,y) = (2x\sin y)\,\mathbf{i} + (x^2\cos y - 3y^2)\,\mathbf{j}$. Then $f(x,y) = x^2\sin y - y^3$ is a potential function
for $\mathbf{F}$, that is, $\nabla f = \mathbf{F}$ so $\mathbf{F}$ is conservative and thus its line integral is independent of path. Hence
$\int_C 2x\sin y\,dx + (x^2\cos y - 3y^2)\,dy = \int_C \mathbf{F} \cdot d\mathbf{r} = f(5,1) - f(-1,0) = 25\sin 1 - 1$.

21. Here $\mathbf{F}(x,y) = x^2y^3\,\mathbf{i} + x^3y^2\,\mathbf{j}$. $W = \int_C \mathbf{F} \cdot d\mathbf{r}$. Since $\partial(x^2y^3)/\partial y = 3x^2y^2 = \partial(x^3y^2)/\partial x$, there
exists a function f such that $\nabla f = \mathbf{F}$. In fact, $f_x = x^2y^3 \Rightarrow f(x,y) = \frac{1}{3}x^3y^3 + g(y) \Rightarrow$
$f_y = x^3y^2 + g'(y) \Rightarrow g'(y) = 0$, so we can take $f(x,y) = \frac{1}{3}x^3y^3$. Thus
$W = \int_C \mathbf{F} \cdot d\mathbf{r} = f(2,1) - f(0,0) = \frac{1}{3}(2^3)(1^3) - 0 = \frac{8}{3}$.

23. We know that if the vector field (call it $\mathbf{F}$) is conservative, then around any closed path C, $\int_C \mathbf{F} \cdot d\mathbf{r} = 0$.
But take C to be some circle centered at the origin, oriented counterclockwise. All of the field vectors
along C oppose motion along C, so the integral around C will be negative. Therefore the field is not
conservative.

25.

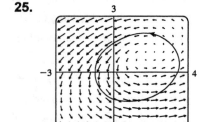

From the graph, it appears that $\mathbf{F}$ is not conservative. For example,
any closed curve containing the point $(2,1)$ seems to have many
field vectors pointing counterclockwise along it, and none pointing
clockwise. So along this path the integral $\int \mathbf{F} \cdot d\mathbf{r} \neq 0$. To confirm
our guess, we calculate

$$\frac{\partial}{\partial y}\left(\frac{x-2y}{\sqrt{1+x^2+y^2}}\right) = (x-2y)\left[\frac{-y}{(1+x^2+y^2)^{3/2}}\right] - \frac{2}{\sqrt{1+x^2+y^2}} = \frac{-2-2x^2-xy}{(1+x^2+y^2)^{3/2}},$$

$$\frac{\partial}{\partial x}\left(\frac{x-2}{\sqrt{1+x^2+y^2}}\right) = (x-2)\left[\frac{-x}{(1+x^2+y^2)^{3/2}}\right] + \frac{1}{\sqrt{1+x^2+y^2}} = \frac{1+y^2+2x}{(1+x^2+y^2)^{3/2}}.$$ These

are not equal, so the field is not conservative, by Theorem 5.

27. Since $\mathbf{F}$ is conservative, there exists a function f such that $\mathbf{F} = \nabla f$, that is, $P = f_x$, $Q = f_y$, and $R = f_z$. Since P, Q and R have continuous first order partial derivatives, Clairaut's Theorem says that $\partial P / \partial y = f_{xy} = f_{yx} = \partial Q / \partial x$, $\partial P / \partial z = f_{xz} = f_{zx} = \partial R / \partial x$, and $\partial Q / \partial z = f_{yz} = f_{zy} = \partial R / \partial y$.

29. $D = \{(x, y) \mid x > 0, y > 0\} =$ the first quadrant (excluding the axes).

(a) D is open because around every point in D we can put a disk that lies in D.

(b) D is connected because the straight line segment joining any two points in D lies in D.

(c) D is simply-connected because it's connected and has no holes.

31. $D = \{(x, y) \mid 1 < x^2 + y^2 < 4\} =$ the annular region between the circles with center $(0, 0)$ and radii 1 and 2.

(a) D is open.

(b) D is connected.

(c) D is not simply-connected. For example, $x^2 + y^2 = (1.5)^2$ is simple and closed and lies within D but encloses points that are not in D. (Or we can say, D has a hole, so is not simply-connected.)

33. (a) $P = -\dfrac{y}{x^2 + y^2}$, $\dfrac{\partial P}{\partial y} = \dfrac{y^2 - x^2}{(x^2 + y^2)^2}$ and $Q = \dfrac{x}{x^2 + y^2}$, $\dfrac{\partial Q}{\partial x} = \dfrac{y^2 - x^2}{(x^2 + y^2)^2}$. Thus $\dfrac{\partial P}{\partial y} = \dfrac{\partial Q}{\partial x}$.

(b) C_1: $x = \cos t$, $y = \sin t$, $0 \le t \le \pi$, C_2: $x = \cos t$, $y = \sin t$, $t = 2\pi$ to $t = \pi$. Then

$$\int_{C_1} \mathbf{F} \cdot d\mathbf{r} = \int_0^\pi \frac{(-\sin t)(-\sin t) + (\cos t)(\cos t)}{\cos^2 t + \sin^2 t}\, dt = \int_0^\pi dt = \pi \text{ and}$$

$\int_{C_2} \mathbf{F} \cdot d\mathbf{r} = \int_{2\pi}^\pi dt = -\pi$. Since these aren't equal, the line integral of $\mathbf{F}$ isn't independent of path. (Or notice that $\int_{C_3} \mathbf{F} \cdot d\mathbf{r} = \int_0^{2\pi} dt = 2\pi$ where C_3 is the circle $x^2 + y^2 = 1$, and apply the contrapositive of Theorem 3.) This doesn't contradict Theorem 6, since the domain of $\mathbf{F}$, which is $\mathbb{R}^2$ except the origin, isn't simply-connected.

Section 13.4 Green's Theorem

1. (a)

$$\oint_C x^2 y\,dx + xy^3\,dy = \oint_{C_1+C_2+C_3+C_4} x^2 y\,dx + xy^3\,dy$$
$$= \int_0^1 0\,dx + \int_0^1 y^3\,dy + \int_1^0 x^2\,dx + \int_1^0 0\,dy$$
$$= \tfrac{1}{4} - \tfrac{1}{3} = -\tfrac{1}{12}$$

(b) $\oint_C x^2 y\,dx + xy^3\,dy = \iint_D \left[\frac{\partial}{\partial x}\left(xy^3\right) - \frac{\partial}{\partial y}\left(x^2 y\right)\right] dA$
$$= \int_0^1 \int_0^1 \left(y^3 - x^2\right) dx\,dy = \int_0^1 \left(y^3 - \tfrac{1}{3}\right) dy = \tfrac{1}{4} - \tfrac{1}{3} = -\tfrac{1}{12}$$

3. We can parametrize C as $x = \cos\theta$, $y = \sin\theta$, $0 \le \theta \le 2\pi$. Then the line integral is

$\oint_C P\,dx + Q\,dy = \int_0^{2\pi} \cos^4\theta \sin^5\theta\,(-\sin\theta)\,d\theta + \int_0^{2\pi} \left(-\cos^7\theta \sin^6\theta\right)\cos\theta\,d\theta = -\frac{29\pi}{1024}$,
according to a CAS. The double integral is

$\iint_D \left(\frac{\partial Q}{\partial x} - \frac{\partial P}{\partial y}\right) dA = \int_{-1}^1 \int_{-\sqrt{1-x^2}}^{\sqrt{1-x^2}} \left(-7x^6 y^6 - 5x^4 y^4\right) dy\,dx = -\frac{29\pi}{1024}$, verifying Green's Theorem
in this case.

5. $\int_C xy\,dx + y^5\,dy = \iint_D \left[\frac{\partial}{\partial x}\left(y^5\right) - \frac{\partial}{\partial y}\left(xy\right)\right] dA = \int_0^2 \int_0^{x/2} (0 - x)\,dy\,dx = \int_0^2 \left(-\tfrac{1}{2}x^2\right) dx = -\tfrac{4}{3}$

7. $\int_C \left(y + e^{\sqrt{x}}\right) dx + \left(2x + \cos y^2\right) dy = \iint_D \left[\frac{\partial}{\partial x}\left(2x + \cos y^2\right) - \frac{\partial}{\partial y}\left(y + e^{\sqrt{x}}\right)\right] dA$
$$= \int_0^1 \int_{y^2}^{\sqrt{y}} (2 - 1)\,dx\,dy = \int_0^1 \left(y^{1/2} - y^2\right) dy = \tfrac{1}{3}$$

9. $\int_C x^2\,dx + y^2\,dy = \iint_D \left[\frac{\partial}{\partial x}\left(y^2\right) - \frac{\partial}{\partial y}\left(x^2\right)\right] dA = \iint_D (0 - 0)\,dA = 0$

11. $\int_C xy\,dx + 2x^2\,dy = \iint_{0 \le x^2 + y^2 \le 4,\,y \ge 0} \left[\frac{\partial}{\partial x}\left(2x^2\right) - \frac{\partial}{\partial y}\left(xy\right)\right] dA$
$$= \iint_{0 \le x^2 + y^2 \le 4,\,y \ge 0} (4x - x)\,dA$$
$$= 3\int_0^\pi \int_0^2 r^2 \cos\theta\,dr\,d\theta = 0 \text{ since } \int_0^\pi \cos\theta\,d\theta = 0 \text{ or } \iint_D 3x\,dA = 3M_y = 0.$$

13. $\int_C \mathbf{F} \cdot d\mathbf{r} = \int_C \left(y^2 - x^2 y\right) dx + xy^2\,dy = \iint_{x^2+y^2 \le 4,\,0 \le y \le x} \left(y^2 - 2y + x^2\right) dA$
$$= \int_0^{\pi/4} \int_0^2 \left(r^2 - 2r\sin\theta\right) r\,dr\,d\theta = \int_0^{\pi/4} \left[4 - \tfrac{16}{3}\sin\theta\right] d\theta$$
$$= \left[4\theta + \tfrac{16}{3}\cos\theta\right]_0^{\pi/4} = \pi + \tfrac{8}{3}\left(\sqrt{2} - 2\right)$$

15. By Green's Theorem, $W = \int_C \mathbf{F} \cdot d\mathbf{r} = \int_C x\,(x + y)\,dx + xy^2\,dy = \iint_D \left(y^2 - x\right) dy\,dx$ where C is
the path described in the question and D is the triangle bounded by C. So

$$W = \int_0^1 \int_0^{1-x} \left(y^2 - x\right) dy\,dx = \int_0^1 \left[\tfrac{1}{3}y^3 - xy\right]_{y=0}^{y=1-x} dx = \int_0^1 \left(\tfrac{1}{3}(1-x)^3 - x\,(1-x)\right) dx$$

$$= \left[-\tfrac{1}{12}(1-x)^4 - \tfrac{1}{2}x^2 + \tfrac{1}{3}x^3\right]_0^1 = \left(-\tfrac{1}{2} + \tfrac{1}{3}\right) - \left(-\tfrac{1}{12}\right) = -\tfrac{1}{12}$$

17. $A = \oint_C x\, dy = \int_0^{2\pi} (\cos^3 t)(3\sin^2 t \cos t)\, dt = 3\int_0^{2\pi} (\cos^4 t \sin^2 t)\, dt$

$= 3\left[-\frac{1}{6}(\sin t \cos^5 t) + \frac{1}{6}\left[\frac{1}{4}(\sin t \cos^3 t) + \frac{3}{8}(\cos t \sin t) + \frac{3}{8}t\right]\right]_0^{2\pi} = 3\left(\frac{1}{6}\right)\left(\frac{6}{8}\pi\right) = \frac{3}{8}\pi$

Or: $3\int_0^{2\pi}(\cos^4 t \sin^2 t)\, dt = 3\int_0^{2\pi}\frac{1}{8}\left[\frac{1}{2}(1 - \cos 4t) + \sin^2 2t \cos 2t\right] dt = \frac{3}{8}\pi$

19. (a) Using Equation 13.2.8, we write parametric equations of the line segment as $x = (1 - t)x_1 + tx_2$, $y = (1 - t)y_1 + ty_2, 0 \le t \le 1$. Then $dx = (x_2 - x_1)\, dt$ and $dy = (y_2 - y_1)\, dt$, so

$\int_C x\, dy - y\, dx = \int_0^1 [(1 - t)x_1 + tx_2](y_2 - y_1)\, dt + [(1 - t)y_1 + ty_2](x_2 - x_1)\, dt$

$= \int_0^1 (x_1(y_2 - y_1) - y_1(x_2 - x_1) + t[(y_2 - y_1)(x_2 - x_1) - (x_2 - x_1)(y_2 - y_1)])\, dt$

$= \int_0^1 (x_1 y_2 - x_2 y_1)\, dt = x_1 y_2 - x_2 y_1$

(b) We apply Green's Theorem to the path $C = C_1 \cup C_2 \cup \cdots \cup C_n$, where C_i is the line segment that joins (x_i, y_i) to (x_{i+1}, y_{i+1}) for $i = 1, 2, \ldots, n - 1$, and C_n is the line segment that joins (x_n, y_n) to (x_1, y_1). From (5), $\frac{1}{2}\int_C x\, dy - y\, dx = \iint_D dA$, where D is the polygon bounded by C. Therefore

area of polygon $= A(D) = \iint_D dA = \frac{1}{2}\int_C x\, dy - y\, dx$

$= \frac{1}{2}\left(\int_{C_1} x\, dy - y\, dx + \int_{C_2} x\, dy - y\, dx + \cdots\right.$

$\left. + \int_{C_{n-1}} x\, dy - y\, dx + \int_{C_n} x\, dy - y\, dx\right)$

To evaluate these integrals we use the formula from (a) to get

$A(D) = \frac{1}{2}[(x_1 y_2 - x_2 y_1) + (x_2 y_3 - x_3 y_2) + \cdots + (x_{n-1} y_n - x_n y_{n-1}) + (x_n y_1 - x_1 y_n)]$.

(c) $A = \frac{1}{2}[(0 \cdot 1 - 2 \cdot 0) + (2 \cdot 3 - 1 \cdot 1) + (1 \cdot 2 - 0 \cdot 3) + (0 \cdot 1 - (-1) \cdot 2) + (-1 \cdot 0 - 0 \cdot 1)]$

$= \frac{1}{2}(0 + 5 + 2 + 2) = \frac{9}{2}$

21. Here $A = \frac{1}{2}(1)(1) = \frac{1}{2}$ and $C = C_1 + C_2 + C_3$, where $C_1: x = x, y = 0, 0 \le x \le 1$; C_2: $x = x, y = 1 - x, x = 1$ to $x = 0$; and C_3: $x = 0, y = 1$ to $y = 0$. Then

$\bar{x} = \frac{1}{2A}\int_C x^2\, dy = \int_{C_1} x^2\, dy + \int_{C_2} x^2\, dy + \int_{C_3} x^2\, dy = 0 + \int_1^0 (x^2)(-dx) + 0 = \frac{1}{3}$. Similarly,

$\bar{y} = -\frac{1}{2A}\int_C y^2\, dx = \int_{C_1} y^2\, dx + \int_{C_2} y^2\, dx + \int_{C_3} y^2\, dx = 0 + \int_1^0 (1 - x)^2(-dx) + 0 = \frac{1}{3}$.

Therefore $(\bar{x}, \bar{y}) = \left(\frac{1}{3}, \frac{1}{3}\right)$.

23. By Green's Theorem, $-\frac{1}{3}\rho\oint_C y^3\, dx = -\frac{1}{3}\rho\iint_D (-3y^2)\, dA = \iint_D y^2\rho\, dA = I_x$ and $\frac{1}{3}\rho\oint_C x^3\, dy = \frac{1}{3}\rho\iint_D (3x^2)\, dA = \iint_D x^2\rho\, dA = I_y$.

25. Since C is a simple closed path which doesn't pass through or enclose the origin, there exists an open region that doesn't contain the origin but does contain D. Thus $P = -y/(x^2 + y^2)$ and $Q = x/(x^2 + y^2)$ have continuous partial derivatives on this open region containing D and we can apply Green's Theorem. But by Exercise 13.3.33(a), $\partial P/\partial y = \partial Q/\partial x$, so $\oint_C \mathbf{F} \cdot d\mathbf{r} = \iint_D 0\, dA = 0$.

27. Using the first part of (5), we have that $\iint_R dx\,dy = A(R) = \int_{\partial R} x\,dy$. But $x = g(u,v)$, and

$dy = \dfrac{\partial h}{\partial u}\,du + \dfrac{\partial h}{\partial v}\,dv$, and we orient ∂S by taking the positive direction to be that which corresponds, under the mapping, to the positive direction along ∂R, so

$$\int_{\partial R} x\,dy = \int_{\partial S} g(u,v)\left(\frac{\partial h}{\partial u}\,du + \frac{\partial h}{\partial v}\,dv\right) = \int_{\partial S} g(u,v)\frac{\partial h}{\partial u}\,du + g(u,v)\frac{\partial h}{\partial v}\,dv$$

$$= \pm\iint_S \left[\frac{\partial}{\partial u}\left(g(u,v)\frac{\partial h}{\partial v}\right) - \frac{\partial}{\partial v}\left(g(u,v)\frac{\partial h}{\partial u}\right)\right] dA \text{ (using Green's Theorem in the } uv\text{-plane)}$$

$$= \pm\iint_S \left(\frac{\partial g}{\partial u}\frac{\partial h}{\partial v} + g(u,v)\frac{\partial^2 h}{\partial u\partial v} - \frac{\partial g}{\partial v}\frac{\partial h}{\partial u} - g(u,v)\frac{\partial^2 h}{\partial v\partial u}\right) dA \text{ (using the Chain Rule)}$$

$$= \pm\iint_S \left(\frac{\partial x}{\partial u}\frac{\partial y}{\partial v} - \frac{\partial x}{\partial v}\frac{\partial y}{\partial u}\right) dA \text{ (by the equality of mixed partials) } = \pm\iint_S \frac{\partial(x,y)}{\partial(u,v)}\,du\,dv$$

The sign is chosen to be positive if the orientation that we gave to ∂S corresponds to the usual positive orientation, and it is negative otherwise. In either case, since $A(R)$ is positive, the sign chosen must be the same as the sign of $\dfrac{\partial(x,y)}{\partial(u,v)}$. Therefore $A(R) = \iint_R dx\,dy = \iint_S \left|\dfrac{\partial(x,y)}{\partial(u,v)}\right| du\,dv$.

Section 13.5 Curl and Divergence

1. (a) $\operatorname{curl}\mathbf{F} = \nabla \times \mathbf{F} = \begin{vmatrix} \mathbf{i} & \mathbf{j} & \mathbf{k} \\ \partial/\partial x & \partial/\partial y & \partial/\partial z \\ yz & xz & xy \end{vmatrix} = (x - x)\,\mathbf{i} + (y - y)\,\mathbf{j} + (z - z)\,\mathbf{k} = \mathbf{0}$

(b) $\operatorname{div}\mathbf{F} = \nabla \cdot \mathbf{F} = \dfrac{\partial}{\partial x}(yz) + \dfrac{\partial}{\partial y}(xz) + \dfrac{\partial}{\partial z}(xy) = 0 + 0 + 0 = 0$

3. (a) $\operatorname{curl}\mathbf{F} = \nabla \times \mathbf{F} = \begin{vmatrix} \mathbf{i} & \mathbf{j} & \mathbf{k} \\ \partial/\partial x & \partial/\partial y & \partial/\partial z \\ 0 & xy & xyz \end{vmatrix} = xz\,\mathbf{i} - yz\,\mathbf{j} + y\,\mathbf{k}$

(b) $\operatorname{div}\mathbf{F} = \nabla \cdot \mathbf{F} = \dfrac{\partial}{\partial x}(0) + \dfrac{\partial}{\partial y}(xy) + \dfrac{\partial}{\partial z}(xyz) = 0 + x + xy = x\,(1 + y)$

5. (a) $\nabla \times \mathbf{F} = \begin{vmatrix} \mathbf{i} & \mathbf{j} & \mathbf{k} \\ \partial/\partial x & \partial/\partial y & \partial/\partial z \\ e^{xz} & -2e^{yz} & 3xe^{y} \end{vmatrix} = (3xe^{y} + 2ye^{yz})\,\mathbf{i} + (xe^{xz} - 3e^{y})\,\mathbf{j}$

(b) $\nabla \cdot \mathbf{F} = \dfrac{\partial}{\partial x}(e^{xz}) + \dfrac{\partial}{\partial y}(-2e^{yz}) + \dfrac{\partial}{\partial z}(3xe^{y}) = ze^{xz} - 2ze^{yz}$

7. If the vector field is $\mathbf{F} = P\mathbf{i} + Q\mathbf{j} + R\mathbf{k}$, then we know $R = 0$. In addition, the x-component of each vector of $\mathbf{F}$ is 0, so $P = 0$, hence $\dfrac{\partial P}{\partial x} = \dfrac{\partial P}{\partial y} = \dfrac{\partial P}{\partial z} = \dfrac{\partial R}{\partial x} = \dfrac{\partial R}{\partial y} = \dfrac{\partial R}{\partial z} = 0$. Q decreases as y

increases, so $\dfrac{\partial Q}{\partial y} < 0$, but Q doesn't change in the x- or z-directions, so $\dfrac{\partial Q}{\partial x} = \dfrac{\partial Q}{\partial z} = 0$.

(a) $\operatorname{div}\mathbf{F} = \dfrac{\partial P}{\partial x} + \dfrac{\partial Q}{\partial y} + \dfrac{\partial R}{\partial z} = 0 + \dfrac{\partial Q}{\partial y} + 0 < 0$

(b) $\operatorname{curl}\mathbf{F} = \left(\dfrac{\partial R}{\partial y} - \dfrac{\partial Q}{\partial z}\right)\mathbf{i} + \left(\dfrac{\partial P}{\partial z} - \dfrac{\partial R}{\partial x}\right)\mathbf{j} + \left(\dfrac{\partial Q}{\partial x} - \dfrac{\partial P}{\partial y}\right)\mathbf{k}$

$= (0 - 0)\,\mathbf{i} + (0 - 0)\,\mathbf{j} + (0 - 0)\,\mathbf{k} = \mathbf{0}$

9. If the vector field is $\mathbf{F} = P\mathbf{i} + Q\mathbf{j} + R\mathbf{k}$, then we know $R = 0$. In addition, the y-component of each vector of $\mathbf{F}$ is 0, so $Q = 0$, hence $\dfrac{\partial Q}{\partial x} = \dfrac{\partial Q}{\partial y} = \dfrac{\partial Q}{\partial z} = \dfrac{\partial R}{\partial x} = \dfrac{\partial R}{\partial y} = \dfrac{\partial R}{\partial z} = 0$. P increases as y

increases, so $\dfrac{\partial P}{\partial y} > 0$, but P doesn't change in the x- or z-directions, so $\dfrac{\partial P}{\partial x} = \dfrac{\partial P}{\partial z} = 0$.

(a) $\operatorname{div}\mathbf{F} = \dfrac{\partial P}{\partial x} + \dfrac{\partial Q}{\partial y} + \dfrac{\partial R}{\partial z} = 0 + 0 + 0 = 0$

(b) $\operatorname{curl}\mathbf{F} = \left(\dfrac{\partial R}{\partial y} - \dfrac{\partial Q}{\partial z}\right)\mathbf{i} + \left(\dfrac{\partial P}{\partial z} - \dfrac{\partial R}{\partial x}\right)\mathbf{j} + \left(\dfrac{\partial Q}{\partial x} - \dfrac{\partial P}{\partial y}\right)\mathbf{k}$

$= (0 - 0)\,\mathbf{i} + (0 - 0)\,\mathbf{j} + \left(0 - \dfrac{\partial P}{\partial y}\right)\mathbf{k} = -\dfrac{\partial P}{\partial y}\mathbf{k}$

Since $\dfrac{\partial P}{\partial y} > 0$, $-\dfrac{\partial P}{\partial y}\mathbf{k}$ is a vector pointing in the negative z-direction.

11. $\text{curl}\,\mathbf{F} = \begin{vmatrix} \mathbf{i} & \mathbf{j} & \mathbf{k} \\ \partial/\partial x & \partial/\partial y & \partial/\partial z \\ y & x & 1 \end{vmatrix} = 0$ and $\mathbf{F}$ is defined on all of $\mathbb{R}^3$ with component functions which have

continuous partial derivatives, so by (4), $\mathbf{F}$ is conservative. Thus there exists f such that $\mathbf{F} = \nabla f$. Then
$f_x\,(x, y, z) = y$ implies $f\,(x, y, z) = xy + g\,(y, z)$ and $f_y\,(x, y, z) = x + g_y\,(y, z)$. But $f_y\,(x, y, z) = x$,
so $g\,(y, z) = h\,(z)$ and $f\,(x, y, z) = xy + h\,(z)$. Thus $f_z\,(x, y, z) = h'\,(z)$ but $f_z\,(x, y, z) = 1$ so
$h\,(z) = z + k$. Hence a potential function for $\mathbf{F}$ is $f\,(x, y, z) = xy + z + k$.

13. $\text{curl}\,\mathbf{F} = \begin{vmatrix} \mathbf{i} & \mathbf{j} & \mathbf{k} \\ \partial/\partial x & \partial/\partial y & \partial/\partial z \\ yz & -z^2 & x^2 \end{vmatrix} = 2z\,\mathbf{i} + (y - 2x)\,\mathbf{j} - z\,\mathbf{k} \neq 0$. Hence $\mathbf{F}$ isn't conservative.

15. Since $\text{curl}\,\mathbf{F} = \begin{vmatrix} \mathbf{i} & \mathbf{j} & \mathbf{k} \\ \partial/\partial x & \partial/\partial y & \partial/\partial z \\ yz & y^2 + xz & xy \end{vmatrix} = (x - x)\,\mathbf{i} + (y - y)\,\mathbf{j} + (z - z)\,\mathbf{k} = 0$, $\mathbf{F}$ is defined on $\mathbb{R}^3$,

and since the partial derivatives of the components of $\mathbf{F}$ are continuous, $\mathbf{F}$ is conservative. Thus there
exists f such that $\nabla f = \mathbf{F}$. Then $f_x\,(x, y, z) = yz$ implies $f\,(x, y, z) = xyz + g\,(y, z)$ and
$f_y\,(x, y, z) = xz + g_y\,(y, z)$. But $f_y\,(x, y, z) = xz + y^2$ so $g\,(y, z) = \frac{1}{3}y^3 + h\,(z)$ and
$f\,(x, y, z) = xyz + \frac{1}{3}y^3 + h\,(z)$. Then $f_z\,(x, y, z) = xy + h'\,(z)$. But $f_z\,(x, y, z) = xy$ so $h\,(z) = k$.
Hence $f\,(x, y, z) = xyz + \frac{1}{3}y^3 + k$ is a potential function for $\mathbf{F}$.

17. No. Assume there is such a $\mathbf{G}$. Then $\text{div}\,(\text{curl}\,\mathbf{G}) = y^2 + z^2 + x^2 \neq 0$, which contradicts Theorem 11.

19. $\text{curl}\,\mathbf{F} = \begin{vmatrix} \mathbf{i} & \mathbf{j} & \mathbf{k} \\ \partial/\partial x & \partial/\partial y & \partial/\partial z \\ f\,(x) & g\,(y) & h\,(z) \end{vmatrix} = (0 - 0)\,\mathbf{i} + (0 - 0)\,\mathbf{j} + (0 - 0)\,\mathbf{k} = 0$.

Hence $\mathbf{F} = f\,(x)\,\mathbf{i} + g\,(y)\,\mathbf{j} + h\,(z)\,\mathbf{k}$ is irrotational.

For **Exercises 21–27**, let $\mathbf{F}\,(x, y, z) = P_1\,\mathbf{i} + Q_1\,\mathbf{j} + R_1\,\mathbf{k}$ and $\mathbf{G}\,(x, y, z) = P_2\,\mathbf{i} + Q_2\,\mathbf{j} + R_2\,\mathbf{k}$.

21. $\text{div}\,(\mathbf{F} + \mathbf{G}) = \dfrac{\partial\,(P_1 + P_2)}{\partial x} + \dfrac{\partial\,(Q_1 + Q_2)}{\partial y} + \dfrac{\partial\,(R_1 + R_2)}{\partial z}$

$= \left(\dfrac{\partial P_1}{\partial x} + \dfrac{\partial Q_1}{\partial y} + \dfrac{\partial R_1}{\partial z} \right) + \left(\dfrac{\partial P_2}{\partial x} + \dfrac{\partial Q_2}{\partial y} + \dfrac{\partial R_3}{\partial z} \right) = \text{div}\,\mathbf{F} + \text{div}\,\mathbf{G}$

23. $\text{div}\,(f\mathbf{F}) = \dfrac{\partial\,(fP_1)}{\partial x} + \dfrac{\partial\,(fQ_1)}{\partial y} + \dfrac{\partial\,(fR_1)}{\partial z}$

$= \left(f\dfrac{\partial P_1}{\partial x} + P_1\dfrac{\partial f}{\partial x} \right) + \left(f\dfrac{\partial Q_1}{\partial y} + Q_1\dfrac{\partial f}{\partial y} \right) + \left(f\dfrac{\partial R_1}{\partial z} + R_1\dfrac{\partial f}{\partial z} \right)$

$= f\left(\dfrac{\partial P_1}{\partial x} + \dfrac{\partial Q_1}{\partial y} + \dfrac{\partial R_1}{\partial z} \right) + \langle P_1, Q_1, R_1 \rangle \cdot \left\langle \dfrac{\partial f}{\partial x}, \dfrac{\partial f}{\partial y}, \dfrac{\partial f}{\partial z} \right\rangle = f\,\text{div}\,\mathbf{F} + \mathbf{F} \cdot \nabla f$

25. $\operatorname{div}(\mathbf{F} \times \mathbf{G}) = \nabla \cdot (\mathbf{F} \times \mathbf{G}) = \begin{vmatrix} \partial/\partial x & \partial/\partial y & \partial/\partial z \\ P_1 & Q_1 & R_1 \\ P_2 & Q_2 & R_2 \end{vmatrix} = \dfrac{\partial}{\partial x} \begin{vmatrix} Q_1 & R_1 \\ Q_2 & R_2 \end{vmatrix} - \dfrac{\partial}{\partial y} \begin{vmatrix} P_1 & R_1 \\ P_2 & R_2 \end{vmatrix} + \dfrac{\partial}{\partial z} \begin{vmatrix} P_1 & Q_1 \\ P_2 & Q_2 \end{vmatrix}$

$= \left[Q_1 \dfrac{\partial R_2}{\partial x} + R_2 \dfrac{\partial Q_1}{\partial x} - Q_2 \dfrac{\partial R_1}{\partial x} - R_1 \dfrac{\partial Q_2}{\partial x} \right]$

$- \left[P_1 \dfrac{\partial R_2}{\partial y} + R_2 \dfrac{\partial P_1}{\partial y} - P_2 \dfrac{\partial R_1}{\partial y} - R_1 \dfrac{\partial P_2}{\partial y} \right]$

$+ \left[P_1 \dfrac{\partial Q_2}{\partial z} + Q_2 \dfrac{\partial P_1}{\partial z} - P_2 \dfrac{\partial Q_1}{\partial z} - Q_1 \dfrac{\partial P_2}{\partial z} \right]$

$= \left[P_2 \left(\dfrac{\partial R_1}{\partial y} - \dfrac{\partial Q_1}{\partial z} \right) + Q_2 \left(\dfrac{\partial P_1}{\partial z} - \dfrac{\partial R_1}{\partial x} \right) + R_2 \left(\dfrac{\partial Q_1}{\partial x} - \dfrac{\partial P_1}{\partial y} \right) \right]$

$- \left[P_1 \left(\dfrac{\partial R_2}{\partial y} - \dfrac{\partial Q_2}{\partial z} \right) + Q_1 \left(\dfrac{\partial P_2}{\partial z} - \dfrac{\partial R_2}{\partial x} \right) + R_1 \left(\dfrac{\partial Q_2}{\partial x} - \dfrac{\partial P_2}{\partial y} \right) \right]$

$= \mathbf{G} \cdot \operatorname{curl} \mathbf{F} - \mathbf{F} \cdot \operatorname{curl} \mathbf{G}$

27. $\operatorname{curl}\operatorname{curl} \mathbf{F} = \nabla \times (\nabla \times \mathbf{F}) = \begin{vmatrix} \mathbf{i} & \mathbf{j} & \mathbf{k} \\ \partial/\partial x & \partial/\partial y & \partial/\partial z \\ \partial R_1/\partial y - \partial Q_1/\partial z & \partial P_1/\partial z - \partial R_1/\partial x & \partial Q_1/\partial x - \partial P_1/\partial y \end{vmatrix}$

$= \left(\dfrac{\partial^2 Q_1}{\partial y \partial x} - \dfrac{\partial^2 P_1}{\partial y^2} - \dfrac{\partial^2 P_1}{\partial z^2} + \dfrac{\partial^2 R_1}{\partial z \partial x} \right) \mathbf{i} + \left(\dfrac{\partial^2 R_1}{\partial z \partial y} - \dfrac{\partial^2 Q_1}{\partial z^2} - \dfrac{\partial^2 Q_1}{\partial x^2} + \dfrac{\partial^2 P_1}{\partial x \partial y} \right) \mathbf{j}$

$+ \left(\dfrac{\partial^2 P_1}{\partial x \partial z} - \dfrac{\partial^2 R_1}{\partial x^2} - \dfrac{\partial^2 R_1}{\partial y^2} + \dfrac{\partial^2 Q_1}{\partial y \partial z} \right) \mathbf{k}$

Now let's consider grad div $\mathbf{F} - \nabla^2 \mathbf{F}$ and compare with the above.

(Note that $\nabla^2 \mathbf{F}$ is defined on page 958.)

$\operatorname{grad} \operatorname{div} \mathbf{F} - \nabla^2 \mathbf{F} = \left[\left(\dfrac{\partial^2 P_1}{\partial x^2} + \dfrac{\partial^2 Q_1}{\partial x \partial y} + \dfrac{\partial^2 R_1}{\partial x \partial z} \right) \mathbf{i} + \left(\dfrac{\partial^2 P_1}{\partial y \partial x} + \dfrac{\partial^2 Q_1}{\partial y^2} + \dfrac{\partial^2 R_1}{\partial y \partial z} \right) \mathbf{j} \right.$

$\left. + \left(\dfrac{\partial^2 P_1}{\partial z \partial x} + \dfrac{\partial^2 Q_1}{\partial z \partial y} + \dfrac{\partial^2 R_1}{\partial z^2} \right) \mathbf{k} \right]$

$- \left[\left(\dfrac{\partial^2 P_1}{\partial x^2} + \dfrac{\partial^2 P_1}{\partial y^2} + \dfrac{\partial^2 P_1}{\partial z^2} \right) \mathbf{i} + \left(\dfrac{\partial^2 Q_1}{\partial x^2} + \dfrac{\partial^2 Q_1}{\partial y^2} + \dfrac{\partial^2 Q_1}{\partial z^2} \right) \mathbf{j} \right.$

$\left. + \left(\dfrac{\partial^2 R_1}{\partial x^2} + \dfrac{\partial^2 R_1}{\partial y^2} + \dfrac{\partial^2 R_1}{\partial z^2} \right) \mathbf{k} \right]$

$= \left(\dfrac{\partial^2 Q_1}{\partial x \partial y} + \dfrac{\partial^2 R_1}{\partial x \partial z} - \dfrac{\partial^2 P_1}{\partial y^2} - \dfrac{\partial^2 P_1}{\partial z^2} \right) \mathbf{i} + \left(\dfrac{\partial^2 P_1}{\partial y \partial x} + \dfrac{\partial^2 R_1}{\partial y \partial z} - \dfrac{\partial^2 Q_1}{\partial x^2} - \dfrac{\partial^2 Q_1}{\partial z^2} \right) \mathbf{j}$

$+ \left(\dfrac{\partial^2 P_1}{\partial z \partial x} + \dfrac{\partial^2 Q_1}{\partial z \partial y} - \dfrac{\partial^2 R_1}{\partial x^2} - \dfrac{\partial^2 R_2}{\partial y^2} \right) \mathbf{k}$

Then applying Clairaut's Theorem to reverse the order of differentiation in the second partial derivatives as needed and comparing, we have $\operatorname{curl} \operatorname{curl} \mathbf{F} = \operatorname{grad} \operatorname{div} \mathbf{F} - \nabla^2 \mathbf{F}$ as desired.

29. (a) $\nabla r = \nabla \sqrt{x^2 + y^2 + z^2} = \dfrac{x}{\sqrt{x^2 + y^2 + z^2}}\,\mathbf{i} + \dfrac{y}{\sqrt{x^2 + y^2 + z^2}}\,\mathbf{j} + \dfrac{z}{\sqrt{x^2 + y^2 + z^2}}\,\mathbf{k}$

$$= \dfrac{x\,\mathbf{i} + y\,\mathbf{j} + z\,\mathbf{k}}{\sqrt{x^2 + y^2 + z^2}} = \dfrac{\mathbf{r}}{r}$$

(b) $\nabla \times \mathbf{r} = \begin{vmatrix} \mathbf{i} & \mathbf{j} & \mathbf{k} \\ \dfrac{\partial}{\partial x} & \dfrac{\partial}{\partial y} & \dfrac{\partial}{\partial z} \\ x & y & z \end{vmatrix}$

$$= \left[\dfrac{\partial}{\partial y}(z) - \dfrac{\partial}{\partial z}(y) \right]\mathbf{i} + \left[\dfrac{\partial}{\partial z}(x) - \dfrac{\partial}{\partial x}(z) \right]\mathbf{j} + \left[\dfrac{\partial}{\partial x}(y) - \dfrac{\partial}{\partial y}(x) \right]\mathbf{k} = 0$$

(c) $\nabla \left(\dfrac{1}{r} \right) = \nabla \left(\dfrac{1}{\sqrt{x^2 + y^2 + z^2}} \right)$

$$= \dfrac{-\dfrac{1}{2\sqrt{x^2+y^2+z^2}}(2x)}{x^2+y^2+z^2}\,\mathbf{i} - \dfrac{\dfrac{1}{2\sqrt{x^2+y^2+z^2}}(2y)}{x^2+y^2+z^2}\,\mathbf{j} - \dfrac{\dfrac{1}{2\sqrt{x^2+y^2+z^2}}(2z)}{x^2+y^2+z^2}\,\mathbf{k}$$

$$= -\dfrac{x\,\mathbf{i} + y\,\mathbf{j} + z\,\mathbf{k}}{(x^2+y^2+z^2)^{3/2}} = -\dfrac{\mathbf{r}}{r^3}$$

(d) $\nabla \ln r = \nabla \ln \left(x^2 + y^2 + z^2 \right)^{1/2} = \tfrac{1}{2} \nabla \ln \left(x^2 + y^2 + z^2 \right)$

$$= \dfrac{x}{x^2+y^2+z^2}\,\mathbf{i} + \dfrac{y}{x^2+y^2+z^2}\,\mathbf{j} + \dfrac{z}{x^2+y^2+z^2}\,\mathbf{k} = \dfrac{x\,\mathbf{i} + y\,\mathbf{j} + z\,\mathbf{k}}{x^2+y^2+z^2} = \dfrac{\mathbf{r}}{r^2}$$

31. By (13), $\oint_C f\,(\nabla g) \cdot \mathbf{n}\,ds = \iint_D \operatorname{div}(f\nabla g)\,dA = \iint_D [f \operatorname{div}(\nabla g) + \nabla g \cdot \nabla f]\,dA$ by Exercise 23. But $\operatorname{div}(\nabla g) = \nabla^2 g$. Hence $\iint_D f\nabla^2 g\,dA = \oint_C f\,(\nabla g) \cdot \mathbf{n}\,ds - \iint_D \nabla g \cdot \nabla f\,dA$.

33. (a) We know that $\omega = v/d$, and from the diagram $\sin\theta = d/r \;\Rightarrow\; v = d\omega = (\sin\theta)\,r\omega = |\mathbf{w} \times \mathbf{r}|$. But $\mathbf{v}$ is perpendicular to both $\mathbf{w}$ and $\mathbf{r}$, so that $\mathbf{v} = \mathbf{w} \times \mathbf{r}$.

(b) From (a),

$$\mathbf{v} = \mathbf{w} \times \mathbf{r} = \begin{vmatrix} \mathbf{i} & \mathbf{j} & \mathbf{k} \\ 0 & 0 & \omega \\ x & y & z \end{vmatrix} = (0 \cdot z - \omega y)\,\mathbf{i} + (\omega x - 0 \cdot z)\,\mathbf{j} + (0 \cdot y - x \cdot 0)\,\mathbf{k} = -\omega y\,\mathbf{i} + \omega x\,\mathbf{j}$$

(c) $\operatorname{curl} \mathbf{v} = \nabla \times \mathbf{v} = \begin{vmatrix} \mathbf{i} & \mathbf{j} & \mathbf{k} \\ \partial/\partial x & \partial/\partial y & \partial/\partial z \\ -\omega y & \omega x & 0 \end{vmatrix}$

$$= \left[\dfrac{\partial}{\partial y}(0) - \dfrac{\partial}{\partial z}(\omega x) \right]\mathbf{i} + \left[\dfrac{\partial}{\partial z}(-\omega y) - \dfrac{\partial}{\partial x}(0) \right]\mathbf{j} + \left[\dfrac{\partial}{\partial x}(\omega x) - \dfrac{\partial}{\partial y}(-\omega y) \right]\mathbf{k}$$

$$= [\omega - (-\omega)]\,\mathbf{k} = 2\omega\mathbf{k} = 2\mathbf{w}$$

Section 13.6 Surface Integrals

1. Each face of the cube has surface area $2^2 = 4$, and the points P_{ij}^* are the points where the cube intersects the coordinate axes. Here, $f(x, y, z) = \sqrt{x^2 + 2y^2 + 3z^2}$, so by Definition 1,

$$\iint_S f(x, y, z)\, dS \approx [f(1,0,0)](4) + [f(-1,0,0)](4) + [f(0,1,0)](4) + [f(0,-1,0)](4)$$
$$+ [f(0,0,1)](4) + [f(0,0,-1)](4)$$
$$= 4\left(1 + 1 + 2\sqrt{2} + 2\sqrt{3}\right) = 8\left(1 + \sqrt{2} + \sqrt{3}\right) \approx 33.170$$

3. We can use the xz- and yz-planes to divide H into four patches of equal size, each with surface area equal to $\frac{1}{8}$ the surface area of a sphere with radius $\sqrt{50}$, so $\Delta S = \frac{1}{8}(4)\pi\left(\sqrt{50}\right)^2 = 25\pi$. Then $(\pm 3, \pm 4, 5)$ are sample points in the four patches, and using a Riemann sum as in Definition 1, we have

$$\iint_H f(x, y, z)\, dS \approx f(3, 4, 5)\,\Delta S + f(3, -4, 5)\,\Delta S + f(-3, 4, 5)\,\Delta S + f(-3, -4, 5)\,\Delta S$$
$$= (7 + 8 + 9 + 12)(25\pi) = 900\pi \approx 2827$$

5. $\mathbf{r}(u, v) = uv\,\mathbf{i} + (u + v)\,\mathbf{j} + (u - v)\,\mathbf{k}$, $u^2 + v^2 \leq 1$ and $|\mathbf{r}_u \times \mathbf{r}_v| = \sqrt{4 + 2u^2 + 2v^2}$ (see Exercise 12.6.9). Then

$$\iint_S yx\, dS = \iint_{u^2 + v^2 \leq 1} (u^2 - v^2)\sqrt{4 + 2u^2 + 2v^2}\, dA$$
$$= \int_0^{2\pi} \int_0^1 r^2 (\cos^2\theta - \sin^2\theta)\sqrt{4 + 2r^2}\, r\, dr\, d\theta$$
$$= \left[\int_0^{2\pi} (\cos^2\theta - \sin^2\theta)\, d\theta\right]\left[\int_0^1 r^3 \sqrt{4 + 2r^2}\, dr\right] = 0$$

since the first integral is 0.

7. $\mathbf{r}(x, y) = x\,\mathbf{i} + y\,\mathbf{j} + (6 - 3x - 2y)\,\mathbf{k}$, $\mathbf{r}_x \times \mathbf{r}_y = 3\,\mathbf{i} + 2\,\mathbf{j} + \mathbf{k}$ (the normal to the plane) and $|\mathbf{r}_x \times \mathbf{r}_y| = \sqrt{14}$. The given plane meets the first octant in the line $3x + 2y = 6$, $z = 0$, $x \geq 0$, $y \geq 0$, so $D = \left\{(x, y) \mid 0 \leq x \leq \frac{1}{3}(6 - 2y), 0 \leq y \leq 3\right\}$. Then

$$\iint_S y\, dS = \int_0^3 \int_0^{(6 - 2y)/3} y\sqrt{14}\, dx\, dy = \sqrt{14}\int_0^3 \left(2y - \tfrac{2}{3}y^2\right) dy = 3\sqrt{14}.$$

9. $\mathbf{r}(x, z) = x\,\mathbf{i} + (x^2 + 4z)\,\mathbf{j} + z\,\mathbf{k}$, $0 \leq x \leq 2$, $0 \leq z \leq 2$, $|\mathbf{r}_x \times \mathbf{r}_z| = \sqrt{4x^2 + 17}$ (see Exercise 12.6.5.) Then $\iint_S x\, dS = \int_0^2 \int_0^2 x\sqrt{4x^2 + 17}\, dx\, dz = 2\left[\frac{1}{12}(4x^2 + 17)^{3/2}\right]_0^2 = \frac{33\sqrt{33} - 17\sqrt{17}}{6}$.

11. Since $z = y + 3$, $|\mathbf{r}_x \times \mathbf{r}_y| = \sqrt{2}$ and

$$\iint_S yz\, dS = \iint_{x^2 + y^2 \leq 1} \sqrt{2}y\,(y + 3)\, dA = \sqrt{2}\int_0^{2\pi} \int_0^1 (r^2 \sin^2\theta + 3r\sin\theta)\, r\, dr\, d\theta$$
$$= \sqrt{2}\int_0^{2\pi} \left[\tfrac{1}{4}\sin^2\theta + \sin\theta\right] d\theta = \frac{\pi}{2\sqrt{2}}$$

13. Using spherical coordinates and Example 1 in Section 12.6 we have
$\mathbf{r}(\phi, \theta) = 2\sin\phi\cos\theta\,\mathbf{i} + 2\sin\phi\sin\theta\,\mathbf{j} + 2\cos\phi\,\mathbf{k}$ and $|\mathbf{r}_\phi \times \mathbf{r}_\theta| = 4\sin\phi$. Then
$$\iint_S (x^2 z + y^2 z)\, dS = \int_0^{2\pi} \int_0^{\pi/2} (4\sin^2\phi)(2\cos\phi)(4\sin\phi)\, d\phi\, d\theta = 16\pi \sin^4\phi\Big]_0^{\pi/2} = 16\pi.$$

15. Using cylindrical coordinates, $\mathbf{r}\,(\theta, z) = 3\cos\theta\,\mathbf{i} + 3\sin\theta\,\mathbf{j} + z\,\mathbf{k}$,

$0 \le \theta \le 2\pi$, $0 \le z \le 2$, and $|\mathbf{r}_\theta \times \mathbf{r}_z| = 3$.

$\iint_S \left(x^2y + z^2\right) dS = \int_0^{2\pi} \int_0^2 \left(27\cos^2\theta\sin\theta + z^2\right) 3\,dz\,d\theta = \int_0^{2\pi} \left(162\cos^2\theta\sin\theta + 8\right) d\theta = 16\pi$

17. $\mathbf{F}\,(\mathbf{r}\,(x, y)) = e^y\,\mathbf{i} + ye^x\,\mathbf{j} + x^2y\,\mathbf{k}$ and $\mathbf{r}_x \times \mathbf{r}_y = -2x\,\mathbf{i} - 2y\,\mathbf{j} + \mathbf{k}$.

Then $\mathbf{F}\,(\mathbf{r}\,(x, y)) \cdot (\mathbf{r}_x \times \mathbf{r}_y) = -2xe^y - 2y^2e^x + x^2y$ and

$\iint_S \mathbf{F} \cdot d\mathbf{S} = \int_0^1 \int_0^1 \left(-2xe^y - 2y^2e^x + x^2y\right) dx\,dy = \int_0^1 \left(-e^y - 2ey^2 + \tfrac{1}{3}y + 2y^2\right) dy = \tfrac{1}{6}\,(11 - 10e)$.

19. As in Exercise 7, $D = \left\{(x, y) \mid 0 \le x \le 2,\, 0 \le y \le \tfrac{1}{2}\,(6 - 3x)\right\}$.

$$\iint_S \mathbf{F} \cdot d\mathbf{S} = \int_0^2 \int_0^{(6-3x)/2} \left[x\,\mathbf{i} + xy\,\mathbf{j} + x\,(6 - 3x - 2y)\,\mathbf{k}\right] \cdot (3\,\mathbf{i} + 2\,\mathbf{j} + \mathbf{k})\,dy\,dx$$
$$= \int_0^2 \int_0^{(6-3x)/2} \left(9x - 3x^2\right) dy\,dx = \int_0^2 \left[27x - \tfrac{45}{2}x^2 + \tfrac{9}{2}x^3\right] dx = 12$$

21. $\mathbf{F}\,(\mathbf{r}\,(\phi, \theta)) = 3\sin\phi\cos\theta\,\mathbf{i} + 3\sin\phi\sin\theta\,\mathbf{j} + 3\cos\phi\,\mathbf{k}$

and $\mathbf{r}_\phi \times \mathbf{r}_\theta = 9\sin^2\phi\cos\theta\,\mathbf{i} + 9\sin^2\phi\sin\theta\,\mathbf{j} + 9\sin\phi\cos\phi\,\mathbf{k}$. Then

$\mathbf{F}\,(\mathbf{r}\,(\phi, \theta)) \cdot (\mathbf{r}_\phi \times \mathbf{r}_\theta) = 27\sin^3\phi\cos^2\theta + 27\sin^3\phi\sin^2\theta + 27\sin\phi\cos^2\phi = 27\sin\phi$ and

$\iint_S \mathbf{F} \cdot d\mathbf{S} = \int_0^{2\pi} \int_0^\pi 27\sin\phi\,d\phi\,d\theta = (2\pi)\,(54) = 108\pi$.

23. Let S_1 be the paraboloid $y = x^2 + z^2$, $0 \le y \le 1$ and S_2 the disk $x^2 + z^2 \le 1$, $y = 1$. Since S is a closed surface, we use the outward orientation. On S_1: $\mathbf{F}\,(\mathbf{r}\,(x, z)) = \left(x^2 + z^2\right)\mathbf{j} - z\,\mathbf{k}$ and $\mathbf{r}_x \times \mathbf{r}_z = 2x\,\mathbf{i} - \mathbf{j} + 2z\,\mathbf{k}$ (since the j-component must be negative on S_1). Then

$$\iint_{S_1} \mathbf{F} \cdot d\mathbf{S} = \iint_{x^2 + z^2 \le 1} \left[-\left(x^2 + z^2\right) - 2z^2\right] dA = -\int_0^{2\pi} \int_0^1 \left(r^2 + 2r^2\cos^2\theta\right) r\,dr\,d\theta$$
$$= -\int_0^{2\pi} \tfrac{1}{4}\left(1 + 2\cos^2\theta\right) d\theta = -\left(\tfrac{\pi}{2} + \tfrac{\pi}{2}\right) = -\pi$$

On S_2: $\mathbf{F}\,(\mathbf{r}\,(x, z)) = \mathbf{j} - z\,\mathbf{k}$ and $\mathbf{r}_z \times \mathbf{r}_x = \mathbf{j}$. Then $\iint_{S_2} \mathbf{F} \cdot d\mathbf{S} = \iint_{x^2 + z^2 \le 1} (1)\,dA = \pi$. Hence $\iint_S \mathbf{F} \cdot d\mathbf{S} = -\pi + \pi = 0$.

25. Here S consists of the six faces of the cube as labeled in the figure. On S_1:

$\mathbf{F} = \mathbf{i} + 2y\mathbf{j} + 3z\mathbf{k}$, $\mathbf{r}_y \times \mathbf{r}_z = \mathbf{i}$ and $\iint_{S_1} \mathbf{F} \cdot d\mathbf{S} = \int_{-1}^1 \int_{-1}^1 dy\,dz = 4$;

S_2: $\mathbf{F} = x\mathbf{i} + 2\mathbf{j} + 3z\mathbf{k}$, $\mathbf{r}_z \times \mathbf{r}_x = \mathbf{j}$ and

$\iint_{S_2} \mathbf{F} \cdot d\mathbf{S} = \int_{-1}^1 \int_{-1}^1 2\,dx\,dz = 8$;

S_3: $\mathbf{F} = x\mathbf{i} + 2y\mathbf{j} + 3\mathbf{k}$, $\mathbf{r}_x \times \mathbf{r}_y = \mathbf{k}$ and

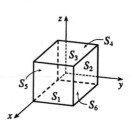

$\iint_{S_3} \mathbf{F} \cdot d\mathbf{S} = \int_{-1}^1 \int_{-1}^1 3\,dx\,dy = 12$;

S_4: $\mathbf{F} = -\mathbf{i} + 2y\mathbf{j} + 3z\mathbf{k}$, $\mathbf{r}_z \times \mathbf{r}_y = -\mathbf{i}$ and $\iint_{S_4} \mathbf{F} \cdot d\mathbf{S} = 4$;

S_5: $\mathbf{F} = x\mathbf{i} - 2\mathbf{j} + 3z\mathbf{k}$, $\mathbf{r}_x \times \mathbf{r}_z = -\mathbf{j}$ and $\iint_{S_5} \mathbf{F} \cdot d\mathbf{S} = 8$;

S_6: $\mathbf{F} = x\mathbf{i} + 2y\mathbf{j} - 3\mathbf{k}$, $\mathbf{r}_y \times \mathbf{r}_x = -\mathbf{k}$ and $\iint_{S_6} \mathbf{F} \cdot d\mathbf{S} = \int_{-1}^1 \int_{-1}^1 3\,dx\,dy = 12$.

Hence $\iint_S \mathbf{F} \cdot d\mathbf{S} = \sum_{i=1}^6 \iint_{S_i} \mathbf{F} \cdot d\mathbf{S} = 48$.

27. We use Formula 4 with $z = 3 - 2x^2 - y^2 \Rightarrow \partial z/\partial x = -4x, \partial z/\partial y = -2y$. The boundaries of the region $3 - 2x^2 - y^2 \geq 0$ are $-\sqrt{\frac{3}{2}} \leq x \leq \sqrt{\frac{3}{2}}$ and $-\sqrt{3 - 2x^2} \leq y \leq \sqrt{3 - 2x^2}$, so we use a CAS (with precision reduced to seven or fewer digits; otherwise the calculation takes a very long time) to calculate

$$\iint_S x^2 y^2 z^2 \, dS = \int_{-\sqrt{3/2}}^{\sqrt{3/2}} \int_{-\sqrt{3-2x^2}}^{\sqrt{3-2x^2}} x^2 y^2 \left(3 - 2x^2 - y^2\right)^2 \sqrt{16x^2 + 4y^2 + 1} \, dy \, dx \approx 3.4895.$$

29. If S is given by $y = h(x, z)$, then S is also the level surface $f(x, y, z) = y - h(x, z) = 0$.

$$\mathbf{n} = \frac{\nabla f(x, y, z)}{|\nabla f(x, y, z)|} = \frac{-h_x \mathbf{i} + \mathbf{j} - h_z \mathbf{k}}{\sqrt{h_x^2 + 1 + h_z^2}}, \text{ and } -\mathbf{n} \text{ is the unit normal that points to the left. Now we}$$

proceed as in the derivation of (10), using Formula 4 to evaluate

$$\iint_S \mathbf{F} \cdot d\mathbf{S} = \iint_S \mathbf{F} \cdot \mathbf{n} \, dS$$

$$= \iint_D (P\mathbf{i} + Q\mathbf{j} + R\mathbf{k}) \frac{\frac{\partial h}{\partial x}\mathbf{i} - \mathbf{j} + \frac{\partial h}{\partial z}\mathbf{k}}{\sqrt{\left(\frac{\partial h}{\partial x}\right)^2 + 1 + \left(\frac{\partial h}{\partial z}\right)^2}} \sqrt{\left(\frac{\partial h}{\partial x}\right)^2 + 1 + \left(\frac{\partial h}{\partial z}\right)^2} \, dA$$

where D is the projection of $f(x, y, z)$ onto the xz- plane. Therefore

$$\iint_S \mathbf{F} \cdot d\mathbf{S} = \iint_D \left(P\frac{\partial h}{\partial x} - Q + R\frac{\partial h}{\partial z}\right) dA.$$

31. $m = \iint_S K \, dS = K \cdot 4\pi \left(\frac{1}{2}a^2\right) = 2\pi a^2 K$; by symmetry $M_{xz} = M_{yz} = 0$, and

$M_{xy} = \iint_S zK \, dS = K \int_0^{2\pi} \int_0^{\pi/2} (a \cos\phi)(a^2 \sin\phi) \, d\phi \, d\theta = 2\pi K a^3 \left[-\frac{1}{4}\cos 2\phi\right]_0^{\pi/2} = \pi K a^3$.

Hence $(\bar{x}, \bar{y}, \bar{z}) = \left(0, 0, \frac{1}{2}a\right)$.

33. (a) $I_z = \iint_S (x^2 + y^2) \rho(x, y, z) \, dS$

(b) $I_z = \iint_S (x^2 + y^2)\left(10 - \sqrt{x^2 + y^2}\right) dS = \iint_{1 \leq x^2 + y^2 \leq 16} (x^2 + y^2)\left(10 - \sqrt{x^2 + y^2}\right) \sqrt{2} \, dA$

$= \int_0^{2\pi} \int_1^4 \sqrt{2} \, (10r^3 - r^4) \, dr \, d\theta = 2\sqrt{2}\pi \left(\frac{4329}{10}\right) = \frac{4329}{5}\sqrt{2}\pi$

35. $\rho(x, y, z) = 1200$, $\mathbf{V} = y\mathbf{i} + \mathbf{j} + z\mathbf{k}$, $\mathbf{F} = \rho\mathbf{V} = (1200)(y\mathbf{i} + \mathbf{j} + z\mathbf{k})$. S is given by $\mathbf{r}(x, y) = x\mathbf{i} + y\mathbf{j} + \left[9 - \frac{1}{4}(x^2 + y^2)\right]\mathbf{k}$, $0 \leq x^2 + y^2 \leq 36$ and $\mathbf{r}_x \times \mathbf{r}_y = \frac{1}{2}x\mathbf{i} + \frac{1}{2}y\mathbf{j} + \mathbf{k}$. Thus the rate of flow is given by

$$\iint_S \mathbf{F} \cdot d\mathbf{S} = \iint_{0 \leq x^2 + y^2 \leq 36} (1200)\left(\frac{1}{2}xy + \frac{1}{2}y + \left[9 - \frac{1}{4}(x^2 + y^2)\right]\right) dA$$

$$= 1200 \int_0^6 \int_0^{2\pi} \left[\frac{1}{2}r^2 \sin\theta \cos\theta + \frac{1}{2}r \sin\theta + 9 - \frac{1}{4}r^2\right] r \, d\theta \, dr$$

$$= 1200 \int_0^6 2\pi \left(9r - \frac{1}{4}r^3\right) dr = (1200)(2\pi)(81) = 194{,}400\pi$$

37. S consists of the hemisphere S_1 given by $z = \sqrt{a^2 - x^2 - y^2}$ and the disk S_2 given by
$0 \le x^2 + y^2 \le a^2$, $z = 0$. On S_1: $\mathbf{E} = a \sin \phi \cos \theta \, \mathbf{i} + a \sin \phi \sin \theta \, \mathbf{j} + 2a \cos \phi \, \mathbf{k}$,
$\mathbf{T}_\phi \times \mathbf{T}_\theta = a^2 \sin^2 \phi \cos \theta \, \mathbf{i} + a^2 \sin^2 \phi \sin \theta \, \mathbf{j} + a^2 \sin \phi \cos \phi \, \mathbf{k}$. Thus

$$\iint_{S_1} \mathbf{E} \cdot d\mathbf{S} = \int_0^{2\pi} \int_0^{\pi/2} \left(a^3 \sin^3 \phi + 2a^3 \sin \phi \cos^2 \phi \right) d\phi \, d\theta$$

$$= \int_0^{2\pi} \int_0^{\pi/2} \left(a^3 \sin \phi + a^3 \sin \phi \cos^2 \phi \right) d\phi \, d\theta = (2\pi) a^3 \left(1 + \tfrac{1}{3} \right) = \tfrac{8}{3} \pi a^3$$

On S_2: $\mathbf{E} = x \, \mathbf{i} + y \, \mathbf{j}$, and $\mathbf{r}_y \times \mathbf{r}_x = -\mathbf{k}$ so $\iint_{S_2} \mathbf{E} \cdot d\mathbf{S} = 0$. Hence the total charge is
$q = \epsilon_0 \iint_S \mathbf{E} \cdot d\mathbf{S} = \tfrac{8}{3} \pi a^3 \epsilon_0$.

39. $K \nabla u = 6.5 \left(4y \, \mathbf{j} + 4z \, \mathbf{k} \right)$. S is given by $\mathbf{r}(x, \theta) = x \, \mathbf{i} + \sqrt{6} \cos \theta \, \mathbf{j} + \sqrt{6} \sin \theta \, \mathbf{k}$ and since we want the
inward heat flow, we use $\mathbf{r}_x \times \mathbf{r}_\theta = -\sqrt{6} \cos \theta \, \mathbf{j} - \sqrt{6} \sin \theta \, \mathbf{k}$. Then the rate of heat flow inward is given
by $\iint_S (-K \nabla u) \cdot d\mathbf{S} = \int_0^{2\pi} \int_0^4 - (6.5)(-24) \, dx \, d\theta = (2\pi)(156)(4) = 1248\pi$.

Section 13.7 Stokes' Theorem

1. Both H and P are oriented piecewise-smooth surfaces that are bounded by the simple, closed, smooth curve $x^2 + y^2 = 4$, $z = 0$ (which we can take to be oriented positively for both surfaces). Then H and P satisfy the hypotheses of Stokes' Theorem, so by (3) we know
$\iint_H \operatorname{curl} \mathbf{F} \cdot d\mathbf{S} = \int_C \mathbf{F} \cdot d\mathbf{r} = \iint_P \operatorname{curl} \mathbf{F} \cdot d\mathbf{S}$ (where C is the boundary curve).

3. The boundary curve is C: $x^2 + y^2 = 1$, $z = 0$ oriented in the counterclockwise direction. The vector equation of C is $\mathbf{r}(t) = \cos t\, \mathbf{i} + \sin t\, \mathbf{j}$, $0 \le t \le 2\pi$.
Then $\mathbf{F}(\mathbf{r}(t)) = \cos t\, \mathbf{j} + e^{\cos t \sin t}\, \mathbf{k}$ and $\mathbf{F}(\mathbf{r}(t)) \cdot \mathbf{r}'(t) = \cos^2 t$. Hence
$\iint_S \operatorname{curl} \mathbf{F} \cdot d\mathbf{S} = \oint_C \mathbf{F} \cdot d\mathbf{r} = \int_0^{2\pi} \cos^2 t\, dt = \int_0^{2\pi} \frac{1}{2}(1 + \cos 2t)\, dt = \pi$.

5. C is the square in the plane $z = -1$. By (3), $\iint_{S_1} \operatorname{curl} \mathbf{F} \cdot d\mathbf{S} = \oint_C \mathbf{F} \cdot d\mathbf{r} = \iint_{S_2} \operatorname{curl} \mathbf{F} \cdot d\mathbf{S}$ where S_1 is the original cube without the bottom and S_2 is the bottom face of the cube.
$\operatorname{curl} \mathbf{F} = x^2 z\, \mathbf{i} + (xy - 2xyz)\, \mathbf{j} + (y - xz)\, \mathbf{k}$. For S_2, we choose $\mathbf{n} = \mathbf{k}$ so that C has the same orientation for both surfaces. Then $\operatorname{curl} \mathbf{F} \cdot \mathbf{n} = y - xz = x + y$ on S_2, where $z = -1$. Thus
$\iint_{S_2} \operatorname{curl} \mathbf{F} \cdot d\mathbf{S} = \int_{-1}^1 \int_{-1}^1 (x + y)\, dx\, dy = 0$ so $\iint_{S_1} \operatorname{curl} \mathbf{F} \cdot d\mathbf{S} = 0$.

7. $\operatorname{curl} \mathbf{F} = 3x\, \mathbf{i} + (x - 3y)\, \mathbf{j} + 2y\, \mathbf{k}$, $\mathbf{n} = \frac{1}{\sqrt{11}}(3\mathbf{i} + \mathbf{j} + \mathbf{k})$ and

$$\oint_C \mathbf{F} \cdot d\mathbf{r} = \iint_S \operatorname{curl} \mathbf{F} \cdot \mathbf{n}\, dS = \int_0^1 \int_0^{3-3x} \frac{1}{\sqrt{11}}[9x + (x - 3y) + 2y] (\sqrt{11})\, dy\, dx$$
$$= \int_0^1 \int_0^{3-3x} (10x - y)\, dy\, dx = \int_0^1 \left[10(3x - 3x^2) - \frac{1}{2}(3 - 3x)^2\right] dx$$
$$= \left[15x^2 - 10x^3 + \frac{3}{2}(1 - x^3)\right]_0^1 = \frac{7}{2}$$

9. The curve of intersection is an ellipse in the plane $z = x + 4$ with unit normal
$\mathbf{n} = \frac{1}{\sqrt{2}}(-\mathbf{i} + \mathbf{k})$ and $\operatorname{curl} \mathbf{F} = 5\mathbf{i} + 2\mathbf{j} + 4\mathbf{k}$ so $\operatorname{curl} \mathbf{F} \cdot \mathbf{n} = -\frac{1}{\sqrt{2}}$. Then
$\oint_C \mathbf{F} \cdot d\mathbf{r} = -\iint_S \frac{1}{\sqrt{2}}\, dS = -\frac{1}{\sqrt{2}}$ (surface area of planar ellipse) $= -\frac{1}{\sqrt{2}}\pi(2)(2\sqrt{2}) = -4\pi$.

11. (a) The curve of intersection is an ellipse in the plane $x + y + z = 1$ with unit normal
$\mathbf{n} = \frac{1}{\sqrt{3}}(\mathbf{i} + \mathbf{j} + \mathbf{k})$, $\operatorname{curl} \mathbf{F} = x^2\, \mathbf{j} + y^2\, \mathbf{k}$ and $\operatorname{curl} \mathbf{F} \cdot \mathbf{n} = \frac{1}{\sqrt{3}}(x^2 + y^2)$. Then
$$\oint_C \mathbf{F} \cdot d\mathbf{r} = \iint_S \frac{1}{\sqrt{3}}(x^2 + y^2)\, dS = \iint_{x^2 + y^2 \le 9} (x^2 + y^2)\, dx\, dy$$
$$= \int_0^{2\pi} \int_0^3 r^3\, dr\, d\theta = 2\pi\left(\frac{81}{4}\right) = \frac{81\pi}{2}$$

(b)

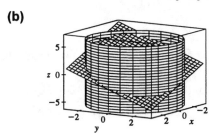

(c) One possible parametrization is $x = 3\cos t$, $y = 3\sin t$, $z = 1 - 3\cos t - 3\sin t$, $0 \le t \le 2\pi$.

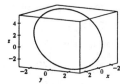

13. The boundary curve C is the circle $x^2 + y^2 = 9$, $z = 0$ oriented in the counterclockwise direction as viewed from $(0,0,1)$. Then $\mathbf{r}(t) = 3\cos t\,\mathbf{i} + 3\sin t\,\mathbf{j}$, $0 \le t \le 2\pi$, so $\mathbf{F}(\mathbf{r}(t)) = 9\sin t\,\mathbf{i} - 18\cos t\,\mathbf{k}$ and $\mathbf{F} \cdot \mathbf{r}'(t) = -27\sin^2 t$. Thus $\oint_C \mathbf{F} \cdot d\mathbf{r} = \int_0^{2\pi} (-27\sin^2 t)\,dt = -27\pi$. Now $\operatorname{curl}\mathbf{F} = -4\,\mathbf{i} + 6\,\mathbf{j} - 3\,\mathbf{k}$, $\mathbf{r}_x \times \mathbf{r}_y = 2x\,\mathbf{i} + 2y\,\mathbf{j} + \mathbf{k}$, so

$$\iint_S \operatorname{curl}\mathbf{F} \cdot d\mathbf{S} = \iint_{x^2+y^2 \le 9} (-8x + 12y - 3)\,dA = \int_0^{2\pi} \int_0^3 (-8r\cos\theta + 12r\sin\theta - 3)\,r\,dr\,d\theta$$

$$= \int_0^3 (-3r)(2\pi)\,dr = -27\pi$$

15. The x-, y-, and z-intercepts of the plane are all 1, so C consists of the three line segments C_1: $\mathbf{r}_1(t) = (1-t)\mathbf{i} + t\mathbf{j}$, $0 \le t \le 1$, C_2: $\mathbf{r}_2(t) = (1-t)\mathbf{j} + t\mathbf{k}$, $0 \le t \le 1$, and C_3: $\mathbf{r}_3(t) = t\mathbf{i} + (1-t)\mathbf{k}$, $0 \le t \le 1$. Then

$$\oint_C \mathbf{F} \cdot d\mathbf{r} = \int_0^1 [t\mathbf{i} + (1-t)\mathbf{k}] \cdot (-\mathbf{i} + \mathbf{j})\,dt + \int_0^1 [(1-t)\mathbf{i} + t\mathbf{j}] \cdot (-\mathbf{j} + \mathbf{k})\,dt$$

$$+ \int_0^1 [(1-t)\mathbf{j} + t\mathbf{k}] \cdot (\mathbf{i} - \mathbf{k})\,dt$$

$$= \int_0^1 (-3t)\,dt = -\tfrac{3}{2}$$

Now $\operatorname{curl}\mathbf{F} = -\mathbf{i} - \mathbf{j} - \mathbf{k}$ and $\mathbf{r}_x \times \mathbf{r}_y = \mathbf{i} + \mathbf{j} + \mathbf{k}$. Hence $\iint_S \operatorname{curl}\mathbf{F} \cdot d\mathbf{S} = \int_0^1 \int_0^{1-x} (-3)\,dy\,dx = -\tfrac{3}{2}$.

17. $\operatorname{curl}\mathbf{F} = \begin{vmatrix} \mathbf{i} & \mathbf{j} & \mathbf{k} \\ \partial/\partial x & \partial/\partial y & \partial/\partial z \\ x^x + z^2 & y^y + x^2 & z^z + y^2 \end{vmatrix} = 2y\,\mathbf{i} + 2z\,\mathbf{j} + 2x\,\mathbf{k}$ and $W = \int_C \mathbf{F} \cdot d\mathbf{r} = \iint_S \operatorname{curl}\mathbf{F} \cdot d\mathbf{S}$.

To parametrize the surface, let $x = 2\cos\theta \sin\phi$, $y = 2\sin\theta \sin\phi$, $z = 2\cos\phi$, so that $\mathbf{r}(\phi,\theta) = 2\sin\phi\cos\theta\,\mathbf{i} + 2\sin\phi\sin\theta\,\mathbf{j} + 2\cos\phi\,\mathbf{k}$, $0 \le \phi \le \tfrac{\pi}{2}$, $0 \le \theta \le \tfrac{\pi}{2}$, and $\mathbf{r}_\phi \times \mathbf{r}_\theta = 4\sin^2\phi\cos\theta\,\mathbf{i} + 4\sin^2\phi\sin\theta\,\mathbf{j} + 4\sin\phi\cos\phi\,\mathbf{k}$. Then $\operatorname{curl}\mathbf{F}(\mathbf{r}(\phi,\theta)) = 4\sin\phi\sin\theta\,\mathbf{i} + 4\cos\phi\,\mathbf{j} + 4\sin\phi\cos\theta\,\mathbf{k}$, and $\operatorname{curl}\mathbf{F} \cdot (\mathbf{r}_\phi \times \mathbf{r}_\theta) = 16\sin^3\phi\sin\theta\cos\theta + 16\cos\phi\sin^2\phi\sin\theta + 16\sin^2\phi\cos\phi\cos\theta$. Therefore

$$\iint_S \operatorname{curl}\mathbf{F} \cdot d\mathbf{S} = \iint_D \operatorname{curl}\mathbf{F} \cdot (\mathbf{r}_\phi \times \mathbf{r}_\theta)\,dA$$

$$= 16\left[\int_0^{\pi/2} \sin\theta\cos\theta\,d\theta\right]\left[\int_0^{\pi/2} \sin^3\phi\,d\phi\right] + 16\left[\int_0^{\pi/2} \sin\theta\,d\theta\right]\left[\int_0^{\pi/2} \sin^2\phi\cos\phi\,d\phi\right]$$

$$+16\left[\int_0^{\pi/2} \cos\theta\,d\theta\right]\left[\int_0^{\pi/2} \sin^2\phi\cos\phi\,d\phi\right]$$

$$= 8\left[-\cos\phi + \tfrac{1}{3}\cos^3\phi\right]_0^{\pi/2} + 16\,(1)\left[\tfrac{1}{3}\sin^3\phi\right]_0^{\pi/2} + 16\,(1)\left[\tfrac{1}{3}\sin^3\phi\right]_0^{\pi/2}$$

$$= 8\left[0 + 1 + 0 - \tfrac{1}{3}\right] + 16\left(\tfrac{1}{3}\right) + 16\left(\tfrac{1}{3}\right) = \tfrac{16}{3} + \tfrac{16}{3} + \tfrac{16}{3} = 16$$

19. Assume S is centered at the origin with radius a and let H_1 and H_2 be the upper and lower hemispheres, respectively, of S. Then $\iint_S \operatorname{curl}\mathbf{F} \cdot d\mathbf{S} = \iint_{H_1} \operatorname{curl}\mathbf{F} \cdot d\mathbf{S} + \iint_{H_2} \operatorname{curl}\mathbf{F} \cdot d\mathbf{S} = \oint_{C_1} \mathbf{F} \cdot d\mathbf{r} + \oint_{C_2} \mathbf{F} \cdot d\mathbf{r}$ by Stokes' Theorem. But C_1 is the circle $x^2 + y^2 = a^2$ oriented in the counterclockwise direction while C_2 is the same circle oriented in the clockwise direction. Hence $\oint_{C_2} \mathbf{F} \cdot d\mathbf{r} = -\oint_{C_1} \mathbf{F} \cdot d\mathbf{r}$ so $\iint_S \operatorname{curl}\mathbf{F} \cdot d\mathbf{S} = 0$ as desired.

Section 13.8 The Divergence Theorem

1. The vectors that end near P_1 are longer than the vectors that start near P_1, so the net flow is inward near P_1 and div $\mathbf{F}\,(P_1)$ is negative. The vectors that end near P_2 are shorter than the vectors that start near P_2, so the net flow is outward near P_2 and div $\mathbf{F}\,(P_2)$ is positive.

3. div $\mathbf{F} = 3 + x + 2x = 3 + 3x$, so

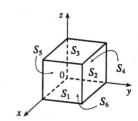

$\iiint_E \operatorname{div}\mathbf{F}\,dV = \int_0^1\int_0^1\int_0^1 (3x+3)\,dx\,dy\,dz = \frac{9}{2}$ (notice the triple

integral is three times the volume of the cube plus three times $\bar{x}$).

To compute $\iint_S \mathbf{F} \cdot d\mathbf{S}$, on S_1: $\mathbf{n} = \mathbf{i}$, $\mathbf{F} = 3\mathbf{i} + y\mathbf{j} + 2z\mathbf{k}$, and

$\iint_{S_1}\mathbf{F}\cdot d\mathbf{S} = \iint_{S_1} 3\,dS = 3$;

S_2: $\mathbf{F} = 3x\,\mathbf{i} + x\mathbf{j} + 2xz\,\mathbf{k}$, $\mathbf{n} = \mathbf{j}$ and $\iint_{S_2}\mathbf{F}\cdot d\mathbf{S} = \iint_{S_2} x\,dS = \frac{1}{2}$;

S_3: $\mathbf{F} = 3x\,\mathbf{i} + xy\,\mathbf{j} + 2x\,\mathbf{k}$, $\mathbf{n} = \mathbf{k}$ and $\iint_{S_3}\mathbf{F}\cdot d\mathbf{S} = \iint_{S_3} 2x\,dS = 1$;

S_4: $\mathbf{F} = 0$, $\iint_{S_4}\mathbf{F}\cdot d\mathbf{S} = 0$; S_5: $\mathbf{F} = 3x\,\mathbf{i} + 2x\,\mathbf{k}$, $\mathbf{n} = -\mathbf{j}$ and $\iint_{S_5}\mathbf{F}\cdot d\mathbf{S} = \iint_{S_5} 0\,dS = 0$;

S_6: $\mathbf{F} = 3x\,\mathbf{i} + xy\,\mathbf{j}$, $\mathbf{n} = -\mathbf{k}$ and $\iint_{S_6}\mathbf{F}\cdot d\mathbf{S} = \iint_{S_6} 0\,dS = 0$. Thus $\iint_S \mathbf{F}\cdot d\mathbf{S} = \frac{9}{2}$.

5. div $\mathbf{F} = \dfrac{\partial}{\partial x}\left(3y^2z^3\right) + \dfrac{\partial}{\partial y}\left(9x^2yz^2\right) + \dfrac{\partial}{\partial z}\left(4xy^2\right) = 9x^2z^2$, so by the Divergence Theorem,

$\iint_S \mathbf{F}\cdot d\mathbf{S} = \iiint_E 9x^2z^2\,dV = \int_{-1}^1\int_{-1}^1\int_{-1}^1 9x^2z^2\,dx\,dy\,dz = 8.$

7. $\iint_S \mathbf{F}\cdot d\mathbf{S} = \iiint_E (-z - z + 2z)\,dV = 0$

9. $\iint_S \mathbf{F}\cdot d\mathbf{S} = \iiint_E 3\left(x^2+y^2+z^2\right)dV = \int_0^{2\pi}\int_0^\pi\int_0^1 3\rho^4 \sin\phi\,d\rho\,d\phi\,d\theta = 2\pi\int_0^\pi \frac{3}{5}\sin\phi\,d\phi = \frac{12}{5}\pi$

11. $\iint_S \mathbf{F}\cdot d\mathbf{S} = \iiint_E 2y\,dV = \iint_{x^2+y^2\le 9}\int_{y-3}^0 2y\,dz\,dA = \int_0^{2\pi}\int_0^3\int_{-3+r\sin\theta}^0 \left(2r^2 \sin\theta\right) dz\,dr\,d\theta$

$= \int_0^{2\pi}\int_0^3\left(6r^2 \sin\theta - 2r^3 \sin^2\theta\right) dr\,d\theta = \int_0^{2\pi}\left[54\sin\theta - \frac{81}{2}\sin^2\theta\right] d\theta = -\frac{81}{2}\pi$

13. $\iint_S \mathbf{F}\cdot d\mathbf{S} = \iiint_E \sqrt{3-x^2}\,dV = \int_{-1}^1\int_{-1}^1\int_0^{2-x^4-y^4}\sqrt{3-x^2}\,dz\,dy\,dx = \frac{341}{60}\sqrt{2} + \frac{81}{20}\sin^{-1}\left(\frac{\sqrt{3}}{3}\right)$

15. For S_1 we have $\mathbf{n} = -\mathbf{k}$, so $\mathbf{F}\cdot\mathbf{n} = \mathbf{F}\cdot(-\mathbf{k}) = -x^2z - y^2 = -y^2$ (since $z = 0$ on S_1). So if D is the unit disk, we get $\iint_{S_1}\mathbf{F}\cdot d\mathbf{S} = \iint_{S_1}\mathbf{F}\cdot\mathbf{n}\,dS = \iint_D (-y^2)\,dA = -\int_0^{2\pi}\int_0^1 r^2 \sin^2\theta\,r\,dr\,d\theta = -\frac{1}{4}\pi$.

Now since S_2 is closed, we can use the Divergence Theorem. Since

div $\mathbf{F} = \dfrac{\partial}{\partial x}\left(z^2x\right) + \dfrac{\partial}{\partial y}\left(\frac{1}{3}y^3 + \tan z\right) + \dfrac{\partial}{\partial z}\left(x^2z + y^2\right) = z^2 + y^2 + x^2$, we use spherical

coordinates to get $\iint_{S_2}\mathbf{F}\cdot d\mathbf{S} = \iiint_E \operatorname{div}\mathbf{F}\,dV = \int_0^{2\pi}\int_0^{\pi/2}\int_0^1 \rho^2\cdot\rho^2\sin\phi\,d\rho\,d\phi\,d\theta = \frac{2}{5}\pi$. Finally

$\iint_S \mathbf{F}\cdot d\mathbf{S} = \iint_{S_2}\mathbf{F}\cdot d\mathbf{S} - \iint_{S_1}\mathbf{F}\cdot d\mathbf{S} = \frac{2}{5}\pi - \left(-\frac{1}{4}\pi\right) = \frac{13}{20}\pi$.

17. Since $\dfrac{\mathbf{x}}{|\mathbf{x}|^3} = \dfrac{x\,\mathbf{i} + y\,\mathbf{j} + z\,\mathbf{k}}{(x^2 + y^2 + z^2)^{3/2}}$ and $\dfrac{\partial}{\partial x}\left(\dfrac{x}{(x^2 + y^2 + z^2)^{3/2}}\right) = \dfrac{(x^2 + y^2 + z^2) - 3x^2}{(x^2 + y^2 + z^2)^{5/2}}$ with similar

expressions for $\dfrac{\partial}{\partial y}\left(\dfrac{y}{(x^2 + y^2 + z^2)^{3/2}}\right)$ and $\dfrac{\partial}{\partial z}\left(\dfrac{z}{(x^2 + y^2 + z^2)^{3/2}}\right)$, we have

$$\operatorname{div}\left(\dfrac{\mathbf{x}}{|\mathbf{x}|^3}\right) = \dfrac{3\,(x^2 + y^2 + z^2) - 3\,(x^2 + y^2 + z^2)}{(x^2 + y^2 + z^2)^{5/2}} = 0,\text{ except at }(0,0,0)\text{ where it is undefined.}$$

19. $\iint_S \mathbf{a} \cdot \mathbf{n}\, dS = \iiint_E \operatorname{div} \mathbf{a}\, dV = 0$ since $\operatorname{div} \mathbf{a} = 0$.

21. $\iint_S \operatorname{curl} \mathbf{F} \cdot d\mathbf{S} = \iiint_E \operatorname{div}(\operatorname{curl} \mathbf{F})\, dV = 0$ by Theorem 13.5.11.

23. $\iint_S (f\nabla g) \cdot \mathbf{n}\, dS = \iiint_E \operatorname{div}(f\nabla g)\, dV = \iiint_E (f\nabla^2 g + \nabla g \cdot \nabla f)\, dV$ by Exercise 13.5.23.

25. If $\mathbf{c} = c_1\,\mathbf{i} + c_2\,\mathbf{j} + c_3\,\mathbf{k}$ is an arbitrary constant vector, we define $\mathbf{F} = f\mathbf{c} = fc_1\,\mathbf{i} + fc_2\,\mathbf{j} + fc_3\,\mathbf{k}$. Then

$$\operatorname{div} \mathbf{F} = \operatorname{div} f\mathbf{c} = \dfrac{\partial f}{\partial x}c_1 + \dfrac{\partial f}{\partial y}c_2 + \dfrac{\partial f}{\partial z}c_3 = \nabla f \cdot \mathbf{c}\text{ and the divergence theorem says}$$

$\iint_S \mathbf{F} \cdot d\mathbf{S} = \iiint_E \operatorname{div} \mathbf{F}\, dV \;\Rightarrow\; \iint_S \mathbf{F} \cdot \mathbf{n}\, dS = \iiint_E \nabla f \cdot \mathbf{c}\, dV$. In particular, if $\mathbf{c} = \mathbf{i}$ then

$\iint_S f\mathbf{i} \cdot \mathbf{n}\, dS = \iiint_E \nabla f \cdot \mathbf{i}\, dV \;\Rightarrow\; \iint_S fn_1\, dS = \iiint_E \dfrac{\partial f}{\partial x}\, dV$ (where $\mathbf{n} = n_1\,\mathbf{i} + n_2\,\mathbf{j} + n_3\,\mathbf{k}$).

Similarly, if $\mathbf{c} = \mathbf{j}$ we have $\displaystyle\iint_S fn_2\, dS = \iiint_E \dfrac{\partial f}{\partial y}\, dV$, and $\mathbf{c} = \mathbf{k}$ gives

$\displaystyle\iint_S fn_3\, dS = \iiint_E \dfrac{\partial f}{\partial z}\, dV$. Then

$$\iint_S f\mathbf{n}\, dS = \left(\iint_S fn_1\, dS\right)\mathbf{i} + \left(\iint_S fn_2\, dS\right)\mathbf{j} + \left(\iint_S fn_3\, dS\right)\mathbf{k}$$

$$= \left(\iiint_E \dfrac{\partial f}{\partial x}\, dV\right)\mathbf{i} + \left(\iiint_E \dfrac{\partial f}{\partial y}\, dV\right)\mathbf{j} + \left(\iiint_E \dfrac{\partial f}{\partial z}\, dV\right)\mathbf{k}$$

$$= \iiint_E \left(\dfrac{\partial f}{\partial x}\mathbf{i} + \dfrac{\partial f}{\partial y}\mathbf{j} + \dfrac{\partial f}{\partial z}\mathbf{k}\right) dV = \iiint_E \nabla f\, dV$$

as desired.

Chapter 13 Review

Concept Check

1. See Definitions 1 and 2 in Section 13.1. A vector field can represent, for example, the wind velocity at any location in space, the speed and direction of the ocean current at any location, or the force vectors of Earth's gravitational field at a location in space.

2. (a) A conservative vector field $\mathbf{F}$ is a vector field which is the gradient of some scalar function f.

   (b) The function f in part (a) is called a potential function for $\mathbf{F}$, that is, $\mathbf{F} = \nabla f$.

3. (a) See Definition 2 in Section 13.2.

   (b) We normally evaluate the line integral using (3) in Section 13.2.

   (c) The mass is $m = \int_C \rho(x, y)\, ds$, and the center of mass is $(\bar{x}, \bar{y})$ where $\bar{x} = \frac{1}{m} \int_C x\rho(x, y)\, ds$, $\bar{y} = \frac{1}{m} \int_C y\rho(x, y)\, ds$.

   (d) See (5) and (6) in Section 13.2 for plane curves; we have similar definitions when C is a space curve [see the equation preceding (10) on page 930].

   (e) For plane curves, see (7) in Section 13.2. We have similar results for space curves [see the equation preceding (10) on page 930].

4. (a) See Definition 13 in Section 13.2.

   (b) If $\mathbf{F}$ is a force field, $\int_C \mathbf{F} \cdot d\mathbf{r}$ represents the work done by $\mathbf{F}$ in moving a particle along the curve C.

   (c) $\int_C \mathbf{F} \cdot d\mathbf{r} = \int_C P\, dx + Q\, dy + R\, dz$

5. See Theorem 2 in Section 13.3.

6. (a) $\int_C \mathbf{F} \cdot d\mathbf{r}$ is independent of path if the line integral has the same value for any two curves that have the same initial and terminal points.

   (b) See Theorem 4 in Section 13.3.

7. See the statement of Green's Theorem on page 946.

8. See (5) in Section 13.4.

9. (a) $\operatorname{curl} \mathbf{F} = \left(\dfrac{\partial R}{\partial y} - \dfrac{\partial Q}{\partial z} \right) \mathbf{i} + \left(\dfrac{\partial P}{\partial z} - \dfrac{\partial R}{\partial x} \right) \mathbf{j} + \left(\dfrac{\partial Q}{\partial x} - \dfrac{\partial P}{\partial y} \right) \mathbf{k} = \nabla \times \mathbf{F}$

   (b) $\operatorname{div} \mathbf{F} = \dfrac{\partial P}{\partial x} + \dfrac{\partial Q}{\partial y} + \dfrac{\partial R}{\partial z} = \nabla \cdot \mathbf{F}$

   (c) For curl $\mathbf{F}$, see the discussion accompanying Figure 1 on page 956 as well as Figure 6 and the accompanying discussion on pages 976–977. For div $\mathbf{F}$, see the discussion at the top of page 958 as well as the discussion preceding (8) on page 983.

10. See Theorem 6 in Section 13.3; see Theorem 4 in Section 13.5.

11. (a) See (1) in Section 13.6.

(b) We normally evaluate the surface integral using (2) in Section 13.6.

(c) See (4) in Section 13.6.

(d) The mass is $m = \iint_S \rho\,(x,y,z)\ dS$ and the center of mass is $(\bar{x}, \bar{y}, \bar{z})$ where

$$\bar{x} = \tfrac{1}{m} \iint_S x\rho\,(x,y,z)\ dS, \bar{y} = \tfrac{1}{m} \iint_S y\rho\,(x,y,z)\ dS, \bar{z} = \tfrac{1}{m} \iint_S z\rho\,(x,y,z)\ dS.$$

12. (a) See Figures 6 and 7 and the accompanying discussion on page 966.

(b) See Definition 8 in Section 13.6.

(c) See (9) in Section 13.6.

(d) See (10) in Section 13.6.

13. See the statement of Stokes' Theorem on page 973.

14. See the statement of the Divergence Theorem on page 979.

15. In each theorem, we have an integral of a "derivative" over a region on the left side, while the right side involves the values of the original function only on the boundary of the region.

True-False Quiz

1. False; div $\mathbf{F}$ is a scalar field.

3. True, by (13.5.3) and the fact that div $\mathbf{0} = 0$.

5. False. See Exercise 13.3.33. [But the assertion is true if D is simply-connected; see (13.3.6).]

7. True. Apply the Divergence Theorem and use the fact that div $\mathbf{F} = 0$.

Exercises

1. (a) Vectors starting on C point in roughly the direction opposite to C, so the tangential component $\mathbf{F} \cdot \mathbf{T}$ is negative. Thus $\int_C \mathbf{F} \cdot d\mathbf{r} = \int_C \mathbf{F} \cdot \mathbf{T}\,ds$ is negative.

(b) The vectors that end near P are shorter than the vectors that start near P, so the net flow is outward near P and div $\mathbf{F}\,(P)$ is positive.

3. $\int_C x^3 z\,ds = \int_0^{\pi/2} (2\sin t)^3\,(2\cos t)\,\sqrt{(2\cos t)^2 + (1)^2 + (-2\sin t)^2}\,dt = \int_0^{\pi/2} (16\sin^3 t \cos t)\,\sqrt{5}\,dt$
$= 4\sqrt{5}\sin^4 t\big]_0^{\pi/2} = 4\sqrt{5}$

5. $x = \cos t \Rightarrow dx = -\sin t\,dt$, $y = \sin t \Rightarrow dy = \cos t\,dt$, $0 \leq t \leq 2\pi$ and
$\int_C x^3 y\,dx - x\,dy = \int_0^{2\pi} (-\cos^3 t \sin^2 t - \cos^2 t)\,dt = \int_0^{2\pi} (-\cos^3 t \sin^2 t - \cos^2 t)\,dt = -\pi$
Or: Since C is a simple closed curve, apply Green's Theorem giving
$\iint_{x^2+y^2\leq 1} (-1 - x^3)\,dA = \int_0^1 \int_0^{2\pi} (-r - r^4 \cos^3\theta)\,d\theta = -\pi$.

7. 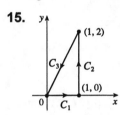 C_1: $x = t$, $y = t$, $z = 2t$, $0 \le t \le 1$;
C_2: $x = 1 + 2t$, $y = 1$, $z = 2 + 2t$, $0 \le t \le 1$. Then
$\int_C y\, dx + z\, dy + x\, dz = \int_0^1 5t\, dt + \int_0^1 (4 + 4t)\, dt = \frac{17}{2}$.

9. $\mathbf{F}(\mathbf{r}(t)) = (2t + t^2)\mathbf{i} + t^4\mathbf{j} + 4t^4\mathbf{k}$, $\mathbf{F} \cdot \mathbf{r}'(t) = 4t + 2t^2 + 2t^5 + 16t^7$ and
$\int_C \mathbf{F} \cdot d\mathbf{r} = \int_0^1 (4t + 2t^2 + 2t^5 + 16t^7)\, dt = 5$.

11. $\dfrac{\partial (\sin y)}{\partial y} = \cos y$ and $\dfrac{\partial (x \cos y + \sin y)}{\partial x} = \cos y$ and the domain of $\mathbf{F}$ is $\mathbb{R}^2$ so $\mathbf{F}$ is conservative.
Hence there exists f such that $\nabla f = \mathbf{F}$. Then $f_x(x, y) = \sin y$ implies $f(x, y) = x \sin y + g(y)$ and
$f_y(x, y) = x \cos y + g'(y)$. But $f_y(x, y) = x \cos y + \sin y$ so $g'(y) = \sin y$ and
$f(x, y) = x \sin y - \cos y + K$ is a potential function for $\mathbf{F}$.

13. Since $\dfrac{\partial (2x + y^2 + 3x^2 y)}{\partial y} = 2y + 3x^2 = \dfrac{\partial (2xy + x^3 + 3y^2)}{\partial x}$ and the domain of $\mathbf{F}$ is $\mathbb{R}^2$, $\mathbf{F}$ is
conservative. Furthermore $f(x, y) = x^2 + xy^2 + x^3 y + y^3 + K$ is a potential function for $\mathbf{F}$. Then
$\int_C \mathbf{F} \cdot d\mathbf{r} = f(\pi, 0) - f(0, 0) = \pi^2$.

15. C_1: $0 \le x \le 1$, $y = 0$; C_2: $x = 1$, $0 \le y \le 2$; C_3: $x = x$, $y = 2x$, $x = 1$ to $x = 0$.
Then $\int_C xy\, dx + x^2\, dy = \int_0^1 0\, dx + \int_0^2 (0 + 1)\, dy + \int_1^0 (2x^2 + 2x^2)\, dx = \frac{2}{3}$.
Using Green's Theorem, we have
$$\int_C xy\, dx + x^2\, dy = \iint_D \left[\tfrac{\partial}{\partial x}(x^2) - \tfrac{\partial}{\partial y}(xy) \right] dA = \iint_D (2x - x)\, dA$$
$$= \int_0^1 \int_0^{2x} x\, dy\, dx = \tfrac{2}{3}$$

17. $\int_C x^2 y\, dx - xy^2\, dy = \iint_{x^2 + y^2 \le 4} \left[\tfrac{\partial}{\partial x}(-xy^2) - \tfrac{\partial}{\partial y}(x^2 y) \right] dA$
$\qquad\qquad = \iint_{x^2 + y^2 \le 4} (-y^2 - x^2)\, dA = -\int_0^{2\pi} \int_0^2 r^3\, dr\, d\theta = -8\pi$

19. If we assume there is such a vector field $\mathbf{G}$, then $\operatorname{div}(\operatorname{curl} \mathbf{G}) = 2 + 3z - 2xz$. But $\operatorname{div}(\operatorname{curl} \mathbf{F}) = 0$ for
all vector fields $\mathbf{F}$. Thus such a $\mathbf{G}$ cannot exist.

21. For any piecewise-smooth simple closed plane curve C bounding a region
D, we can apply Green's Theorem to $\mathbf{F}(x, y) = f(x)\mathbf{i} + g(y)\mathbf{j}$ to get
$$\int_C f(x)\, dx + g(y)\, dy = \iint_D \left[\tfrac{\partial}{\partial x} g(y) - \tfrac{\partial}{\partial y} f(x) \right] dA = \iint_D 0\, dA = 0.$$

23. $\nabla^2 f = 0$ means that $\dfrac{\partial^2 f}{\partial x^2} + \dfrac{\partial^2 f}{\partial y^2} = 0$. Now if $\mathbf{F} = f_y\,\mathbf{i} - f_x\,\mathbf{j}$ and C is any closed path in D, then applying Green's Theorem, we get

$$\oint_C \mathbf{F} \cdot d\mathbf{r} = \int_C f_y\,dx - f_x\,dy = \iint_D \left[\frac{\partial}{\partial x}(-f_x) - \frac{\partial}{\partial y}(f_y) \right] dA = -\iint_D (f_{xx} + f_{yy})\,dA$$

$$= -\iint_D 0\,dA = 0$$

Therefore the line integral is independent of path, by Theorem 13.3.3.

25. $z = f(x, y) = x^2 + y^2$ with $0 \le x^2 + y^2 \le 4$ so $\mathbf{r}_x \times \mathbf{r}_y = -2x\,\mathbf{i} - 2y\,\mathbf{j} + \mathbf{k}$ (using upward orientation). Then

$$\iint_S z\,dS = \iint_{x^2 + y^2 \le 4} (x^2 + y^2)\,\sqrt{4x^2 + 4y^2 + 1}\,dA = \int_0^{2\pi} \int_0^2 r^3 \sqrt{1 + 4r^2}\,dr\,d\theta$$

$$= \tfrac{1}{60}\pi\left(391\sqrt{17} + 1\right)$$

(Substitute $u = 1 + 4r^2$ and use tables.)

27. Since the sphere bounds a simple solid region, the Divergence Theorem applies and

$$\iint_S \mathbf{F} \cdot d\mathbf{S} = \iiint_E (z - 2)\,dV = \iiint_E z\,dV - 2\iiint_E dV = m\bar{z} - 2\left(\tfrac{4}{3}\pi 2^3\right) = -\tfrac{64}{3}\pi.$$

Alternate Solution: $\mathbf{F}(\mathbf{r}(\phi, \theta)) = 4\sin\phi\cos\theta\cos\phi\,\mathbf{i} - 4\sin\phi\sin\theta\,\mathbf{j} + 6\sin\phi\cos\theta\,\mathbf{k}$,
$\mathbf{r}_\phi \times \mathbf{r}_\theta = 4\sin^2\phi\cos\theta\,\mathbf{i} + 4\sin^2\phi\sin\theta\,\mathbf{j} + 4\sin\phi\cos\phi\,\mathbf{k}$, and
$\mathbf{F} \cdot (\mathbf{r}_\phi \times \mathbf{r}_\theta) = 16\sin^3\phi\cos^2\theta\cos\phi - 16\sin^3\phi\sin^2\theta + 24\sin^2\phi\cos\phi\cos\theta$. Then

$$\iint_S \mathbf{F} \cdot d\mathbf{S} = \int_0^{2\pi} \int_0^\pi \left(16\sin^3\phi\cos\phi\cos^2\theta - 16\sin^3\phi\sin^2\theta + 24\sin^2\phi\cos\phi\cos\theta\right) d\phi\,d\theta$$

$$= \int_0^{2\pi} \tfrac{4}{3}\left(-16\sin^2\theta\right) d\theta = -\tfrac{64}{3}\pi$$

29. Since $\operatorname{curl}\mathbf{F} = \mathbf{0}$, $\iint_S (\operatorname{curl}\mathbf{F}) \cdot d\mathbf{S} = 0$. We parametrize C: $\mathbf{r}(t) = \cos t\,\mathbf{i} + \sin t\,\mathbf{j}$, $0 \le t \le 2\pi$ and

$$\oint_C \mathbf{F} \cdot d\mathbf{r} = \int_0^{2\pi} \left(-\cos^2 t\sin t + \sin^2 t\cos t\right) dt = \tfrac{1}{3}\cos^3 t + \tfrac{1}{3}\sin^3 t \Big]_0^{2\pi} = 0.$$

31. The surface is given by $x + y + z = 1$ or $z = 1 - x - y$, $0 \le x \le 1$, $0 \le y \le 1 - x$ and $\mathbf{r}_x \times \mathbf{r}_y = \mathbf{i} + \mathbf{j} + \mathbf{k}$. Then

$$\oint_C \mathbf{F} \cdot d\mathbf{r} = \iint_S \operatorname{curl}\mathbf{F} \cdot d\mathbf{S} = \iint_D (-y\,\mathbf{i} - z\,\mathbf{j} - x\,\mathbf{k}) \cdot (\mathbf{i} + \mathbf{j} + \mathbf{k})\,dA$$

$$= \iint_D (-1)\,dA = -(\text{area of } D) = -\tfrac{1}{2}$$

33. $\iiint_E \operatorname{div}\mathbf{F}\,dV = \iiint_{x^2 + y^2 + z^2 \le 1} 3\,dV = 3\,(\text{volume of sphere}) = 4\pi$. Then
$\mathbf{F}(\mathbf{r}(\phi, \theta)) \cdot (\mathbf{r}_\phi \times \mathbf{r}_\theta) = \sin^3\phi\cos^2\theta + \sin^3\phi\sin^2\theta + \sin\phi\cos^2\phi = \sin\phi$ and
$\iint_S \mathbf{F} \cdot d\mathbf{S} = \int_0^{2\pi} \int_0^\pi \sin\phi\,d\phi\,d\theta = (2\pi)(2) = 4\pi$.

35. Because $\operatorname{curl}\mathbf{F} = \mathbf{0}$, $\mathbf{F}$ is conservative, and if $f(x, y, z) = x^3 yz - 3xy + z^2$, then $\nabla f = \mathbf{F}$. Hence
$\int_C \mathbf{F} \cdot d\mathbf{r} = \int_C \nabla f \cdot d\mathbf{r} = f(0, 3, 0) - f(0, 0, 2) = 0 - 4 = -4$.

37. By the Divergence Theorem, $\iint_S \mathbf{F} \cdot \mathbf{n}\,dS = \iiint_E \operatorname{div}\mathbf{F}\,dV = 3\,(\text{volume of } E) = 3\,(8 - 1) = 21$.

Focus on Problem Solving

1. Let S_1 be the portion of $\Omega\,(S)$ between $S\,(a)$ and S, and let ∂S_1 be its boundary. Also let S_L be the lateral surface of S_1 [that is, the surface of S_1 except S and $S\,(a)$]. Applying the Divergence Theorem we have $\iint_{\partial S_1} \dfrac{\mathbf{r}\cdot\mathbf{n}}{r^3}\,dS = \iiint_{S_1} \nabla\cdot\dfrac{\mathbf{r}}{r^3}\,dV$. But

$$\nabla\cdot\frac{\mathbf{r}}{r^3} = \left\langle \frac{\partial}{\partial x}, \frac{\partial}{\partial y}, \frac{\partial}{\partial z} \right\rangle \cdot \left\langle \frac{x}{(x^2+y^2+z^2)^{3/2}}, \frac{y}{(x^2+y^2+z^2)^{3/2}}, \frac{z}{(x^2+y^2+z^2)^{3/2}} \right\rangle$$

$$= \frac{\left(x^2+y^2+z^2-3x^2\right)+\left(x^2+y^2+z^2-3y^2\right)+\left(x^2+y^2+z^2-3z^2\right)}{\left(x^2+y^2+z^2\right)^{5/2}} = 0$$

$$\Rightarrow \iint_{\partial S_1} \frac{\mathbf{r}\cdot\mathbf{n}}{r^3}\,dS = \iiint_{S_1} 0\,dV = 0.$$ On the other hand, notice that for the surfaces of ∂S_1 other than $S\,(a)$ and S, $\mathbf{r}\cdot\mathbf{n} = 0 \;\Rightarrow$

$$0 = \iint_{\partial S_1} \frac{\mathbf{r}\cdot\mathbf{n}}{r^3}\,dS = \iint_{S} \frac{\mathbf{r}\cdot\mathbf{n}}{r^3}\,dS + \iint_{S(a)} \frac{\mathbf{r}\cdot\mathbf{n}}{r^3}\,dS + \iint_{S_L} \frac{\mathbf{r}\cdot\mathbf{n}}{r^3}\,dS$$

$$= \iint_{S} \frac{\mathbf{r}\cdot\mathbf{n}}{r^3}\,dS + \iint_{S(a)} \frac{\mathbf{r}\cdot\mathbf{n}}{r^3}\,dS$$

$$\Rightarrow \iint_{S} \frac{\mathbf{r}\cdot\mathbf{n}}{r^3}\,dS = -\iint_{S(a)} \frac{\mathbf{r}\cdot\mathbf{n}}{r^3}\,dS.$$ Notice that on $S\,(a)$,

$$r = a \;\Rightarrow\; \mathbf{n} = -\frac{\mathbf{r}}{r} = -\frac{\mathbf{r}}{a} \text{ and } \mathbf{r}\cdot\mathbf{r} = r^2 = a^2, \text{ so that}$$

$$-\iint_{S(a)} \frac{\mathbf{r}\cdot\mathbf{n}}{r^3}\,dS = \iint_{S(a)} \frac{\mathbf{r}\cdot\mathbf{r}}{a^4}\,dS = \iint_{S(a)} \frac{a^2}{a^4}\,dS = \frac{1}{a^2}\iint_{S(a)} dS = \frac{\text{area of } S\,(a)}{a^2} = |\Omega\,(S)|.$$

Therefore $|\Omega\,(S)| = \displaystyle\iint_{S} \frac{\mathbf{r}\cdot\mathbf{n}}{r^3}\,dS.$

3. The given line integral $\frac{1}{2}\int_C (bz - cy)\,dx + (cx - az)\,dy + (ay - bx)\,dz$ can be expressed as $\int_C \mathbf{F}\cdot d\mathbf{r}$ if we define the vector field $\mathbf{F}$ by
$\mathbf{F}\,(x,y,z) = P\mathbf{i} + Q\mathbf{j} + R\mathbf{k} = \frac{1}{2}(bz - cy)\,\mathbf{i} + \frac{1}{2}(cx - az)\,\mathbf{j} + \frac{1}{2}(ay - bx)\,\mathbf{k}$. Then define S to be the planar interior of C, so S is an oriented, smooth surface. Stokes' Theorem says
$\int_C \mathbf{F}\cdot d\mathbf{r} = \iint_S \operatorname{curl}\mathbf{F}\cdot d\mathbf{S} = \iint_S \operatorname{curl}\mathbf{F}\cdot\mathbf{n}\,dS$. Now

$$\operatorname{curl}\mathbf{F} = \left(\frac{\partial R}{\partial y} - \frac{\partial Q}{\partial z}\right)\mathbf{i} + \left(\frac{\partial P}{\partial z} - \frac{\partial R}{\partial x}\right)\mathbf{j} + \left(\frac{\partial Q}{\partial x} - \frac{\partial P}{\partial y}\right)\mathbf{k}$$

$$= \left(\tfrac{1}{2}a + \tfrac{1}{2}a\right)\mathbf{i} + \left(\tfrac{1}{2}b + \tfrac{1}{2}b\right)\mathbf{j} + \left(\tfrac{1}{2}c + \tfrac{1}{2}c\right)\mathbf{k} = a\mathbf{i} + b\mathbf{j} + c\mathbf{k} = \mathbf{n}$$

so $\operatorname{curl}\mathbf{F}\cdot\mathbf{n} = \mathbf{n}\cdot\mathbf{n} = |\mathbf{n}|^2 = 1$, hence $\iint_S \operatorname{curl}\mathbf{F}\cdot\mathbf{n}\,dS = \iint_S dS$ which is simply the surface area of S. Thus, $\int_C \mathbf{F}\cdot d\mathbf{r} = \frac{1}{2}\int_C (bz - cy)\,dx + (cx - az)\,dy + (ay - bx)\,dz$ is the plane area enclosed by C.

Appendixes

Appendix D   Precise Definitions of Limits

1. (*Note:* This is Exercise 21 in the full version of the text.)

Let $\varepsilon > 0$. We want to find $\delta > 0$ such that

$$\left| \frac{xy}{\sqrt{x^2 + y^2}} - 0 \right| < \varepsilon \qquad \text{whenever} \qquad 0 < \sqrt{x^2 + y^2} < \delta$$

that is,

$$\frac{|xy|}{\sqrt{x^2 + y^2}} < \varepsilon \qquad \text{whenever} \qquad 0 < \sqrt{x^2 + y^2} < \delta$$

But $|x| = \sqrt{x^2} \le \sqrt{x^2 + y^2}$ and $|y| = \sqrt{y^2} \le \sqrt{x^2 + y^2}$, so

$$\frac{|xy|}{\sqrt{x^2 + y^2}} \le \frac{\left(\sqrt{x^2 + y^2}\right)^2}{\sqrt{x^2 + y^2}} = \sqrt{x^2 + y^2}$$

Thus, if we choose $\delta = \varepsilon$ and let $0 < \sqrt{x^2 + y^2} < \delta$, then

$$\left| \frac{xy}{\sqrt{x^2 + y^2}} - 0 \right| \le \sqrt{x^2 + y^2} < \delta = \varepsilon$$

Hence, by Definition 1,

$$\lim_{(x,y) \to (0,0)} \frac{xy}{\sqrt{x^2 + y^2}} = 0$$

Appendix G Polar Coordinates

Section G.1   Curves in Polar Coordinates

1. (a) By adding 2π to $\frac{\pi}{2}$, we obtain the point $\left(1, \frac{5\pi}{2}\right)$. The direction opposite $\frac{\pi}{2}$ is $\frac{3\pi}{2}$, so $\left(-1, \frac{3\pi}{2}\right)$ is a point that satisfies the $r < 0$ requirement.

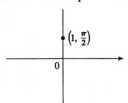

(b) $\left(-1, \frac{\pi}{5}\right)$

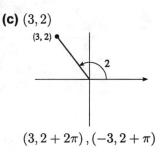

$\left(1, \frac{6\pi}{5}\right), \left(-1, \frac{11\pi}{5}\right)$

(c) $(3, 2)$

$(3, 2 + 2\pi), (-3, 2 + \pi)$

3. (a)

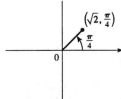

Using Equations 1 with $r = \sqrt{2}$ and $\theta = \frac{\pi}{4}$, we have
$x = r\cos\theta = \sqrt{2}\cos\frac{\pi}{4} = 1$ and
$y = r\sin\theta = \sqrt{2}\sin\frac{\pi}{4} = 1$. Thus, the point has Cartesian coordinates $(1, 1)$.

(b)

$x = 1.5\cos\frac{3\pi}{2} = 0$ and
$y = 1.5\sin\frac{3\pi}{2} = -1.5$
give us $\left(0, -\frac{3}{2}\right)$.

(c)

$x = -1\cos\frac{\pi}{3} = -\frac{1}{2}$ and
$y = -1\sin\frac{\pi}{3} = -\frac{\sqrt{3}}{2}$
give us $\left(-\frac{1}{2}, -\frac{\sqrt{3}}{2}\right)$.

5. (a) $(x, y) = (-1, 1), r = \sqrt{(-1)^2 + 1^2} = \sqrt{2}$, $\tan\theta = y/x = -1$ and (x, y) is in quadrant II, so $\theta = \frac{3\pi}{4}$. The polar coordinates are $\left(\sqrt{2}, \frac{3\pi}{4}\right)$.

(b) $(x, y) = (2\sqrt{3}, -2), r = \sqrt{12 + 4} = 4$, $\tan\theta = y/x = -\frac{1}{\sqrt{3}}$ and (x, y) is in quadrant IV, so $\theta = \frac{11\pi}{6}$. The polar coordinates are $\left(4, \frac{11\pi}{6}\right)$.

7. $r > 1$

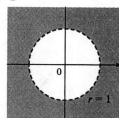

9. $0 \le r \le 2, \frac{\pi}{2} \le \theta \le \pi$

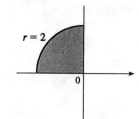

11. $3 < r < 4, -\frac{\pi}{2} \le \theta \le \pi$

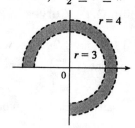

13. Since $y = r \sin \theta$, the equation $r \sin \theta = 2$ becomes $y = 2$.

15. $r = \dfrac{1}{1 - \cos \theta} \quad \Rightarrow \quad r - r \cos \theta = 1 \quad \Rightarrow \quad r = 1 + r \cos \theta \quad \Rightarrow \quad r^2 = (1 + r \cos \theta)^2 \quad \Rightarrow$

$x^2 + y^2 = (1 + x)^2 = 1 + 2x + x^2 \quad \Leftrightarrow \quad y^2 = 1 + 2x$

17. $y = 5 \quad \Leftrightarrow \quad r \sin \theta = 5$

19. $x^2 + y^2 = 25 \quad \Leftrightarrow \quad r^2 = 25 \quad \Rightarrow \quad r = 5$

21. As in Example 4, $r = 5$ represents the circle with center O and radius 5.

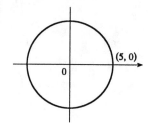

23. $r = 2 \sin \theta \quad \Rightarrow \quad r^2 = 2r \sin \theta \quad \Rightarrow$
$x^2 + y^2 = 2y \quad \Rightarrow$
$x^2 + (y^2 - 2y + 1) = 1 \quad \Rightarrow$
$x^2 + (y - 1)^2 = 1$, a circle with center
$(0, 1)$ and radius 1.

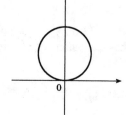

25. $r = \theta, \theta \ge 0$

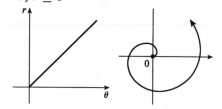

27. $r = 1 - 2 \cos \theta$

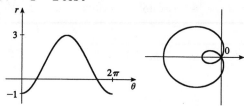

29. $r = 2\cos 4\theta$

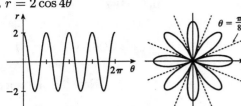

31. $r^2 = 4\cos 2\theta$

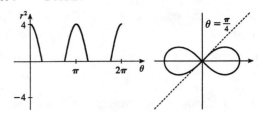

33. $x = (r)\cos\theta = (4 + 2\sec\theta)\cos\theta = 4\cos\theta + 2$. Now, $r \to \infty \Rightarrow$ $(4 + 2\sec\theta) \to \infty \Rightarrow \theta \to \left(\frac{\pi}{2}\right)^- $ or $\theta \to \left(\frac{3\pi}{2}\right)^+$ (since we need only consider $0 \le \theta < 2\pi$), so $\displaystyle\lim_{r\to\infty} x = \lim_{\theta\to\pi/2-} (4\cos\theta + 2) = 2$.

Also, $r \to -\infty \Rightarrow (4 + 2\sec\theta) \to -\infty \Rightarrow \theta \to \left(\frac{\pi}{2}\right)^+$ or

$\theta \to \left(\frac{3\pi}{2}\right)^-$, so $\displaystyle\lim_{r\to-\infty} x = \lim_{\theta\to\pi/2+} (4\cos\theta + 2) = 2$. Therefore,

$\displaystyle\lim_{r\to\pm\infty} x = 2 \Rightarrow x = 2$ is a vertical asymptote.

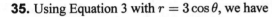

35. Using Equation 3 with $r = 3\cos\theta$, we have

$$\frac{dy}{dx} = \frac{dy/d\theta}{dx/d\theta} = \frac{(dr/d\theta)(\sin\theta) + r\cos\theta}{(dr/d\theta)(\cos\theta) - r\sin\theta} = \frac{-3\sin\theta\sin\theta + 3\cos\theta\cos\theta}{-3\sin\theta\cos\theta - 3\cos\theta\sin\theta} = \frac{3\left(\cos^2\theta - \sin^2\theta\right)}{-3\left(2\sin\theta\cos\theta\right)}$$

$$= -\frac{\cos 2\theta}{\sin 2\theta} = -\cot 2\theta = \frac{1}{\sqrt{3}} \text{ when } \theta = \frac{\pi}{3}.$$

Another Solution: $r = 3\cos\theta \Rightarrow x = r\cos\theta = 3\cos^2\theta, y = r\sin\theta = 3\sin\theta\cos\theta \Rightarrow$

$$\frac{dy}{dx} = \frac{dy/d\theta}{dx/d\theta} = \frac{-3\sin^2\theta + 3\cos^2\theta}{-6\cos\theta\sin\theta} = \frac{\cos 2\theta}{-\sin 2\theta} = -\cot 2\theta = \frac{1}{\sqrt{3}} \text{ when } \theta = \frac{\pi}{3}.$$

37. $r = \theta \Rightarrow x = r\cos\theta = \theta\cos\theta, y = r\sin\theta = \theta\sin\theta \Rightarrow$

$$\frac{dy}{dx} = \frac{dy/d\theta}{dx/d\theta} = \frac{\sin\theta + \theta\cos\theta}{\cos\theta - \theta\sin\theta} = \frac{1}{-\pi/2} = -\frac{2}{\pi} \text{ when } \theta = \frac{\pi}{2}.$$

39. $r = \cos 2\theta \Rightarrow x = r\cos\theta = \cos 2\theta\cos\theta, y = r\sin\theta = \cos 2\theta\sin\theta \Rightarrow$

$dy/d\theta = -2\sin 2\theta\sin\theta + \cos 2\theta\cos\theta = -4\sin^2\theta\cos\theta + \left(\cos^3\theta - \sin^2\theta\cos\theta\right)$

$= \cos\theta\left(\cos^2\theta - 5\sin^2\theta\right) = \cos\theta\left(1 - 6\sin^2\theta\right) = 0 \Rightarrow$

$\cos\theta = 0$ or $\sin\theta = \pm\frac{1}{\sqrt{6}} \Rightarrow \theta = \frac{\pi}{2}, \frac{3\pi}{2}, \alpha, \pi - \alpha, \pi + \alpha$, or $2\pi - \alpha$ (where $\alpha = \sin^{-1}\frac{1}{\sqrt{6}}$).

So the tangent is horizontal at $\left(-1, \frac{\pi}{2}\right), \left(-1, \frac{3\pi}{2}\right), \left(\frac{2}{3}, \alpha\right), \left(\frac{2}{3}, \pi - \alpha\right), \left(\frac{2}{3}, \pi + \alpha\right)$, and $\left(\frac{2}{3}, 2\pi - \alpha\right)$.

$dx/d\theta = -2\sin 2\theta\cos\theta - \cos 2\theta\sin\theta = -4\sin\theta\cos^2\theta - \left(2\cos^2\theta - 1\right)\sin\theta$

$= \sin\theta\left(1 - 6\cos^2\theta\right) = 0 \Rightarrow$

$\sin\theta = 0$ or $\cos\theta = \pm\frac{1}{\sqrt{6}} \Rightarrow \theta = 0, \pi, \frac{\pi}{2} - \alpha, \frac{\pi}{2} + \alpha, \frac{3\pi}{2} - \alpha$, or $\frac{3\pi}{2} + \alpha$ (where $\alpha = \cos^{-1}\frac{1}{\sqrt{6}}$).

So the tangent is vertical at $(1, 0), (1, \pi), \left(\frac{2}{3}, \frac{3\pi}{2} - \alpha\right), \left(\frac{2}{3}, \frac{3\pi}{2} + \alpha\right), \left(\frac{2}{3}, \frac{\pi}{2} - \alpha\right)$, and $\left(\frac{2}{3}, \frac{\pi}{2} + \alpha\right)$.

41. $r = 1 + \cos\theta$ $\Rightarrow$ $x = r\cos\theta = \cos\theta\,(1 + \cos\theta)$, $y = r\sin\theta = \sin\theta\,(1 + \cos\theta)$ $\Rightarrow$
$dy/d\theta = (1 + \cos\theta)\cos\theta - \sin^2\theta = 2\cos^2\theta + \cos\theta - 1 = (2\cos\theta - 1)(\cos\theta + 1) = 0$ $\Rightarrow$
$\cos\theta = \frac{1}{2}$ or -1 $\Rightarrow$ $\theta = \frac{\pi}{3}, \pi,$ or $\frac{5\pi}{3}$ $\Rightarrow$ horizontal tangents at $\left(\frac{3}{2}, \frac{\pi}{3}\right)$, $(0, \pi)$, and $\left(\frac{3}{2}, \frac{5\pi}{3}\right)$.
$dx/d\theta = -(1 + \cos\theta)\sin\theta - \cos\theta\sin\theta = -\sin\theta\,(1 + 2\cos\theta) = 0$ $\Rightarrow$ $\sin\theta = 0$ or $\cos\theta = -\frac{1}{2}$
$\Rightarrow$ $\theta = 0, \pi, \frac{2\pi}{3},$ or $\frac{4\pi}{3}$ $\Rightarrow$ vertical tangents at $(2, 0)$, $\left(\frac{1}{2}, \frac{2\pi}{3}\right)$, and $\left(\frac{1}{2}, \frac{4\pi}{3}\right)$. Note that the tangent is
horizontal, not vertical, when $\theta = \pi$, since $\displaystyle\lim_{\theta \to \pi} \frac{dy/d\theta}{dx/d\theta} = 0$ [by l'Hospital's Rule].

43. $r = a\sin\theta + b\cos\theta$ $\Rightarrow$ $r^2 = ar\sin\theta + br\cos\theta$ $\Rightarrow$ $x^2 + y^2 = ay + bx$ $\Rightarrow$
$\left(x - \frac{1}{2}b\right)^2 + \left(y - \frac{1}{2}a\right)^2 = \frac{1}{4}\left(a^2 + b^2\right)$, and this is a circle with center $\left(\frac{1}{2}b, \frac{1}{2}a\right)$ and radius $\frac{1}{2}\sqrt{a^2 + b^2}$.

45. (a) We see that the curve $r = 1 + c\sin\theta$ crosses itself at the origin, where $r = 0$ (in fact the inner loop
corresponds to negative r-values), so we solve the equation of the limaçon for $r = 0$ $\Leftrightarrow$
$c\sin\theta = -1$ $\Leftrightarrow$ $\sin\theta = -1/c$. Now if $|c| < 1$, then this equation has no solution and hence
there is no inner loop. But if $c < -1$, then on the interval $(0, 2\pi)$ the equation has the two solutions
$\theta = \sin^{-1}(-1/c)$ and $\theta = \pi - \sin^{-1}(-1/c)$, and if $c > 1$, the solutions are $\theta = \pi + \sin^{-1}(1/c)$
and $\theta = 2\pi - \sin^{-1}(1/c)$. In each case, $r < 0$ for θ between the two solutions, indicating a loop.

(b) For $0 < c < 1$, the dimple (if it exists) is characterized by the fact that y has a local maximum at
$\theta = \frac{3\pi}{2}$. So we determine for what c-values $d^2y/d\theta^2$ is negative at $\theta = \frac{3\pi}{2}$, since by the Second
Derivative Test this indicates a maximum: $y = r\sin\theta = \sin\theta + c\sin^2\theta$ $\Rightarrow$
$dy/d\theta = \cos\theta + 2c\sin\theta\cos\theta = \cos\theta + c\sin 2\theta$ $\Rightarrow$ $d^2y/d\theta^2 = -\sin\theta + 2c\cos 2\theta$. At $\theta = \frac{3\pi}{2}$,
this is equal to $-(-1) + 2c\,(-1) = 1 - 2c$, which is negative only for $c > \frac{1}{2}$. A similar argument
shows that for $-1 < c < 0$, y only has a local minimum at $\theta = \frac{\pi}{2}$ (indicating a dimple) for $c < -\frac{1}{2}$.

Exercises 47-50: Maple is able to plot polar curves using the `polarplot` command, or using the
`coords=polar` option in a regular `plot` command. In Mathematica, use `PolarPlot`. In
Derive, change to `Polar` under `Options State`. If your graphing device cannot plot polar
equations, you must convert to parametric equations. For example, in Exercise 47,
$x = r\cos\theta = [1 + 2\sin(\theta/2)]\cos\theta$, $y = r\sin\theta = [1 + 2\sin(\theta/2)]\sin\theta$.

47. $r = 1 + 2\sin(\theta/2)$. The correct parameter
interval is $[0, 4\pi]$.

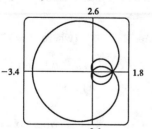

49. $r = \sin(9\theta/4)$. The correct parameter
interval is $[0, 8\pi]$.

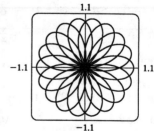

51.

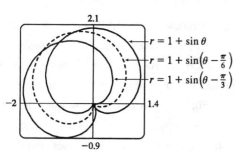

It appears that the graph of $r = 1 + \sin\left(\theta - \frac{\pi}{6}\right)$ is the same shape as the graph of $r = 1 + \sin\theta$, but rotated counterclockwise about the origin by $\frac{\pi}{6}$. Similarly, the graph of $r = 1 + \sin\left(\theta - \frac{\pi}{3}\right)$ is rotated by $\frac{\pi}{3}$. In general, the graph of $r = f\left(\theta - \alpha\right)$ is the same shape as that of $r = f\left(\theta\right)$, but rotated counterclockwise through α about the origin. That is, for any point (r_0, θ_0) on the curve $r = f\left(\theta\right)$, the point $(r_0, \theta_0 + \alpha)$ is on the curve the curve $r = f\left(\theta - \alpha\right)$, since

$$r_0 = f\left(\theta_0\right) = f\left(\left(\theta_0 + \alpha\right) - \alpha\right).$$

53. (a) $r = \sin n\theta$. From the graphs, it seems that when n is even, the number of loops in the curve (called a rose) is $2n$, and when n is odd, the number of loops is simply n.

This is because in the case of n odd, every point on the graph is traversed twice, due to the fact

that $r\left(\theta + \pi\right) = \sin\left[n\left(\theta + \pi\right)\right] = \sin n\theta \cos n\pi + \cos n\theta \sin n\pi = \begin{cases} \sin n\theta & \text{if } n \text{ is even} \\ -\sin n\theta & \text{if } n \text{ is odd} \end{cases}$

| $n = 2$ | $n = 3$ | $n = 4$ | $n = 5$ |

(b) The graph of $r = |\sin n\theta|$ has $2n$ loops whether n is odd or even, since $r\left(\theta + \pi\right) = r\left(\theta\right)$.

| $n = 2$ | $n = 3$ | $n = 4$ | $n = 5$ |

55. $r = \dfrac{1 - a\cos\theta}{1 + a\cos\theta}$. We start with $a = 0$, since in this case the curve is simply the circle $r = 1$.

As a increases, the graph moves to the left, and its right side becomes flattened. As a increases through about 0.4, the right side seems to grow a dimple, which upon closer investigation (with narrower θ-ranges) seems to appear at $a \approx 0.42$ (the actual value is $\sqrt{2} - 1$). As $a \to 1$, this dimple becomes more pronounced, and the curve begins to stretch out horizontally, until at $a = 1$ the denominator vanishes at $\theta = \pi$, and the dimple becomes an actual cusp. For $a > 1$ we must choose our parameter interval carefully, since $r \to \infty$ as $1 + a\cos\theta \to 0 \iff \theta \to \pm\cos^{-1}(-1/a)$. As a increases from 1, the curve splits into two parts. The left part has a loop, which grows larger as a increases, and the right part grows broader vertically, and its left tip develops a dimple when $a \approx 2.42$ (actually, $\sqrt{2} + 1$). As a increases, the dimple grows more and more pronounced.

If $a < 0$, we get the same graph as we do for the corresponding positive a-value, but with a rotation through π about the pole, as happened when c was replaced with $-c$ in Exercise 54.

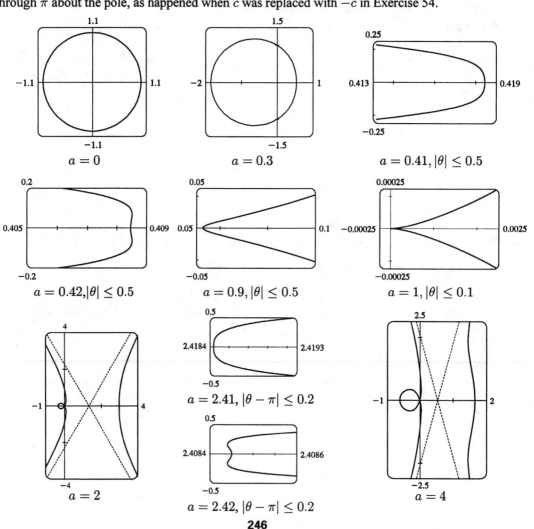

57.

$$\tan \psi = \tan (\phi - \theta) = \frac{\tan \phi - \tan \theta}{1 + \tan \phi \tan \theta} = \frac{\dfrac{dy}{dx} - \tan \theta}{1 + \dfrac{dy}{dx} \tan \theta} = \frac{\dfrac{dy/d\theta}{dx/d\theta} - \tan \theta}{1 + \dfrac{dy/d\theta}{dx/d\theta} \tan \theta}$$

$$= \frac{\dfrac{dy}{d\theta} - \dfrac{dx}{d\theta} \tan \theta}{\dfrac{dx}{d\theta} + \dfrac{dy}{d\theta} \tan \theta} = \frac{\left(\dfrac{dr}{d\theta} \sin \theta + r \cos \theta\right) - \tan \theta \left(\dfrac{dr}{d\theta} \cos \theta - r \sin \theta\right)}{\left(\dfrac{dr}{d\theta} \cos \theta - r \sin \theta\right) + \tan \theta \left(\dfrac{dr}{d\theta} \sin \theta + r \cos \theta\right)}$$

$$= \frac{r \cos \theta + r \cdot \dfrac{\sin^2 \theta}{\cos \theta}}{\dfrac{dr}{d\theta} \cos \theta + \dfrac{dr}{d\theta} \cdot \dfrac{\sin^2 \theta}{\cos \theta}} = \frac{r \cos^2 \theta + r \sin^2 \theta}{\dfrac{dr}{d\theta} \cos^2 \theta + \dfrac{dr}{d\theta} \sin^2 \theta} = \frac{r}{dr/d\theta}$$

Section G.2 Areas and Lengths in Polar Coordinates

1. $A = \int_0^\pi \frac{1}{2}r^2 d\theta = \int_0^\pi \frac{1}{2}\theta^2 d\theta = \left[\frac{1}{6}\theta^3\right]_0^\pi = \frac{1}{6}\pi^3$

3. $A = \int_0^{\pi/6} \frac{1}{2}\left(2\cos\theta\right)^2 d\theta = \int_0^{\pi/6} 2\cos^2\theta d\theta = \int_0^{\pi/6}\left(1+\cos 2\theta\right)d\theta$ [since $\cos^2\theta = \frac{1}{2}\left(1+\cos 2\theta\right)$]
$= \left[\theta + \frac{1}{2}\sin 2\theta\right]_0^{\pi/6} = \frac{\pi}{6} + \frac{\sqrt{3}}{4}$

5. $A = 4\int_0^{\pi/4} \frac{1}{2}r^2 d\theta = 2\int_0^{\pi/4}\left(4\cos 2\theta\right)d\theta = 8\int_0^{\pi/4}\cos 2\theta d\theta$
$= 4\left[\sin 2\theta\right]_0^{\pi/4} = 4$

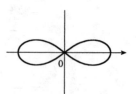

7. $A = 2\int_{-\pi/2}^{\pi/2} \frac{1}{2}\left(4-\sin\theta\right)^2 d\theta = \int_{-\pi/2}^{\pi/2}\left(16 - 8\sin\theta + \sin^2\theta\right)d\theta$
$= \int_{-\pi/2}^{\pi/2}\left(16 + \sin^2\theta\right)d\theta$ [by Theorem 5.5.6b]
$= 2\int_0^{\pi/2}\left(16 + \sin^2\theta\right)d\theta$ [by Theorem 5.5.6a]
$= 2\int_0^{\pi/2}\left[16 + \frac{1}{2}\left(1-\cos 2\theta\right)\right]d\theta = 2\left[\frac{33}{2}\theta - \frac{1}{4}\sin 2\theta\right]_0^{\pi/2}$
$= \frac{33\pi}{2}$

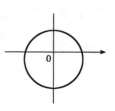

9. By symmetry, the total area is twice the area enclosed above the
polar axis, so
$A = 2\int_0^\pi \frac{1}{2}r^2 d\theta = \int_0^\pi \left[2 + \cos 6\theta\right]^2 d\theta$
$= \int_0^\pi \left(4 + 4\cos 6\theta + \cos^2 6\theta\right)d\theta$
$= \left[4\theta + 4\left(\frac{1}{6}\sin 6\theta\right) + \left(\frac{1}{24}\sin 12\theta + \frac{1}{2}\theta\right)\right]_0^\pi$
$= 4\pi + \frac{\pi}{2} = \frac{9\pi}{2}$.

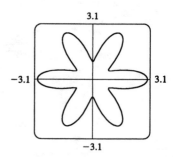

11. $A = \int_0^{\pi/5} \frac{1}{2}\sin^2 5\theta d\theta = \frac{1}{4}\int_0^{\pi/5}\left(1-\cos 10\theta\right)d\theta = \frac{1}{4}\left[\theta - \frac{1}{10}\sin 10\theta\right]_0^{\pi/5} = \frac{\pi}{20}$

13.

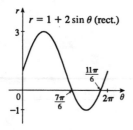

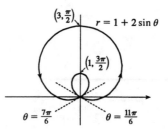

This is a limaçon, with inner loop traced out between $\theta = \frac{7\pi}{6}$ and $\frac{11\pi}{6}$ [found by solving $r = 0$].

$A = 2\int_{7\pi/6}^{3\pi/2} \frac{1}{2}\left(1 + 2\sin\theta\right)^2 d\theta = \int_{7\pi/6}^{3\pi/2}\left(1 + 4\sin\theta + 4\sin^2\theta\right)d\theta$

$= \left[\theta - 4\cos\theta + 2\theta - \sin 2\theta\right]_{7\pi/6}^{3\pi/2} = \left(\frac{9\pi}{2}\right) - \left(\frac{7\pi}{2} + 2\sqrt{3} - \frac{\sqrt{3}}{2}\right) = \pi - \frac{3\sqrt{3}}{2}$

15. $1 - \cos\theta = \frac{3}{2} \implies \cos\theta = -\frac{1}{2} \implies \theta = \frac{2\pi}{3}$ or $\frac{4\pi}{3} \implies$

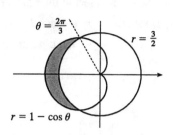

$$A = 2\int_{2\pi/3}^{\pi} \frac{1}{2}\left[(1-\cos\theta)^2 - \left(\frac{3}{2}\right)^2\right] d\theta$$
$$= \int_{2\pi/3}^{\pi}\left(-\frac{5}{4} - 2\cos\theta + \cos^2\theta\right) d\theta$$
$$= \left[-\frac{5}{4}\theta - 2\sin\theta\right]_{2\pi/3}^{\pi} + \frac{1}{2}\int_{2\pi/3}^{\pi}(1+\cos 2\theta)\, d\theta$$
$$= -\frac{5}{12}\pi + \sqrt{3} + \frac{1}{2}\left[\theta + \frac{1}{2}\sin 2\theta\right]_{2\pi/3}^{\pi}$$
$$= -\frac{5}{12}\pi + \sqrt{3} + \frac{1}{6}\pi + \frac{\sqrt{3}}{8} = \frac{9\sqrt{3}}{8} - \frac{1}{4}\pi$$

17. $4\sin\theta = 2 \iff \sin\theta = \frac{1}{2} \implies \theta = \frac{\pi}{6}$ or $\frac{5\pi}{6} \implies$

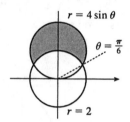

$$A = 2\int_{\pi/6}^{\pi/2} \frac{1}{2}\left[(4\sin\theta)^2 - 2^2\right] d\theta = \int_{\pi/6}^{\pi/2}(16\sin^2\theta - 4)\, d\theta$$
$$= \int_{\pi/6}^{\pi/2}[8(1-\cos 2\theta) - 4]\, d\theta = [4\theta - 4\sin 2\theta]_{\pi/6}^{\pi/2}$$
$$= \frac{4}{3}\pi + 2\sqrt{3}$$

19. $A = 2\int_0^{\pi/4} \frac{1}{2}\sin^2\theta\, d\theta = \int_0^{\pi/4} \frac{1}{2}(1-\cos 2\theta)\, d\theta$

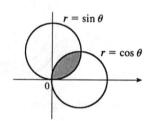

$$= \left[\frac{1}{2}\theta - \frac{1}{4}\sin 2\theta\right]_0^{\pi/4}$$
$$= \frac{1}{8}\pi - \frac{1}{4}$$

21. $\sin 2\theta = \cos 2\theta \implies \tan 2\theta = 1 \implies 2\theta = \frac{\pi}{4} \implies \theta = \frac{\pi}{8} \implies$

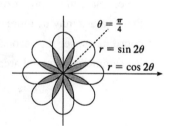

$$A = 16\int_0^{\pi/8} \frac{1}{2}\sin^2 2\theta\, d\theta = 4\int_0^{\pi/8}(1-\cos 4\theta)\, d\theta$$
$$= 4\left[\theta - \frac{1}{4}\sin 4\theta\right]_0^{\pi/8} = \frac{1}{2}\pi - 1$$

23. $A = 2\left[\int_0^{2\pi/3} \frac{1}{2}\left(\frac{1}{2} + \cos\theta\right)^2 d\theta - \int_{2\pi/3}^{\pi} \frac{1}{2}\left(\frac{1}{2} + \cos\theta\right)^2 d\theta\right]$

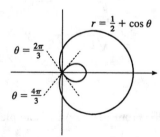

$$= \int_0^{2\pi/3}\left(\frac{1}{4} + \cos\theta + \cos^2\theta\right) d\theta - \int_{2\pi/3}^{\pi}\left(\frac{1}{4} + \cos\theta + \cos^2\theta\right) d\theta$$
$$= \left[\frac{\theta}{4} + \sin\theta + \frac{\theta}{2} + \frac{\sin 2\theta}{4}\right]_0^{2\pi/3} - \left[\frac{\theta}{4} + \sin\theta + \frac{\theta}{2} + \frac{\sin 2\theta}{4}\right]_{2\pi/3}^{\pi}$$
$$= \left(\frac{\pi}{2} + \frac{3\sqrt{3}}{8}\right) - \left(\frac{3\pi}{4}\right) + \left(\frac{\pi}{2} + \frac{3\sqrt{3}}{8}\right) = \frac{1}{4}\left(\pi + 3\sqrt{3}\right)$$

25. The curves intersect at the pole since $\left(0, \frac{\pi}{2}\right)$ satisfies $r = \cos\theta$ and

$(0, 0)$ satisfies $r = 1 - \cos\theta$. $\cos\theta = 1 - \cos\theta$ $\Rightarrow$ $\cos\theta = \frac{1}{2}$ $\Rightarrow$

$\theta = \frac{\pi}{3}$ or $\frac{5\pi}{3}$ $\Rightarrow$ the other intersection points are $\left(\frac{1}{2}, \frac{\pi}{3}\right)$ and

$\left(\frac{1}{2}, \frac{5\pi}{3}\right)$.

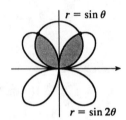

27. The pole is a point of intersection. $\sin\theta = \sin 2\theta = 2\sin\theta\cos\theta$ $\Leftrightarrow$

$\sin\theta(1 - 2\cos\theta) = 0$ $\Leftrightarrow$ $\sin\theta = 0$ or $\cos\theta = \frac{1}{2}$ $\Rightarrow$ $\theta = 0, \pi,$

$\frac{\pi}{3}, -\frac{\pi}{3}$ $\Rightarrow$ $\left(\frac{\sqrt{3}}{2}, \frac{\pi}{3}\right)$ and $\left(\frac{\sqrt{3}}{2}, \frac{2\pi}{3}\right)$ (by symmetry) are the other

intersection points.

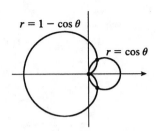

29.

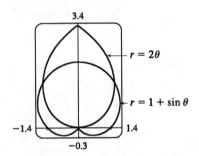

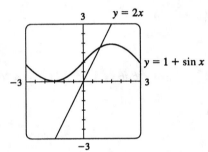

From the first graph, we see that the pole is one point of intersection. By zooming in or using the cursor, we estimate the θ-values of the intersection points to be about 0.89 and $\pi - 0.89 \approx 2.25$. (The first of these values may be more easily estimated by plotting $y = 1 + \sin x$ and $y = 2x$ in rectangular coordinates; see the second graph.)

By symmetry, the total area contained is twice the area contained in the first quadrant, that is,

$$A \approx 2\int_0^{0.89} \frac{1}{2}(2\theta)^2\,d\theta + 2\int_{0.89}^{\pi/2} \frac{1}{2}(1 + \sin\theta)^2\,d\theta$$

$$= \left[\frac{4}{3}\theta^3\right]_0^{0.89} + \left[\theta - 2\cos\theta + \left(\frac{1}{2}\theta - \frac{1}{4}\sin 2\theta\right)\right]_{0.89}^{\pi/2} \approx 3.46$$

31. $L = \int_a^b \sqrt{r^2 + (dr/d\theta)^2}\,d\theta = \int_0^{2\pi}\sqrt{(2^\theta)^2 + [(\ln 2)\,2^\theta]^2}\,d\theta = \int_0^{2\pi} 2^\theta\sqrt{1 + \ln^2 2}\,d\theta$

$$= \left[\sqrt{1 + \ln^2 2}\left(\frac{2^\theta}{\ln 2}\right)\right]_0^{2\pi} = \frac{\sqrt{1 + \ln^2 2}\,(2^{2\pi} - 1)}{\ln 2}$$

33. $L = \int_0^{2\pi}\sqrt{(\theta^2)^2 + (2\theta)^2}\,d\theta = \int_0^{2\pi}\theta\sqrt{\theta^2 + 4}\,d\theta = \frac{1}{2}\cdot\frac{2}{3}\left[(\theta^2 + 4)^{3/2}\right]_0^{2\pi} = \frac{8}{3}\left[(\pi^2 + 1)^{3/2} - 1\right]$

35. From Figure 4 in Example 1,

$$L = \int_{-\pi/4}^{\pi/4}\sqrt{r^2 + (r')^2}\,d\theta = 2\int_0^{\pi/4}\sqrt{\cos^2 2\theta + 4\sin^2 2\theta}\,d\theta \approx 2\,(1.211056) \approx 2.4221.$$

Appendix H   Complex Numbers

1. $(3 + 2i) + (7 - 3i) = (3 + 7) + (2 - 3)i = 10 - i$

3. $(3 - i)(4 + i) = 12 + 3i - 4i - (-1) = 13 - i$

5. $\overline{12 + 7i} = 12 - 7i$

7. $\dfrac{2 + 3i}{1 - 5i} = \dfrac{2 + 3i}{1 - 5i} \cdot \dfrac{1 + 5i}{1 + 5i} = \dfrac{2 + 10i + 3i + 15(-1)}{1 - 25(-1)} = \dfrac{-13 + 13i}{26} = -\frac{1}{2} + \frac{1}{2}i$

9. $\dfrac{1}{1 + i} = \dfrac{1}{1 + i} \cdot \dfrac{1 - i}{1 - i} = \dfrac{1 - i}{1 - (-1)} = \dfrac{1 - i}{2} = \frac{1}{2} - \frac{1}{2}i$

11. $i^3 = i^2 \cdot i = (-1)i = -i$

13. $\sqrt{-25} = \sqrt{25}\,i = 5i$

15. $\overline{3 + 4i} = 3 - 4i,\ |3 + 4i| = \sqrt{3^2 + 4^2} = \sqrt{25} = 5$

17. $\overline{-4i} = \overline{0 - 4i} = 0 + 4i = 4i,\ |-4i| = \sqrt{0^2 + (-4)^2} = 4$

19. $4x^2 + 9 = 0 \ \Leftrightarrow\ 4x^2 = -9 \ \Leftrightarrow\ x^2 = -\frac{9}{4} \ \Leftrightarrow\ x = \pm\sqrt{-\frac{9}{4}} = \pm\sqrt{\frac{9}{4}}\,i = \pm\frac{3}{2}i.$

21. By the quadratic formula, $x^2 - 8x + 17 = 0 \ \Leftrightarrow$

$$x = \frac{8 \pm \sqrt{(-8)^2 - 4(1)(17)}}{2(1)} = \frac{8 \pm \sqrt{-4}}{2} = \frac{8 \pm 2i}{2} = 4 \pm i.$$

23. By the quadratic formula, $z^2 + z + 2 = 0 \ \Leftrightarrow\ z = \dfrac{-1 \pm \sqrt{1^2 - 4(1)(2)}}{2(1)} = \dfrac{-1 \pm \sqrt{-7}}{2} = -\frac{1}{2} \pm \frac{\sqrt{7}}{2}i.$

25. For $z = -3 + 3i,\ r = \sqrt{(-3)^2 + 3^2} = 3\sqrt{2}$ and $\tan\theta = \frac{3}{-3} = -1 \ \Rightarrow\ \theta = \frac{3}{4}\pi$ (since z lies in the second quadrant). Therefore, $-3 + 3i = 3\sqrt{2}\left(\cos\frac{3\pi}{4} + i\sin\frac{3\pi}{4}\right)$.

27. For $z = 3 + 4i,\ r = \sqrt{3^2 + 4^2} = 5$ and $\tan\theta = \frac{4}{3} \ \Rightarrow\ \theta = \tan^{-1}\frac{4}{3}$ (since z lies in the second quadrant). Therefore, $3 + 4i = 5\left[\cos\left(\tan^{-1}\frac{4}{3}\right) + i\sin\left(\tan^{-1}\frac{4}{3}\right)\right]$.

29. For $z = \sqrt{3} + i,\ r = \sqrt{\left(\sqrt{3}\right)^2 + 1^2} = 2$ and $\tan\theta = \frac{1}{\sqrt{3}} \ \Rightarrow\ \theta = \frac{\pi}{6} \ \Rightarrow\ z = 2\left(\cos\frac{\pi}{6} + i\sin\frac{\pi}{6}\right)$.
For $w = 1 + \sqrt{3}i,\ r = 2$ and $\tan\theta = \sqrt{3} \ \Rightarrow\ \theta = \frac{\pi}{3} \ \Rightarrow\ w = 2\left(\cos\frac{\pi}{3} + i\sin\frac{\pi}{3}\right)$. Therefore,
$zw = 2 \cdot 2\left[\cos\left(\frac{\pi}{6} + \frac{\pi}{3}\right) + i\sin\left(\frac{\pi}{6} + \frac{\pi}{3}\right)\right] = 4\left(\cos\frac{\pi}{2} + i\sin\frac{\pi}{2}\right)$,
$z/w = \frac{2}{2}\left[\cos\left(\frac{\pi}{6} - \frac{\pi}{3}\right) + i\sin\left(\frac{\pi}{6} - \frac{\pi}{3}\right)\right] = \cos\left(-\frac{\pi}{6}\right) + i\sin\left(-\frac{\pi}{6}\right)$, and
$1 = 1 + 0i = 1\left(\cos 0 + i\sin 0\right) \ \Rightarrow$
$1/z = \frac{1}{2}\left[\cos\left(0 - \frac{\pi}{6}\right) + i\sin\left(0 - \frac{\pi}{6}\right)\right] = \frac{1}{2}\left[\cos\left(-\frac{\pi}{6}\right) + i\sin\left(-\frac{\pi}{6}\right)\right]$. For $1/z$, we could also use the formula that precedes Example 5 to obtain $1/z = \frac{1}{8}\left(\cos\frac{\pi}{6} - i\sin\frac{\pi}{6}\right)$.

31. For $z = 2\sqrt{3} - 2i$, $r = \sqrt{\left(2\sqrt{3}\right)^2 + (-2)^2} = 4$ and $\tan\theta = \frac{-2}{2\sqrt{3}} = -\frac{1}{\sqrt{3}}$ $\Rightarrow$

$\theta = -\frac{\pi}{6}$ $\Rightarrow$ $z = 4\left[\cos\left(-\frac{\pi}{6}\right) + i\sin\left(-\frac{\pi}{6}\right)\right]$. For $w = -1 + i$, $r = \sqrt{2}$,

$\tan\theta = \frac{1}{-1} = -1$ $\Rightarrow$ $\theta = \frac{3\pi}{4}$ $\Rightarrow$ $z = \sqrt{2}\left(\cos\frac{3\pi}{4} + i\sin\frac{3\pi}{4}\right)$. Therefore,

$zw = 4\sqrt{2}\left[\cos\left(-\frac{\pi}{6} + \frac{3\pi}{4}\right) + i\sin\left(-\frac{\pi}{6} + \frac{3\pi}{4}\right)\right] = 4\sqrt{2}\left(\cos\frac{7\pi}{12} + i\sin\frac{7\pi}{12}\right)$,

$z/w = \frac{4}{\sqrt{2}}\left[\cos\left(-\frac{\pi}{6} - \frac{3\pi}{4}\right) + i\sin\left(-\frac{\pi}{6} - \frac{3\pi}{4}\right)\right] = \frac{4}{\sqrt{2}}\left[\cos\left(-\frac{11\pi}{12}\right) + i\sin\left(-\frac{11\pi}{12}\right)\right]$

$\qquad = 2\sqrt{2}\left(\cos\frac{13\pi}{12} + i\sin\frac{13\pi}{12}\right)$, and

$1/z = \frac{1}{4}\left[\cos\left(-\frac{\pi}{6}\right) - i\sin\left(-\frac{\pi}{6}\right)\right] = \frac{1}{4}\left(\cos\frac{\pi}{6} + i\sin\frac{\pi}{6}\right)$.

33. For $z = 1 + i$, $r = \sqrt{2}$ and $\tan\theta = \frac{1}{1} = 1$ $\Rightarrow$ $\theta = \frac{\pi}{4}$ $\Rightarrow$ $z = \sqrt{2}\left(\cos\frac{\pi}{4} + i\sin\frac{\pi}{4}\right)$. So by
De Moivre's Theorem,

$(1+i)^{20} = \left[\sqrt{2}\left(\cos\frac{\pi}{4} + i\sin\frac{\pi}{4}\right)\right]^{20} = \left(2^{1/2}\right)^{20}\left(\cos\frac{20\cdot\pi}{4} + i\sin\frac{20\cdot\pi}{4}\right)$

$\qquad = 2^{10}\left(\cos 5\pi + i\sin 5\pi\right) = 2^{10}\left[-1 + i\left(0\right)\right] = -2^{10} = -1024$.

35. For $z = 2\sqrt{3} + 2i$, $r = 4$ and $\tan\theta = \frac{2}{2\sqrt{3}} = \frac{1}{\sqrt{3}}$ $\Rightarrow$ $\theta = \frac{\pi}{6}$ $\Rightarrow$ $z = 4\left(\cos\frac{\pi}{6} + i\sin\frac{\pi}{6}\right)$. So by
De Moivre's Theorem,

$\left(2\sqrt{3} + 2i\right)^5 = \left[4\left(\cos\frac{\pi}{6} + i\sin\frac{\pi}{6}\right)\right]^5 = 4^5\left(\cos\frac{5\pi}{6} + i\sin\frac{5\pi}{6}\right) = 1024\left[-\frac{\sqrt{3}}{2} + \frac{1}{2}i\right]$

$\qquad = -512\sqrt{3} + 512i$.

37. $1 = 1 + 0i = 1\left(\cos 0 + i\sin 0\right)$. Using Equation 3 with $r = 1$, $n = 8$, and $\theta = 0$, we have

$w_k = 1^{1/8}\left[\cos\left(\frac{0 + 2k\pi}{8}\right) + i\sin\left(\frac{0 + 2k\pi}{8}\right)\right] = \cos\frac{k\pi}{4} + i\sin\frac{k\pi}{4}$, where $k = 0, 1, 2, \ldots, 7$.

$w_0 = 1\left(\cos 0 + i\sin 0\right) = 1$, $w_1 = 1\left(\cos\frac{\pi}{4} + i\sin\frac{\pi}{4}\right) = \frac{1}{\sqrt{2}} + \frac{1}{\sqrt{2}}i$,

$w_2 = 1\left(\cos\frac{\pi}{2} + i\sin\frac{\pi}{2}\right) = i$, $w_3 = 1\left(\cos\frac{3\pi}{4} + i\sin\frac{3\pi}{4}\right) = -\frac{1}{\sqrt{2}} + \frac{1}{\sqrt{2}}i$,

$w_4 = 1\left(\cos\pi + i\sin\pi\right) = -1$, $w_5 = 1\left(\cos\frac{5\pi}{4} + i\sin\frac{5\pi}{4}\right) = -\frac{1}{\sqrt{2}} - \frac{1}{\sqrt{2}}i$,

$w_6 = 1\left(\cos\frac{3\pi}{2} + i\sin\frac{3\pi}{2}\right) = -i$, $w_7 = 1\left(\cos\frac{7\pi}{4} + i\sin\frac{7\pi}{4}\right) = \frac{1}{\sqrt{2}} - \frac{1}{\sqrt{2}}i$

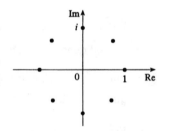

39. $0 = 0 + i = 1\left(\cos\frac{\pi}{2} + i\sin\frac{\pi}{2}\right)$. Using Equation 3 with $r = 1$, $n = 3$, and $\theta = \frac{\pi}{2}$, we have

$w_k = 1^{1/3}\left[\cos\left(\frac{\frac{\pi}{2} + 2k\pi}{3}\right) + i\sin\left(\frac{\frac{\pi}{2} + 2k\pi}{3}\right)\right]$, where $k = 0, 1, 2$.

$w_0 = \left(\cos\frac{\pi}{6} + i\sin\frac{\pi}{6}\right) = \frac{\sqrt{3}}{2} + \frac{1}{2}i$

$w_1 = \left(\cos\frac{5\pi}{6} + i\sin\frac{5\pi}{6}\right) = -\frac{\sqrt{3}}{2} + \frac{1}{2}i$

$w_2 = \left(\cos\frac{9\pi}{6} + i\sin\frac{9\pi}{6}\right) = -i$

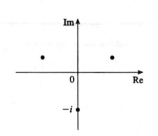

41. Using Euler's formula (6) with $y = \frac{\pi}{2}$, we have $e^{i\pi/2} = \cos\frac{\pi}{2} + i\sin\frac{\pi}{2} = 0 + 1i = i$.

43. Using Euler's formula with $y = \frac{3\pi}{4}$, we have $e^{i3\pi/4} = \cos \frac{3\pi}{4} + i \sin \frac{3\pi}{4} = -\frac{1}{\sqrt{2}} + \frac{1}{\sqrt{2}} i$.

45. Using Equation 7 with $x = 2$ and $y = \pi$, we have

$$e^{2+i\pi} = e^2 e^{i\pi} = e^2 \left(\cos \pi + i \sin \pi \right) = e^2 \left(-1 + 0 \right) = -e^2.$$

47. Take $r = 1$ and $n = 3$ in De Moivre's Theorem to get

$$[1 \left(\cos \theta + i \sin \theta \right)]^3 = 1^3 \left(\cos 3\theta + i \sin 3\theta \right) \quad \Rightarrow \quad \left(\cos \theta + i \sin \theta \right)^3 = \cos 3\theta + i \sin 3\theta \quad \Rightarrow$$

$$\cos^3 \theta + 3 \left(\cos^2 \theta \right) \left(i \sin \theta \right) + 3 \left(\cos \theta \right) \left(i \sin \theta \right)^2 + \left(i \sin \theta \right)^3 = \cos 3\theta + i \sin 3\theta \quad \Rightarrow$$

$$\left(\cos^3 \theta - 3 \sin^2 \theta \cos \theta \right) + \left(3 \sin \theta \cos^2 \theta - \sin^3 \theta \right) i = \cos 3\theta + i \sin 3\theta. \text{ Equating real and imaginary}$$

parts gives $\cos 3\theta = \cos^3 \theta - 3 \sin^2 \theta \cos \theta$ and $\sin 3\theta = 3 \sin \theta \cos^2 \theta - \sin^3 \theta$.

49. $F(x) = e^{rx} = e^{(a+bi)x} = e^{ax+bxi} = e^{ax} \left(\cos bx + i \sin bx \right) = e^{ax} \cos bx + i \left(e^{ax} \sin bx \right) \quad \Rightarrow$

$$F'(x) = \left(e^{ax} \cos bx \right)' + i \left(e^{ax} \sin bx \right)' = \left(ae^{ax} \cos bx - be^{ax} \sin bx \right) + i \left(ae^{ax} \sin bx + be^{ax} \cos bx \right)$$

$$= a \left[e^{ax} \left(\cos bx + i \sin bx \right) \right] + b \left[e^{ax} \left(-\sin bx + i \cos bx \right) \right] = ae^{rx} + b \left[e^{ax} \left(i^2 \sin bx + i \cos bx \right) \right]$$

$$= ae^{rx} + bi \left[e^{ax} \left(\cos bx + i \sin bx \right) \right] = ae^{rx} + bie^{rx} = (a + bi) e^{rx} = re^{rx}$$